Electrical
Wiring
Commercial

Based on the 1999
NATIONAL ELECTRICAL CODE®

Electrical
Wiring
Commercial

Tenth Edition

Ray C. Mullin   Robert L. Smith

Delmar Publishers

an International Thomson Publishing company I(T)P®

Albany • Bonn • Boston • Cincinnati • Detroit • London • Madrid
Melbourne • Mexico City • New York • Pacific Grove • Paris • San Francisco
Singapore • Tokyo • Toronto • Washington

NOTICE TO THE READER

Cover Design: Courtesy of Brucie Rosch

Delmar Staff:
Publisher: Alar Elken
Acquisitions Editor: Mark Huth
Developmental Editor: Jeanne Mesick
Project Editor: Megeen Mulholland
Production Coordinator: Toni Hansen
Art and Design Coordinator: Cheri Plasse
Editorial Assistant: Dawn Daugherty
Production Manager: Mary Ellen Black

COPYRIGHT © 1999
By Delmar Publishers
an International Thomson Publishing company

The ITP logo is a trademark under license
Printed in the United States of America

For more information, contact:

Delmar Publishers
3 Columbia Circle, Box 15015
Albany, New York 12212-5015

International Thomson Publishing Europe
Berkshire House 168-173
168-173 High Holborn
London, WC1V 7AA
United Kingdom

Nelson ITP, Australia
102 Dodds Street
South Melbourne
Victoria, 3205 Australia

Nelson Canada
1120 Birchmont Road
Scarborough, Ontario
M1K 5G4, Canada

International Thomson Publishing France
Tour Maine-Montparnasse
33 Avenue du Maine
75755 Paris Cedex 15, France

International Thomson Editores
Seneca 53
Colonia Polanco
11560 Mexico D. F. Mexico

International Thomson Publishing GmbH
Königswinterer Straße 418
53227 Bonn
Germany

International Thomson Publishing Asia
60 Albert Street
#15-01 Albert Complex
Singapore 189969

International Thomson Publishing Japan
Hirakawa-cho Kyowa Building, 3F
2-2-1 Hirakawa-cho, Chiyoda-ku,
Tokyo 102, Japan

ITE Spain/Parninfo
Calle Magallanes, 25
28015-Madrid, Espana

1 2 3 4 5 6 7 8 9 10 XXX 03 02 01 00 99 98

Library of Congress Cataloging-in-Publication Data

Mullin, Ray C.
    Electrical wiring commercial / Ray C. Mullin, Robert L. Smith. —
10th ed.
      p.   cm.
    "Based on the 1999 National Electrical Code."
    Includes bibliographical references and index.
    ISBN 0-7668-0179-9 (alk. paper)
    1. Electric wiring, Interior. 2. Commercial buildings — Electric
equipment. 3. Electric wiring — Insurance requirements. I. Smith,
Robert L., 1926–   . II. Title.
TK3284.M85 1998
621.319'24 — dc21
                                                    98-30058
                                                  CIP

CONTENTS

PREFACE

INTRODUCTION

The tenth edition of ELECTRICAL WIRING — COMMERCIAL is based on the 1999 *National Electrical Code.** The new edition thoroughly and clearly explains the *NEC*® changes that relate to commercial wiring.

The *National Electrical Code*® is used as the basic standard for the layout and construction of electrical systems. To gain the greatest benefit from this text, the learner must use the *National Electrical Code*® on a continuing basis.

State and local codes may contain modifications of the *National Electrical Code*® to meet local requirements. The instructor is encouraged to furnish students with any variations from the *NEC*® as they affect this commercial installation in a specific area.

This text takes the learner through the essential minimum requirements as set forth in the *National Electrical Code*® for commercial installations. In addition to Code minimums, the reader will find such information above and beyond the minimum requirements.

The commercial electrician is required to work in three common situations: where the work is planned in advance; where there is no advance planning; and where repairs are needed. The first situation exists when the work is designed by a consulting engineer. In this case, the electrician must know the installation procedures, must be able to read plans, and must be able to understand and interpret specifications. The second situation occurs either during or after construction when changes or remodeling are required. The third situation arises any time after a system is installed. Whenever a problem occurs with an installation, the electrician must understand the operation of all equipment included in the installation in order to solve the problem.

When the electrician is working on the initial installation or is modifying an existing installation, the circuit loads must be determined. Thorough explanations and numerous examples of calculating these loads help prepare the reader for similar problems on the job. The text and assignments make frequent reference to the commercial building drawings at the back of the text.

Readers should be aware that many of the electrical loads used as *examples in the text were contrived in order to create Code problems*. The authors' purpose in putting this building together is to demonstrate, and thus enhance learning, as many Code problems as possible. As an example, there is a single-phase feeder to the Doctor's office. This could have been a three-phase feeder similar to those in the other occupancies. However, using the single-phase feeder allows us to demonstrate many additional Code applications.

---

* *National Electrical Code*® and *NEC*® are registered trademarks of the National Fire Protection Association, Inc., Quincy, MA 02269. Applicable tables and section references are reprinted with permission from NFPA – 1999, the *National Electrical Code,*® Copyright © 1998, National Fire Protection Association, Quincy, Massachusetts 02269. This reprinted material is not the complete and official position of the National Fire Protection Associaion on the referenced subject, which is represented only by the standard in its entirety.

This is not a typical or an ideal design for a commercial building. It is a composite to demonstrate a range of Code applications. The authors also carry many calculations to a higher level of accuracy as compared to the accuracy required in many actual job situations. This is done to demonstrate the correct method. Then, if the reader wants to back off from this level, based upon installation requirements, it can be done intelligently.

THE TENTH EDITION

The unchanging goal of this text is to explain and demonstrate how, why, and where the *National Electrical Code®* is applied to the wiring of a commercial type building. Three completely rewritten units provide an expanded and simplified insight into the planning of an electrical installation, carefully demonstrating how the load requirements are converted into branch circuits, then into the feeders and finally into the building service. New tables and drawings enhance this progression. This edition totally reflects the reorganization of the 1999 *National Electrical Code.* The authors believe that these changes will result in a Code that will be recognized as a document that is easier to interpret and apply. Every Code reference in tenth edition of *Electrical Wiring Commercial* is the result of carefully comparing the past Code references with the 1999 *NEC®*.

CHANGES FOR THIS EDITION AS A RESULT OF CHANGES IN THE 1999 *NATIONAL ELECTRICAL CODE®*

- Over 1,000 *NEC®* references have been reviewed and updated as dictated by the revision and reorganization of the 1999 *NEC.*

- An expansion of Unit 2 to include working drawing review of the entry level occupancies.

- A complete rewrite of Unit 4 to clarify and simplify the creation of branch circuits. Included is a worksheet that facilitates the selection of branch circuit overcurrent protection and conductors.

- A completely new Unit 8 explains and demonstrates each step necessary in the selection of feeder phase and neutral conductors and feeder overcurrent protection.

- Unit 10 has been revised to include all three of the upper level occupancies, including an extended discussion of the merits of 3-wire single-phase and 4-wire three-phase feeders.

- A major modification of Unit 13 incorporates an entirely new procedure for making decisions related to the service entrance equipment.

- Several branch circuits have been revised and rerouted to provide more examples of *NEC®* requirements.

- An updating of the sizing of neutrals where there is significant nonlinear loading.

ABOUT THE AUTHORS

This text was prepared by Ray C. Mullin and Robert L. Smith. Mr. Mullin is a former electrical circuit instructor for the Electrical Trades, Wisconsin Schools of Vocational, Technical and Adult Education. A former member of the International Brotherhood of Electrical Workers, Mr. Mullin is presently a member of the International Association of Electrical Inspectors, the Institute of Electrical and Electronic Engineers, and the National Fire Protection Association, Electrical Section. He served on Code Making Panel 4 for the *National Electrical Code®*, NFPA-70 for the National Fire Protection Association.

Mr. Mullin completed his apprenticeship training and has worked as a journeyman and supervisor. He has taught both day and night electrical apprentice and journeyman courses and has conducted engineering seminars. Mr. Mullin has contributed to and assisted other authors in their writing of texts and articles relating to overcurrent protection and conductor withstand ratings. He has had many articles relating to overcurrent protection published in various trade magazines.

Mr. Mullin attended the University of Wisconsin, Colorado State University, and Milwaukee School of Engineering.

He serves on the Executive Board of the Western Section, International Association of Electrical Inspectors. He also serves on their National Electrical Code Committee and on their Code Clearing Committee. He also serves on the Electrical Commission in his home-town. Mr. Mullin has conducted many technical Code workshops and seminars at state chapter and section meetings of the International Association of Electrical Inspectors and serves on their Code panels.

Mr. Mullin is past Director, Technical Liaison, and Code Coordinator for a large electrical manufacturer.

Robert L. Smith, P. E., is an Emeritus Professor of Architecture at the University of Illinois where he taught courses in electrical systems, lighting design, acoustics and energy-conscious design. He worked in electrical construction for more than 25 years, and has taught electrical apprentice classes and courses in industrial electricity.

Professor Smith has been affiliated with a number of professional organizations, including the Illinois Society of Professional Engineers, the International Association of Electrical Inspectors, and the National Academy of Code Administration. He is a life-long honorary member of the International Brotherhood of Electrical Workers. He gives frequent public lectures on the application of the *National Electrical Code®* and was a regular presenter at the University of Wisconsin Extension.

Delmar Publishers Is Your Electrical Book Source!

Whether you're a beginning student or a master electrician, Delmar Publishers has the right book for you. Our complete selection of proven best-sellers and all-new titles is designed to bring you the most up-to-date, technically accurate information available.

NATIONAL ELECTRICAL CODE

National Electrical Code® 1999/NFPA

Revised every three years, the *National Electrical Code®* is the basis of all U.S. electrical codes.
Order # 0-8776-5432-8

Loose-leaf version in binder   *Order # 0-8776-5433-6*

National Electrical Code® Handbook 1999/NFPA

This essential resource pulls together all the extra facts, figures, and explanations you need to interpret the 1999 *NEC®* It includes the entire text of the Code, plus expert commentary, real-world examples, diagrams, and illustrations that clarify requirements.   *Order # 0-8776-5437-9*

Illustrated Changes in the 1999 National Electrical Code®/O'Riley

This book provides an abundantly illustrated and easy-to-understand analysis of the changes made to the 1999 *NEC®*.   *Order # 0-7668-0763-0*

Understanding the National Electrical Code,® 3E/Holt

This book gives users at every level the ability to understand what the *NEC®* requires, and simplifies this sometimes intimidating and confusing code.   *Order # 0-7668-0350-3*

Illustrated Guide to the National Electrical Code®/Miller

Highly detailed illustrations offer insight into Code requirements, and are further enhanced through clearly written, concise blocks of text that can be read very quickly and understood with ease. Organized by classes of occupancy.   *Order # 0-7668-0529-8*

Interpreting the National Electrical Code, *5E*/Surbrook

This updated resource provides a process for understanding and applying the *National Electrical Code*® to electrical contracting, plan development, and review.   *Order # 0-7668-0187-X*

Electrical Grounding, 5E/O'Riley

Electrical Grounding is a highly illustrated, systematic approach for understanding grounding principles and their application to the 1999 *NEC.*®   *Order # 0-7668-0486-0*

ELECTRICAL WIRING

Electrical Raceways and Other Wiring Methods, 3E/Loyd

The most authoritative resource on metallic and nonmetallic raceways, provides users with a concise, easy-to-understand guide to the specific design criteria and wiring methods and materials required by the 1999 *NEC.*®   *Order # 0-7668-0266-3*

Electrical Wiring—Residential, 13E/Mullin

Now in full color! Users can learn all aspects of residential wiring and how to apply them to the wiring of a typical house from this, the most widely used residential wiring book in the country.
Softcover Order #0-8273-8607-9
Hardcover Order # 0-8273-8610-9

House Wiring with the NEC®/Mullin

The focus of this new book is the applications of the *NEC*® to house wiring.
Order # 0-8273-8350-9

Electrical Wiring—Commercial, 10E/Mullin and Smith

Users can learn commercial wiring in accordance with the *NEC*® from this comprehensive guide to applying the newly revised 1999 *NEC.*®   *Order # 0-7668-0179-9*

Electrical Wiring—Industrial, 10E/Smith and Herman

This practical resource has users work their way through an entire industrial building—wiring the branch-circuits, feeders, service entrances, and many of the electrical appliances and subsystems found in commercial buildings.   *Order # 0-7668-0193-4*

Cables and Wiring, 2E/AVO

This concise, easy-to-use book is your single-source guide to electrical cables—it's a "must-have" reference for journeyman electricians, contractors, inspectors, and designers.

Order # 0-7668-0270-1

ELECTRICAL MACHINES AND CONTROLS

Industrial Motor Control, 4E/Herman and Alerich

This newly revised and expanded book, now in full color, provides easy-to-follow instructions and essential information for controlling industrial motors. Also available are a new lab manual and an interactive CD-ROM.   *Order # 0-8273-8640-0*

Electric Motor Control, 6E/Alerich and Herman

Fully updated in this new sixth edition, this book has been a long-standing leader in the area of electric motor controls.   *Order # 0-8273-8456-4*

Introduction to Programmable Logic Controllers/Dunning

This book offers an introduction to Programmable Logic Controllers.

Order # 0-8273-7866-1

Technician's Guide to Programmable Controllers, 3E/Cox

Uses a plain, easy-to-understand approach and covers the basics of programmable controllers.
Order # 0-8273-6238-2

Programmable Controller Circuits/Bertrand

This book is a project manual designed to provide practical laboratory experience for one studying industrial controls.   *Order # 0-8273-7066-0*

Electronic Variable Speed Drives/Brumbach

Aimed squarely at maintenance and troubleshooting, *Electronic Variable Speed Drives* is the only book devoted exclusively to this topic.   *Order # 0-8273-6937-9*

Electrical Controls for Machines, 5E/Rexford

State-of-the-art process and machine control devices, circuits, and systems for all types of industries are explained in detail in this comprehensive resource.   *Order # 0-8273-7644-8*

Electrical Transformers and Rotating Machines/Herman

This new book is an excellent resource for electrical students and professionals in the electrical trade.   *Order # 0-7668-0579-4*

Delmar's Standard Guide to Transformers/Herman

Delmar's Standard Guide to Transformers was developed from the best-seller *Standard Textbook of Electricity* with expanded transformer coverage not found in any other book.
Order # 0-8273-7209-4

DATA AND VOICE COMMUNICATION CABLING AND FIBER OPTICS

Complete Guide to Fiber Optic Cable System Installation/Pearson

This book offers comprehensive, unbiased, state-of-the-art information and procedures for installing fiber optic cable systems.   *Order # 0-8273-7318-X*

Fiber Optics Technician's Manual/Hayes

Here's an indispensable tool for all technicians and electricians who need to learn about optimal fiber optic design and installation as well as the latest troubleshooting tips and techniques.
Order # 0-8273-7426-7

A Guide for Telecommunications Cable Splicing/Highhouse

A "how-to" guide for splicing all types of telecommunications cables.
Order # 0-8273-8066-6

Premises Cabling/Sterling

This reference is ideal for electricians, electrical contractors, and inspectors needing specific information on the principles of structured wiring systems.   *Order # 0-8273-7244-2*

ELECTRICAL THEORY

Delmar's Standard Textbook of Electricity, 2E/Herman

This exciting full-color book is the most comprehensive book on DC/AC circuits and machines for those learning the electrical trades.   *Order # 0-8273-8550-1*

Industrial Electricity, 6E/Nadon, Gelmine and Brumbach

This revised, illustrated book offers broad coverage of the basics of electrical theory and is perfect for those who wish to be industrial maintenance technicians.   *Order # 0-7668-0101-2*

EXAM PREPARATION

Journeyman Electrician's Exam Preparation, 2E/Holt

This comprehensive exam prep guide includes all of the topics on the journeyman electrician competency exams.   *Order # 0-7668-0375-9*

Master Electrician's Exam Preparation, 2E/Holt

This comprehensive exam prep guide includes all of the topics on the master electrician's competency exams.   *Order # 0-7668-0376-7*

REFERENCE

ELECTRICAL REFERENCE SERIES

This series of technical reference books is written by experts and designed to provide the electrician, electrical contractor, industrial maintenance technician, and other electrical workers with a source of reference information about virtually all of the electrical topics that they encounter.

*Electrician's Technical Reference—**Motor Controls***/Carpenter

Electrical Reference—Motor Controls is a source of comprehensive information on understanding the controls that start, stop, and regulate the speed of motors.   *Order # 0-8273-8514-5*

*Electrician's Technical Reference—**Motors***/Carpenter

Electrician's Technical Reference—Motors builds an understanding of the operation, theory, and applications of motors.   *Order # 0-8273-8513-7*

*Electrician's Technical Reference—**Theory and Calculations***/Herman

Electrician's Technical Reference—Theory and Calculations provides detailed examples of problem-solving for different kinds of DC and AC circuits.   *Order # 0-8273-7885-8*

*Electrician's Technical Reference—**Transformers***/Herman

Electrician's Technical Reference—Transformers focuses on the theoretical and practical aspects of single-phase and 3-phase transformers and transformer connections.

Order # 0-8273-8496-3

*Electrician's Technical Reference—**Hazardous Locations***/Loyd

Electrician's Technical Reference—Hazardous Locations covers electrical wiring methods and basic electrical design considerations for hazardous locations.   *Order # 0-8273-8380-0*

*Electrician's Technical Reference—**Wiring Methods***/Loyd

Electrician's Technical Reference—Wiring Methods covers electrical wiring methods and basic electrical design considerations for all locations, and shows how to provide efficient, safe, and economical applications of various types of available wiring methods.   *Order # 0-8273-8379-7*

*Electrician's Technical Reference—**Industrial Electronics***/Herman

Electrician's Technical Reference—Industrial Electronics covers components most used in heavy industry, such as silicon control rectifiers, triacs, and more. It also includes examples of common rectifiers and phase-shifting circuits.   *Order # 0-7668-0347-3*

RELATED TITLES

Common Sense Conduit Bending and Cable Tray Techniques/Simpson

Now geared especially for students, this manual remains the only complete treatment of the topic in the electrical field.   *Order # 0-8273-7110-1*

Practical Problems in Mathematics for Electricians, 5E/Herman

This book details the mathematics principles needed by electricians.   *Order # 0-8273-6708-2*

Electrical Estimating/Holt

This book provides a comprehensive look at how to estimate electrical wiring for residential and commercial buildings with extensive discussion of manual versus computer-assisted estimating.

Order # 0-8273-8100-X

Electrical Studies for Trades/Herman

Based on Delmar's *Standard Textbook of Electricity*, this new book provides non-electrical trades students with the basic information they need to understand electrical systems.

Order # 0-8273-7845-9

To request examination copies or a catalog of all our titles, call or write to:
Delmar Publishers
3 Columbia Circle
P.O. Box 15015
Albany, NY 12212-5015
Phone: 1-800-347-7707 • 1-518-464-3500 • FAX: 1-518-464-0301

ACKNOWLEDGMENTS

ELECTRICAL WIRING — COMMERCIAL, tenth edition, has been thoroughly reviewed and tested to ensure its accuracy and usefulness as a learning tool.

The authors would like to thank Mel Sanders for his many contributions to the new edition of this text. His suggestions were very helpful in updating the text. Mr. Sanders teaches Code seminars in all areas of the country and contributes regularly to the *Electrical Construction and Maintenance* magazine and to the International Association of Electrical Inspectors.

The following companies provided technical information and figures used in the text.

American National Standards Institute, Inc., 1430 Broadway, New York, NY 10018.

Appleton Electrical Company, 1747 W. Wellington Avenue, Chicago, IL 60657.

Boltswitch, Inc., 6107 West Lou Avenue, Crystal Lake, IL 60014.

Bussmann Division, Cooper Industries, P.O. Box 14460, St. Louis, MO 63178-9977.

Canadian Standards Association, 178 Rexdale Blvd., Rexdale, Ontario, Canada, M9W 1R3.

Carlon Electrical Products, 25701 Science Park Drive, Cleveland, OH 44122.

Cutler Industries, Inc., 7425 Croname Road, Niles, IL 60648.

General Electric Co., 225 Service Avenue, Warick, RI 02886.

Cooper Lighting, 400 Busse Road, Elk Grove Village, IL 60007.

Honeywell, Inc., Commercial Div., 2701 Fourth Ave. S., Minneapolis, MN 55408.

Harvey Hubbell, Inc., Wiring Device Div., P.O. Box 3999, Bridgeport, CT 06605.

Illuminating Engineering Society, 120 Wall Street, 17th Floor, New York, NY 10005-4001.

International Association of Electrical Inspectors, 930 Busse Highway, Park Ridge, IL 60068.

Jefferson Electric Co., Div. Litton Industries, 840 South 25th Ave., Bellwood, IL 60104.

Juno Lighting, Inc., 2001 S. Mt. Prospect Road, Des Plaines, IL 60618.

Kurt Versen Company, 10 Charles Street, P.O. Box 677, Westwood, NJ 07675.

National Electrical Code, National Fire Protection Association, Batterymarch Park, Quincy, MA 02269.

National Electrical Manufacturers Association, 2101 L Street NW, Washington, DC 20037.

Onan Division, Onan Corporation, 1400 73rd Ave. NE, Minneapolis, MN 55432.

Square D Co., Distribution Equip., Group Headquarters, 1601 Mercer Rd., Lexington, KY 40505.

The Trane Co., 3600 Pammel Creek Rd., LaCrosse, WI 54601.

Underwriters Laboratories, Inc., 333 Pfingsten Road, Northbrook, IL 60062.

The authors and Delmar Publishers gratefully acknowledge the time and efforts of those who reviewed *Electrical Wiring Commercial.* Their comments and suggestions were invaluable. Special thanks to:

Greg Fletcher
Kennebec Valley Technical College
Fairfield, ME

John Penley
Riverland Community College
Albert Lea, MN

David Gehlauf
Tri-County Vocational School
Glouster, OH

Larry Catron
Scott County Vocational School
Gate City, VA

Dan Kubala
Triangle Technical College
Pittsburgh, PA

Lanny McMahill
City of Phoenix, DSD
Phoenix, AZ

Larry A. Catron
Scott County Vocational Center

Eric David
Tennessee State Area Vocational-Technical School

Robert Hayden
Atlantic County Vocational Technical School

UNIT 1

Commercial Building Plans and Specifications

OBJECTIVES

After studying this unit, the student will be able to

- understand the basic safety rules for working on electrical systems.
- define the project requirements from the contract documents.
- demonstrate the application of building plans and specification.
- locate specific information on the building plans.
- obtain information from industry-related organizations.
- apply and interchange SI and English measurements.

SAFETY IN THE WORKPLACE

Before we get started on our venture into the wiring of a typical commercial building, let us talk about safety.

Electricity is dangerous! Working on electrical equipment with the power turned on can result in death or serious injury, either as a direct result of electricity or from an indirect secondary reaction such as falling off a ladder or falling into moving parts of equipment. Dropping a metal tool onto live parts or allowing metal shavings from a drilling operation to fall onto live parts of electrical equipment generally results in an *arc blast* that can cause deadly burns. The heat of an electrical arc has been determined to be hotter than the sun. Pressures developed during an arc blast can blow a person clear across the room. Dirt, debris, and moisture can also set the stage for catastrophic equipment failures and personal injury. Neatness and cleanliness in the workplace are a must.

The Federal Regulations in the Occupational Safety and Health Act (OSHA) Number 29, Subpart S, in Part 1910.332 discusses the training needed for those who face a risk of electrical injury. Proper training means "trained in and familiar with the safety-related work practices required by paragraphs 1910.331 through 1910.335." Numerous texts are available that delve into the OSHA requirements in great detail.

The *National Electrical Code®* defines a *qualified person* as "one familiar with the construction and operation of the equipment and the hazards involved." Merely telling someone or being told "be careful" does not meet the definition of proper training, and does not make the person qualified.

Only qualified persons are permitted to work on or near exposed energized equipment. To become qualified, a person must have the skill and technique necessary to distinguish exposed live parts from other parts of electrical equipment, be able to determine the voltage of exposed live parts, and be trained in the use of special precautionary techniques such as personal protective equipment, insulations, shielding material, and insulated tools.

In Subpart S, paragraph 1910.333 requires that safety-related work practices shall be employed to prevent electric shock or other injuries resulting from either direct or indirect electrical contact. Live parts to which an employee may be exposed shall be de-energized before the employee works on or near them, unless the employer can demonstrate that de-energizing introduces additional or increased hazards.

Working on equipment "live" is acceptable only if there would be a greater hazard if the system were de-energized. Examples of this would be life support systems, some alarm systems, certain ventilation systems in hazardous locations, and the power for critical illumination circuits. Working on energized equipment requires properly insulated tools, proper nonflammable clothing, rubber gloves, protective shields and goggles, and in some cases, rubber blankets.

OSHA regulations allow only qualified personnel to work on or near electrical circuits or equipment that has not been de-energized. The OSHA regulations provide rules regarding "lockout and tagging" to make sure that the electrical equipment being worked on will not inadvertently be turned on while someone is working on the supposedly "dead" equipment. As the OSHA regulations state, "a lock and a tag shall be placed on each disconnecting means used to de-energize circuits and equipment. . . ."

Some electricians' contractual agreements require that as a safety measure, two or more qualified electricians must work together when working on energized circuits. They do not allow untrained apprentices to work on "live" equipment, but do allow apprentices to stand back and observe.

The National Fire Protection Association's publications *Safety Related Work Practices* NFPA 70E and *Electrical Equipment Maintenance* NFPA 70B present much of the same material regarding electrical safety as do the OSHA regulations.

Safety cannot be compromised.

Follow this rule: turn off and lock off the power, then properly tag the disconnect with a description as to exactly what that particular disconnect controls.

With safety the utmost concern in our minds, let us begin our course on the wiring of a typical commercial building.

COMMERCIAL BUILDING SPECIFICATIONS

When a building project contract is awarded, the electrical contractor is given the plans and specifications for the building. These two contract documents govern the construction of the building. It is very important that the electrical contractor and the electricians employed by the contractor to perform the electrical construction follow the specifications exactly. The electrical contractor will be held responsible for any deviations from the specifications and may be required to correct such deviations or variations at personal expense. Thus, it is important that any changes or deviations be verified — in writing. Avoid verbal change orders.

It is suggested that the electrician assigned to a new project first read the specifications carefully. These documents provide the detailed information that will simplify the task of studying the plans. The specifications are usually prepared in book form and may consist of a few pages to as many as several hundred pages covering all phases of the construction. This text presents in detail only that portion of the specifications that directly affects the electrician; however, summaries of the other specification sections are presented to acquaint the electrician with the full scope of the document.

The specification is a book of rules governing all of the material to be used and the work to be performed on a construction project. The specification is usually divided into several sections.

GENERAL CLAUSES AND CONDITIONS

The first section of the specification, titled *General Clauses and Conditions*, deals with the legal requirements of the project. The index to this section may include the following headings:

> Notice to Bidders
> Schedule of Drawings
> Instruction to Bidders
> Proposal
> Agreement
> General Conditions

Some of these items will affect the electrician on the job and others will be of primary concern to the electrical contractor. The following paragraphs give a brief, general description of each item and how it affects either the electrician on the job or the contractor.

Notice to Bidders. This item is of value to the contractor and his estimator only. The notice describes the project, its location, the time and place of the bid opening, and where and how the plans and specifications can be obtained.

Schedule of Drawings. The schedule is a list, by number and title, of all of the drawings related to the project. The contractor, estimator, and electrician will each use this schedule prior to preparing the bid for the job: the contractor to determine if all the drawings required are at hand, the estimator to do a take-off and formulate a bid, and the electrician to determine if all of the drawings necessary to do the installation are available.

Instructions to Bidders. This section provides the contractor with a brief description of the project, its location, and how the job is to be bid (lump sum, one contract, or separate contracts for the various construction trades, such as plumbing, heating, electrical, and general). In addition, bidders are told where and how the plans and specifications can be obtained prior to the preparation of the bid, how to make out the proposal form, where and when to deliver the proposal, the amount of any bid deposits required, any performance bonds required, and bidders' qualifications. Other specific instructions may be given, depending upon the particular job.

Proposal. The proposal is a form that is filled out by the contractor and submitted at the proper time and place. The proposal is the contractor's bid on a project. The form is the legal instrument that binds the contractor to the owner if: (a) the contractor completes the proposal properly, (b) the contractor does not forfeit the bid bond, (c) the owner accepts the proposal, and (d) the owner signs the agreement. Generally, only the contractor will be using this section.

The proposal may show that alternate bids were requested by the owner. In this case, the electrician on the job should study the proposal and consult with the contractor to learn which of the alternate bids has been accepted in order to determine the extent of the work to be completed.

On occasion, the proposal may include a specified time for the completion of the project. This information is important to the electrician on the job since the work must be scheduled to meet the completion date.

Agreement. The agreement is the legal binding portion of the proposal. The contractor and the owner sign the agreement and the result is a legal contract. Once the agreement is signed, both parties are bound by the terms and conditions given in the specification.

General Conditions. The following items are normally included under the General Conditions heading of the *General Clauses and Conditions*. A brief description is presented for each item.

- General Note: includes the general conditions as part of the contract documents.
- Definition: as used in the contract documents, this item defines the owner, contractor, architect, engineer, and other people and objects involved in the project.
- Contract Documents: a listing of the documents involved in the contract, including plans, specifications, and agreement.
- Insurance: specifies the insurance a contractor must carry on all employees, and on the materials involved in the project.
- Workmanship and Materials: specifies that the work must be done by skilled workers and that the materials must be new and of good quality.
- Substitutions: materials used must be as specified or equivalent materials must be shown to have the required properties.
- Shop Drawings: this item identifies the drawings that must be submitted by the contractor to show how the specific pieces of equipment are to be installed.
- Payments: the method of paying the contractor during the construction is specified.
- Coordination of Work: each contractor on the job must cooperate with every other contractor to insure that the final product is complete and functional.
- Correction Work: describes how work must be corrected, at no cost to the owner, if any part of the job is installed improperly by the contractor.

- Guarantee: the contractor guarantees the work for a certain length of time, usually one year.

- Compliance with All Laws and Regulations: specifies that the contractor will perform all work in accordance with all required laws, ordinances, and codes such as the *National Electrical Code®* and city codes.

- Others: these sections are added as necessary by the owner, architect, and engineer when the complexity of the job and other circumstances require them. None of the items listed in the General Conditions has precedence over another item in terms of its effect on the contractor or the electrician on the job. The electrician must study each of the items before taking a position and assuming responsibilities with respect to the job.

SUPPLEMENTARY GENERAL CONDITIONS

The second main section of the specifications is titled *Supplementary General Conditions*. These conditions usually are more specific than the General Conditions. While the General Conditions can be applied to any job or project in almost any location with little change, the *Supplementary General Conditions* are rewritten for each project. The following list covers the items normally specified by the *Supplementary General Conditions*.

- The contractor must instruct all crews to exercise caution while digging; any utilities damaged during the digging must be replaced by the contractor responsible.

- The contractor must verify the existing conditions and measurements.

- The contractor must employ qualified individuals to lay out the work site accurately. A registered land surveyor or engineer may be part of the crew responsible for the layout work.

- Job offices are to be maintained as specified on the site by the contractor; this office space may include space for owner representatives.

- The contractor may be required to provide telephones at the project site for use by the architect, engineer, subcontractor, or owner.

- Temporary toilet facilities and water are to be provided by the contractor for the construction personnel.

- Temporary light and power. The contractor must supply an electrical service of a specified size to provide temporary light and power at the site.

- It may be necessary for the contractor to supply a specified type of temporary heating to keep the temperature at the level specified for the structure.

- According to the terms of the guarantee, the contractor agrees to replace faulty equipment and correct construction errors for a period of one year.

The above listing is by no means a complete catalog of all of the items that can be included in the section on *Supplementary General Conditions*.

Other names may be applied to the *Supplementary General Conditions* section; these names include *Special Conditions* and *Special Requirements*. Regardless of the name used, these sections contain the same types of information. All sections of the specifications must be read and studied by all of the construction trades involved. In other words, the electrician must study the heating, plumbing, ventilating, air-conditioning, and general construction specifications to determine if there is any equipment furnished by the other trades, where the contract specifies that such equipment is to be installed and wired by the electrical contractor. The electrician must also study the general construction specifications since the roughing in of the electrical system will depend on the types of construction that will be encountered in the building.

This overview of the *General* and *Supplementary General Conditions* of a specification is intended to show the student that the construction worker on the job is affected by parts of the specification other than the part designated for each particular trade.

Contractor Specification

In addition to the sections of the specification that apply to all contractors, there are separate sections for each of the contractors, such as the general contractor who constructs the building proper, the plumbing contractor who installs the water and sewage systems, the heating and air-conditioning contractor, and the electrical contractor. The contract documents usually do not make one contractor responsible for work specified in another section of the specifications. However, it is always considered good practice for each contractor to be aware of being involved in each of the other contracts in the total job.

WORKING DRAWINGS

The construction plans for a building are often called *blueprints*. This term is a carry-over from the days when the plans were blue with white lines. Today, a majority of the plans used have black lines on white since this combination is considered easier to read. The terms plans and working drawings will be commonly used in this text.

A set of 10 plan sheets is included at the back of the text showing the general and electrical portions of the work specified.

- *Sheet A1 — Plot Plan, East Elevation, West Elevation, Index to Drawings:* The plot plan shows the location of the commercial building and gives needed elevations. The east elevation is the street view of the building and the west elevation is the back of the building. The index lists the content of all the plan sheets.

- *Sheet A2 — Architectural Floor Plan; Basement*

- *Sheet A3 — Architectural Floor Plan; First Floor*

- *Sheet A4 — Architectural Floor Plan; Second Floor*
  The architectural floor plans give the wall and partition details for the building. These sheets are dimensioned; the electrician can find exact locations by referring to these sheets. The electrician should also check the plans for the materials used in the general construction as these will affect when and how the system will be installed.

- *Sheet A5 — Elevations; North and South:* The electrician must study the elevation dimensions, which are given in feet and hundredths of a foot above sea level. For example, the finished second floor, which is shown at 218.33', is 218 feet 4 inches above sea level.

- *Sheet A6 — Sections; Longitudinal, Transverse:* This sheet gives detail drawings of the more important sections of the building. The location of the section is indicated on the floor plans. When looking at a section, imagine that you are looking in the direction of the arrows at a building that is cut in two at the place indicated. You should see the section exactly as you view the imaginary building from this point.

- *Sheet E1 Electrical Working Drawing; Basement, Owner — Emergency Panelboard Directory, Telephone Riser Diagram*

- *Sheet E2 — Electrical Working Drawing; First Floor, Drug Store Panelboard Directory, Bakery Panelboard Directory*

- *Sheet E3 — Electrical Working Drawing; Second Floor, Insurance Office Panelboard Directory, Beauty Salon Panelboard Directory, Doctor's Office Panelboard Directory*
  These sheets show the detailed electrical work on an outline of the building. Since dimensions usually are not shown on the electrical plans, the electrician must consult the other sheets for this information. It is recommended that the electrician refer frequently to the other plan sheets to insure that the electrical installation does not conflict with the work of the other construction trades.

- *Sheet E4 — Luminaire — Lamp Schedule, Electrical Symbol Schedule, Electrical Power Distribution Diagram, Detail of Typical Roof-type Cooling System Unit, Panelboard Summary*

To assist the electrician in recognizing components used by other construction trades, the following illustrations are included: Figure 1-1A and Figure 1-1B, Architectural Drafting Symbols; Figure 1-2, Standard Symbols for Plumbing, Piping, and Valves; Figure 1-3, Sheet Metal Ductwork Symbols; and Figure 1-4, Generic Symbols for Electrical Plans. However, the electrician should be aware that variations of these symbols may be used and the specification and/or plans for a specific project must always be consulted.

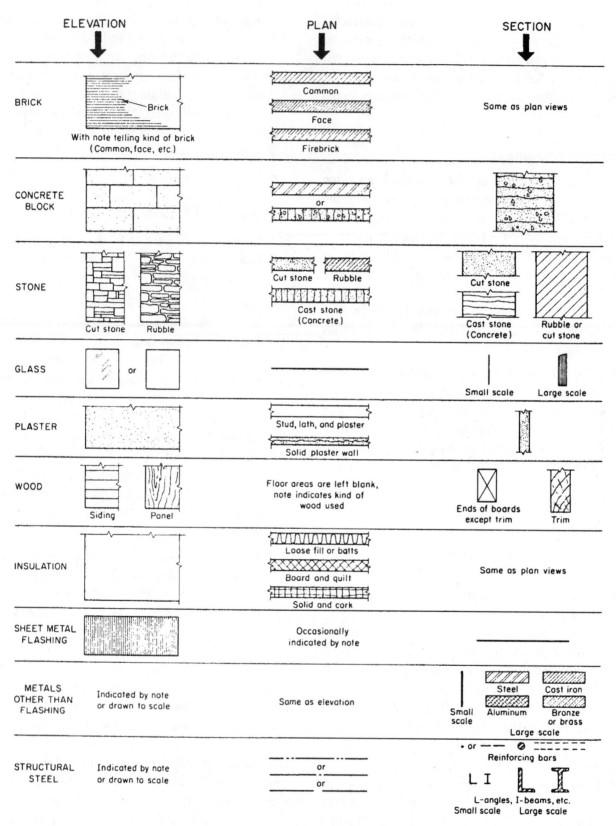

Fig. 1-1A   Architectural drafting symbols.

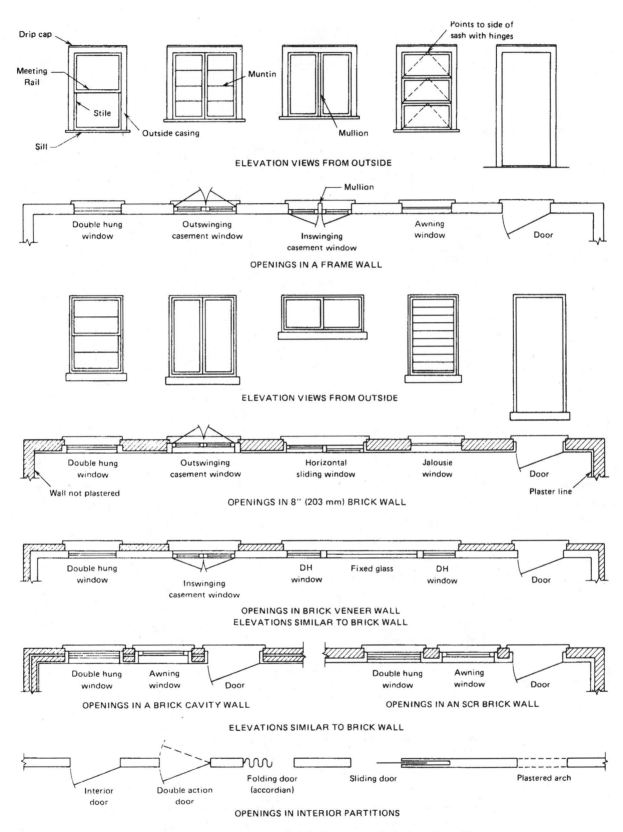

Fig. 1-1B   Architectural drafting symbols (continued).

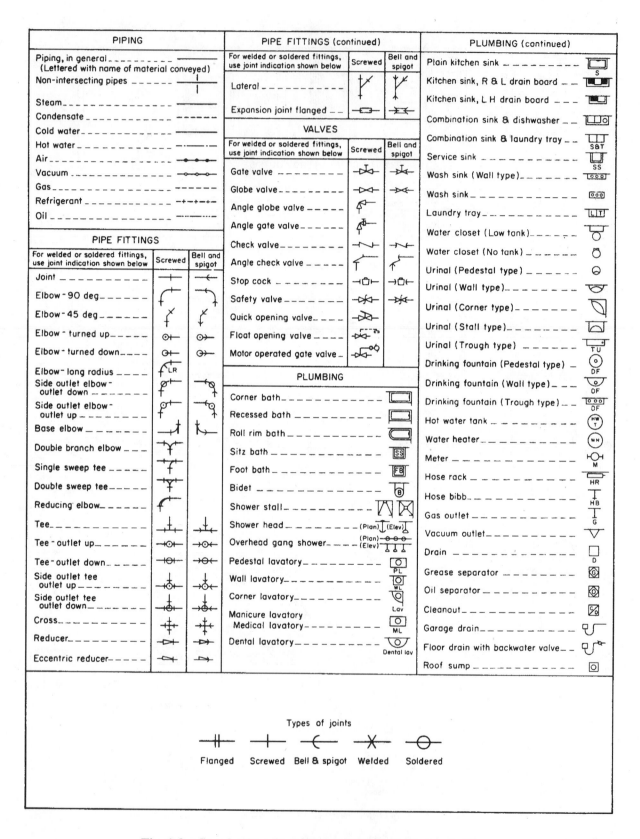

Fig. 1-2   Standard symbols for plumbing, piping, and valves.

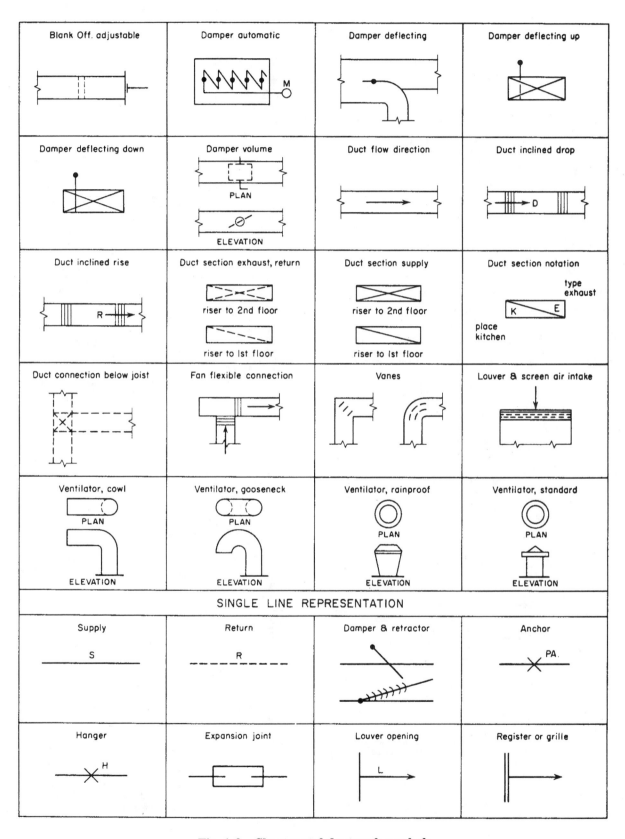

Fig. 1-3   Sheet metal ductwork symbols.

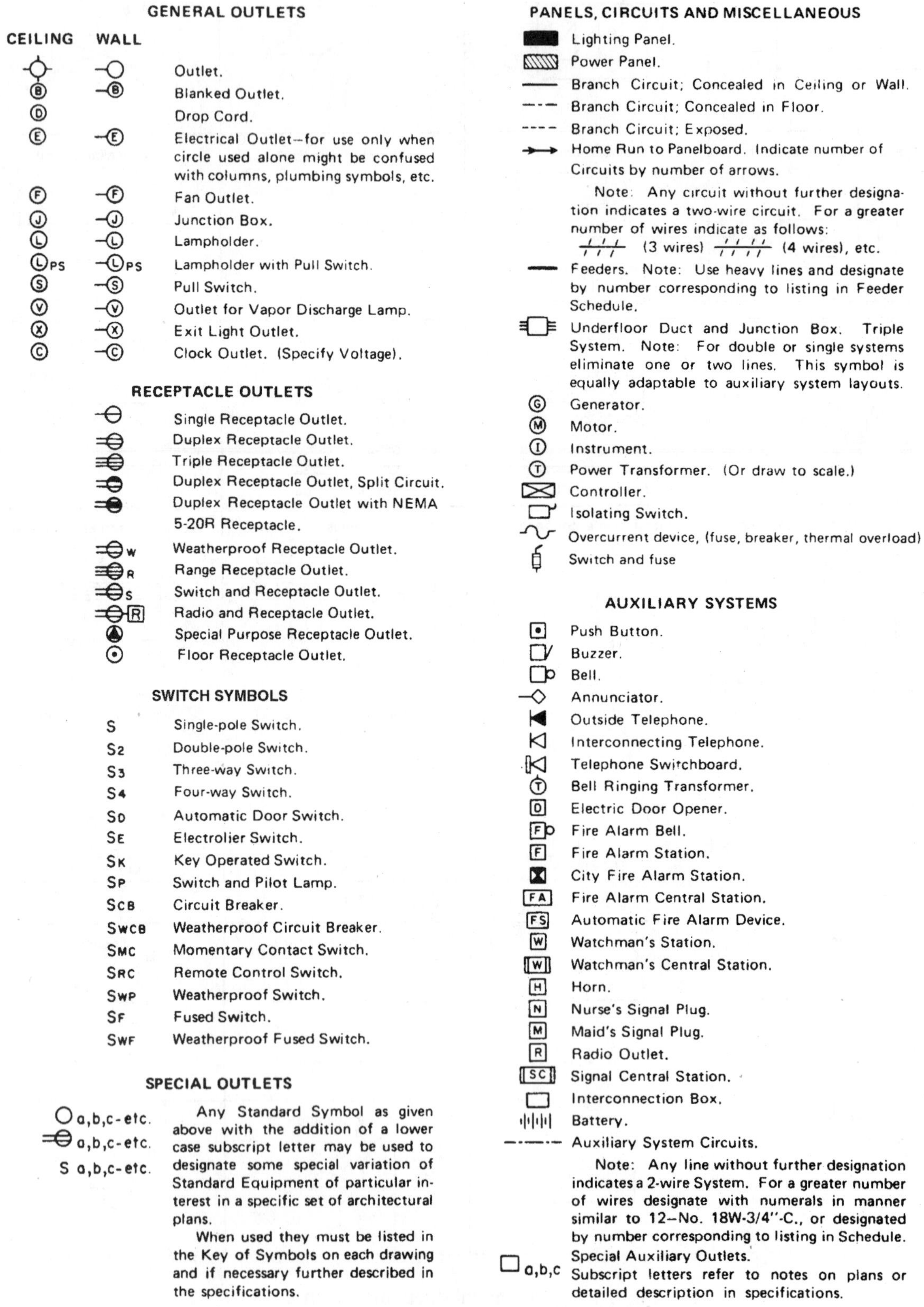

GENERAL OUTLETS

CEILING   WALL

Outlet.

Blanked Outlet.

Drop Cord.

Electrical Outlet—for use only when circle used alone might be confused with columns, plumbing symbols, etc.

Fan Outlet.

Junction Box.

Lampholder.

Lampholder with Pull Switch.

Pull Switch.

Outlet for Vapor Discharge Lamp.

Exit Light Outlet.

Clock Outlet. (Specify Voltage).

RECEPTACLE OUTLETS

Single Receptacle Outlet.

Duplex Receptacle Outlet.

Triple Receptacle Outlet.

Duplex Receptacle Outlet, Split Circuit.

Duplex Receptacle Outlet with NEMA 5-20R Receptacle.

Weatherproof Receptacle Outlet.

Range Receptacle Outlet.

Switch and Receptacle Outlet.

Radio and Receptacle Outlet.

Special Purpose Receptacle Outlet.

Floor Receptacle Outlet.

SWITCH SYMBOLS

S   Single-pole Switch.

S₂   Double-pole Switch.

S₃   Three-way Switch.

S₄   Four-way Switch.

S_D   Automatic Door Switch.

S_E   Electrolier Switch.

S_K   Key Operated Switch.

S_P   Switch and Pilot Lamp.

S_CB   Circuit Breaker.

S_WCB   Weatherproof Circuit Breaker.

S_MC   Momentary Contact Switch.

S_RC   Remote Control Switch.

S_WP   Weatherproof Switch.

S_F   Fused Switch.

S_WF   Weatherproof Fused Switch.

SPECIAL OUTLETS

O a,b,c - etc.

⊖ a,b,c - etc.

S a,b,c - etc.

Any Standard Symbol as given above with the addition of a lower case subscript letter may be used to designate some special variation of Standard Equipment of particular interest in a specific set of architectural plans.

When used they must be listed in the Key of Symbols on each drawing and if necessary further described in the specifications.

PANELS, CIRCUITS AND MISCELLANEOUS

Lighting Panel.

Power Panel.

Branch Circuit; Concealed in Ceiling or Wall.

Branch Circuit; Concealed in Floor.

Branch Circuit; Exposed.

Home Run to Panelboard. Indicate number of Circuits by number of arrows.

Note: Any circuit without further designation indicates a two-wire circuit. For a greater number of wires indicate as follows: ─/// ─ (3 wires) ─//// ─ (4 wires), etc.

Feeders. Note: Use heavy lines and designate by number corresponding to listing in Feeder Schedule.

Underfloor Duct and Junction Box. Triple System. Note: For double or single systems eliminate one or two lines. This symbol is equally adaptable to auxiliary system layouts.

G   Generator.

M   Motor.

I   Instrument.

T   Power Transformer. (Or draw to scale.)

Controller.

Isolating Switch.

Overcurrent device, (fuse, breaker, thermal overload)

Switch and fuse

AUXILIARY SYSTEMS

Push Button.

Buzzer.

Bell.

Annunciator.

Outside Telephone.

Interconnecting Telephone.

Telephone Switchboard.

T   Bell Ringing Transformer.

D   Electric Door Opener.

F   Fire Alarm Bell.

F   Fire Alarm Station.

City Fire Alarm Station.

FA   Fire Alarm Central Station.

FS   Automatic Fire Alarm Device.

W   Watchman's Station.

W   Watchman's Central Station.

H   Horn.

N   Nurse's Signal Plug.

M   Maid's Signal Plug.

R   Radio Outlet.

SC   Signal Central Station.

Interconnection Box.

Battery.

Auxiliary System Circuits.

Note: Any line without further designation indicates a 2-wire System. For a greater number of wires designate with numerals in manner similar to 12—No. 18W-3/4''-C., or designated by number corresponding to listing in Schedule. Special Auxiliary Outlets.

□ a,b,c   Subscript letters refer to notes on plans or detailed description in specifications.

Fig. 1-4   Generic symbols for electrical plans.

CODES AND ORGANIZATIONS

Local Codes

Many organizations such as cities and power companies develop electrical codes that they enforce within their areas of influence. These codes generally are concerned with the design and installation of electrical systems. In all cases, the latest *National Electrical Code®* is used as the basis for the local code. It is always advisable to consult these organizations before work is started on any project. The local codes may contain special requirements that apply to specific and particular installations. Additionally, the contractor may be required to obtain special permits and/or licenses before construction work can begin.

National Fire Protection Association

Organized in 1896, the National Fire Protection Association (NFPA) is an international, non-profit organization dedicated to the twin goals of promoting the science of fire protection and improving fire protection methods. The NFPA publishes an 11-volume series covering the national fire codes. The *National Electrical Code® (NEC)®** is a part of volume 3 of this series. The purpose and scope of this code are set forth in *NEC®* Article 90.

Although the NFPA is an advisory organization, the recommended practices contained in its published codes are widely used as a basis for local codes. Additional information concerning the publications of the NFPA and membership in the organization can be obtained by writing to:

National Fire Protection Association
Batterymarch Park
Quincy, Massachusetts 02269

National Electrical Code®

The original *National Electrical Code®* was developed in 1897. Sponsorship of the Code was assumed by the NFPA in 1911.

The *National Electrical Code®* generally is the "bible" for the electrician. However, the *NEC®* does not have a legal status until the appropriate authori-

ties adopt it as a legal standard. In May 1971, the Department of Labor, through the Occupational Safety and Health Administration (OSHA), adopted the *NEC®* as a national consensus standard. Therefore, in the areas where OSHA is enforced, the *NEC®* is the law.

Throughout this text, references are made to articles and sections of the *National Electrical Code®.* It is suggested that the student, and any other person interested in electrical construction, obtain and use a copy of the latest edition of the *NEC®.* To help the user of this text, relevant Code sections are paraphrased where appropriate. However, the *NEC®* must be consulted before any decision related to electrical installation is made.

The *NEC®* is revised and updated every three years.

Code Terms. The following terms are used throughout the Code. It is important to understand the meanings of these terms.

APPROVED: Acceptable to the authority having jurisdiction.

AUTHORITY HAVING JURISDICTION: An organization, office, or individual responsible for "approving" equipment, an installation, or a procedure.

IDENTIFIED: (As applied to equipment.) Recognizable as suitable for a specific purpose, function, use, environment, or application, where described in a particular Code requirement. For example, "Identified for use in a wet location."

LABELED: Equipment or materials to which has been attached a label, symbol, or other identifying mark of an organization acceptable to the authority having jurisdiction and concerned with product evaluation, that maintains periodic inspection of production of labeled equipment or materials and by whose labeling the manufacturer indicates compliance with appropriate standards or performance in a specified manner.

LISTED: Equipment, materials, or services included in a list published by an organization that is acceptable to the authority having jurisdiction and concerned with evaluation of products or services, that maintains periodic inspection of production of listed equipment or materials or periodic inspection of services, and whose listing states either that the

---

*National Electrical Code® and *NEC®* are Registered Trademarks of the National Fire Protection Association, Inc., Quincy, MA.

equipment, material or services meets identified standards or has been tested and found suitable for use in a specified manner.

SHALL: Indicates a mandatory requirement.

FPN: *Fine Print Notes* are found throughout the *National Electrical Code*. *FPNs* are explanatory in nature, in that they make reference to other sections of the Code. *FPNs* also define things where further description is necessary. *FPNs*, by themselves, are NOT Code rules.

Copies of the *NEC®* are available from the National Fire Protection Association, the International Association of Electrical Inspectors, Delmar Publishers Inc., and from many bookstores.

Citing Code References

Every time that an electrician makes a decision concerning the electrical wiring, the decision should be checked by reference to the Code. Usually this is done from memory without actually using the Code book. If there is any doubt in the electrician's mind, then the Code should be referenced directly — just to make sure. When the Code is referenced it is a good idea to record the location of the information in the code book — this is referred to as citing the code reference. There is a very exact way that the location of a code item is to be cited. The various levels of Code referencing are shown in Table 1-1. Starting at the top of the table, each step becomes a more specific reference. If a person references Chapter 1, this reference includes all the information and requirements that are set forth in several pages. When citing a Section or an Exception, this is being

very specific and may be including only a few words in the citation. The electrician wants to be as specific as possible when citing the Code.

Underwriters Laboratories, Inc.

Founded in 1894, Underwriters Laboratories (UL) is a nonprofit organization that operates laboratories to investigate materials, devices, products, equipment, and construction methods and systems to define any hazards that may affect life and property. The organization provides a listing service to manufacturers, Figure 1-5. Any product authorized to carry an Underwriters' listing has been evaluated with respect to all reasonable foreseeable hazards to life and property, and it has been determined that the product provides safeguards to these hazards to an acceptable degree. A listing by the Underwriters Laboratories does not mean that a product is *approved* by the *National Electrical Code*. However, many local agencies do make this distinction.

The *National Electrical Code®* defines *approved* as ". . . acceptable to the authority having jurisdiction . . ." *NEC® Section 110-3(b)* stipulates that, *"Listed or labeled equipment shall be installed and used in accordance with any instructions included in the listing or labeling."*

TABLE 1-1 Citing the *NEC®*		
Division	**Designation**	**Example**
Chapter	1 through 9	*Chapter 1*
Article	90 through 820	*Article 310*
Part	Uppercase letter	*Article 230, Part H*
Section	Article number, a hyphen, plus one, two or three digits	*Section 250-50*
Paragraph	Section designation, plus lower case letter in ( ) followed by digit in ( ) if required	*Section 250-61(a)(1)*
Exception to	Precedes or follows a Section designation	*Exception No. 3a to Section 250-61(b) or Section 250-61(b) Exception No. 3a*

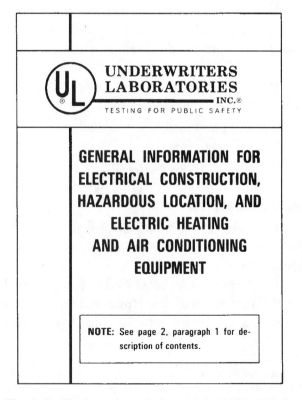

Fig. 1-5   Underwriters Laboratories *White Book*.

Useful Underwriters Laboratories (UL) publications are:

- *Electrical Construction Materials Directory* (Green Book)
- *Electrical Appliances and Utilization Equipment Directory* (Orange Book)
- *Hazardous Location Equipment Directory* (Red Book)
- *General Information for Electrical Construction, Hazardous Location, and Electrical Heating and Air-Conditioning Equipment* (White Book)
- *Recognized Component Directories* (Yellow Books)

Many inspection authorities continually refer to these books in addition to the *National Electrical Code.* If the answer to a question cannot be found readily in the *National Electrical Code,* then it generally can be found in the UL publications listed.

An index of publications and information concerning the Underwriters Laboratories can be obtained by writing to:

Public Information
Underwriters Laboratories, Inc.
333 Pfingsten Road
Northbrook, IL 60062

Recent "harmonizing" of standards has resulted in the fact that Underwriters Laboratories can now evaluate equipment for a manufacturer under both a UL Standard as well as a Canadian Standard. This can save a manufacturer time and money in that the manufacturer can have the testing and evaluation done in one location, acceptable to both US and Canadian authorities enforcing the Code.

Nationally Recognized Testing Agencies

When legal issues arise regarding electrical equipment, the courts must rely on the testing, evaluation, and listing of products by qualified testing agencies. The Occupational Safety and Health Act (OSHA) legislation lists a number of agencies that they consider to be nationally recognized testing agencies. These agencies are referred to in the industry as a "NRTL."

Intertek Testing Services (ITS)

Formerly known as Electrical Testing Laboratories, ITS is a nationally recognized testing laboratory. They provide testing, evaluation, labeling, listing, and follow-up service for the safety testing of electrical products. This is done in conformance to nationally recognized safety standards, or to specifically designated requirements of jurisdictional authorities. Information can be obtained by writing to:

Intetek Testing Services
3933 U.S. Route 11
Cortland, NY 13045
Phone: 1-800-245-3851

National Electrical Manufacturers Association

The National Electrical Manufacturers Association (NEMA) is a nonprofit organization supported by the manufacturers of electrical equipment and supplies. NEMA develops standards that are designed to assist the purchaser in selecting and obtaining the correct product for specific applications, Figure 1-6. Information concerning NEMA standards may be obtained by writing to:

National Electrical Manufacturers Association
11300 North 17th Street
Suite 1847
Rosslyn, Virginia 22209
Phone: 703-841-3200

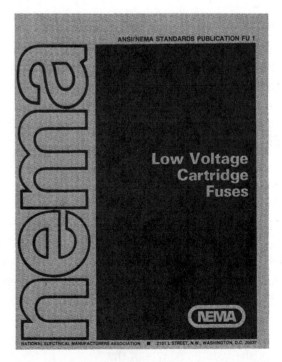

Fig. 1-6   National Electrical Manufacturers Association. (This material is reproduced by permission of the National Electrical Manufacturers Association from NEMA Standards Publication No. FU1, *Low Voltage Cartridge Fuses,* **copyright by NEMA.)**

American National Standards Institute, Inc.

The American National Standards Institute, Inc. (ANSI) is located at 11 West 42nd Street, New York, New York 10036. Phone: 212-642-4900. Fax: 212-398-0023. Various working groups in the organization study the numerous codes and standards. An American National Standard implies "a consensus of those concerned with its scope and provisions." The *National Electrical Code®* is approved by ANSI and is numbered ANSI/NFPA 70-1999.

Canadian Standards Association

The Canadian Electrical Code is significantly different than the *National Electrical Code®.* Those using this text in Canada must follow the Canadian Electrical Code. The Code, which was in effect at the time of preparing this text, is known as the "Safety Standards for Electrical Installations," Standard C22.1.

The Canadian Code is a voluntary code suitable for adoption and enforcement by electrical inspection authorities. The Canadian Electrical Code is published by and available from:

Canadian Standards Association
178 Rexdale Blvd.
Rexdale, Ontario, Canada
M9W 1R3

Enforcement of the Canadian Electrical Code is generally under the jurisdiction of the utility company providing electrical power in the particular province. In Ontario, for instance, electrical inspection is done by field inspectors employed by Ontario Hydro.

International Association of Electrical Inspectors

The International Association of Electrical Inspectors (IAEI) is a nonprofit organization. The IAEI membership consists of electrical inspectors, electricians, contractors, and manufacturers throughout the United States and Canada. One goal of the IAEI is to improve the understanding of the *NEC®.* Representatives of this organization serve as members of the various panels of the *National Electrical Code®* Committee and share equally with other members in the task of reviewing and revising the *NEC®.* The IAEI publishes a bimonthly magazine, the *IAEI*

News. Additional information concerning the organization may be obtained by writing to:

International Association of Electrical
  Inspectors
901 Waterfall Way
Suite 602
Richardson, TX 75080-7702

Illuminating Engineering Society of North America

The Illuminating Engineering Society of North America (IESNA) was formed more than 65 years ago. The objective of this group is to communicate information about all facets of good lighting practice to its members and to consumers. The IESNA produces numerous publications that are concerned with illumination.

The *IESNA Lighting Handbooks* are regarded as the standard for the illumination industry and contain essential information about light, lighting, and luminaires. Information about publications or membership may be obtained by writing to:

Illuminating Engineering Society
  of North America
120 Wall St., 17th Floor
New York, NY 10005-4001
Phone: 212-248-5000

National Electrical Installation Standards

The National Electrical Installation Standards (NEIS) in cooperation with the National Electrical Contractors Association (NECA) and The Illuminating Engineering Society of North America (IESNA) has developed **ANSI/NECA/IESNA 500-1998, Recommended Practice for Installing Indoor Commercial Lighting**. The document is available from either NECA or the IESNA.

Registered Professional Engineer (PE)

Although the requirements may vary slightly from state to state, the general statement can be made that a registered professional engineer has demonstrated his or her competence by graduating from college and passing a difficult licensing examination. Following the successful completion of the examination, the engineer is authorized to practice engineering under the laws of the state. A requirement is usually made that a registered professional engineer must supervise the design of any building that is to be used by the public. The engineer must indicate approval of the design by affixing a seal to the plans.

Information concerning the procedure for becoming a registered professional engineer and a definition of the duties of the professional engineer can be obtained by writing the state government department that supervises licensing and registration.

NEC® USE OF SI (METRIC) MEASUREMENTS

The 1999 *National Electrical Code*® includes both English and metric measurements. The metric system is known as the *International System of Units* (SI). See *NEC*® *Section 90-9.*

Metric (SI) measurements appear in the Code as follows:

- In the Code paragraphs, the approximate metric (SI) measurement appears in parentheses following the English measurement.

- In the Code tables, a footnote shows the SI conversion factors.

A metric (SI) measurement is not shown for conduit size, box size, wire size, horsepower designation for motors, and other "trade sizes" that do not reflect actual measurements.

Guide to Metric (SI) Usage

This system is international and many compromises were made to accommodate regional practice. In the United States system of measurement, it is the practice to use the period as the decimal marker and the comma to separate a string of numbers into groups of three for easier reading. In many countries, the comma is used as the decimal marker and spaces are left to separate the string of numbers into groups of three. The SI system, taking something from both, uses the period as the decimal marker and the space to separate a string of numbers into groups of three starting from the decimal point and counting in either direction (e.g., 12 345.789 99). The only

TABLE 1-2	Metric (SI) Prefixes and Their Values.	
mega	1 000 000	(one million)
kilo	1 000	(one thousand)
hecto	100	(one hundred)
deka	10	(ten)
the unit	1	(one)
deci	0.1	(one-tenth) (1/10)
centi	0.01	(one-hundredth) (1/100)
milli	0.001	(one-thousandth) (1/1 000)
micro	0.000 001	(one-millionth) (1/1 000 000)
nano	0.000 000 001	(one-billionth) (1/1 000 000 000)

exception to this is when there are four numbers, on either side of the decimal point, in the string. In this case, the third and fourth numbers from the decimal point are not separated (e.g., 2015.1415). In this text the English method of separating digits, a comma, will be used with English units and the SI method, a space, with SI units.

In the metric (SI) system, the units increase or decrease in multiples of 10, 100, 1000, and so on. For instance, one megawatt (1 000 000 watts) is 1000 times greater than one kilowatt (1000 watts).

By assigning a name to a measurement, such as a *watt*, the name becomes the unit. Adding a prefix to the unit, such as *kilo-*, forms the new name *kilowatt*, meaning 1000 watts. Refer to Table 1-2 for prefixes used in the metric (SI) system.

Certain of the prefixes shown in Table 1-2 have a preference in usage. These prefixes are *mega-*, *kilo-*, the unit itself, *centi-*, *milli-*, *micro-*, and *nano-*. Consider that the basic unit is a meter (one). Therefore, a kilometer is 1000 meters, a centimeter is 0.01 meter, and a millimeter is 0.001 meter.

The advantage of the metric (SI) system is that recognizing the meaning of the proper prefix lessens the possibility of confusion. For example, a four-foot fluorescent lamp is approximately 1200 millimeters, or 1.2 meters in length.

Some common measurements of length in the English system are shown with their metric (SI) equivalents in Table 1-3.

Electricians will find it useful to refer to the conversion factors and their abbreviations shown in Table 1-4.

Following the Specifications in the Appendix of this text is a comprehensive metric-to-English conversion multiplier table.

TABLE 1-3 Measurements of Length and Their Metric (SI) Equivalents.			
one inch	=	2.54	centimeters
	=	25.4	millimeters
	=	0.025 4	meter
one foot	=	12	inches
	=	0.304 8	meter
	=	30.48	centimeters
	=	304.8	millimeters
one yard	=	3	feet
	=	36	inches
	=	0.914 4	meter
	=	914.4	millimeters
one meter	=	100	centimeters
	=	1 000	millimeters
	=	1.093	yards
	=	3.281	feet
	=	39.370	inches

TABLE 1-4   Useful Conversions (English/SI-SI/English) and Their Abbreviations.
inches (in) X 0.025 4 = meter (m)
inches (in) X 0.254 = decimeters (dm)
inches (in) X 2.54 = centimeters (cm)
centimeters (cm) X 0.393 7 = inches (in)
millimeters (mm) = inches (in) X 25.4
millimeters (mm) X 0.039 37 = inches (in)
feet (ft) X 0.304 8 = meters (m)
meters (m) X 3.280 8 = feet (ft)
square inches (in²) X 6.452 = square centimeters (cm²)
square centimeters (cm²) X 0.155 = square inches (in²)
square feet (ft²) X 0.093 = square meter (m²)
square meters (m²) X 10.764 = square feet (ft²)
square yards (yd²) X 0.836 1 = square meters (m²)
square meters (m²) X 1.196 = square yards (yd²)
kilometers (km) X 1 000 = meters (m)
kilometers (km) X 0.621 = miles (mi)
miles (mi) X 1.609 = kilometers (km)

REVIEW QUESTIONS

Refer to the *National Electrical Code®* or the working drawings when necessary. Where applicable, responses should be written in complete sentences. Write units using unit names, do not use abbreviations or symbols. (1 foot, not 1' or 1 ft).

1. What section of the specification contains a list of contract documents?

   _____

2. The requirement for temporary light and power at the job site will be found in what portion of the specification?

   _____

3. The electrician uses the Schedule of Working Drawings for what purpose?

   _____

Complete the following items by indicating the letter(s) designating the correct source(s) of information for:

4. _____ Ceiling height   A. Architectural floor plan

5. _____ Electrical receptacle style   B. Details

6. _____ Electrical outlet location   C. Electrical layout drawings

7. _____ Exterior wall finishes   D. Electrical symbol schedule

8. _____ Grading elevations   E. Elevations

9. _____ Panelboard Schedules   F. Sections

10. _____ Room width   G. Site plan

11. _____ Swing of door   H. Specification

12. _____ View of interior wall

Match the items on the left with those on the right by writing the letter designation of the appropriate organization from the list on the right.

13. _____ Electrical code   A. IAEI

14. _____ Electrical inspectors   B. IESNA

15. _____ Fire codes   C. *NEC®*

16. _____ Lighting information   D. NEMA

17. _____ Listing service   E. NFPA

18. _____ Manufacturers' standards   F. PE

19. _____ Seal   G. UL

Match the items on the left with those on the right by writing the letter designation of the proper level of *NEC®* interpretation from the list on the right.

20. _____ Allowed by the Code   A. never

21. _____ May be done   B. shall

22. _____ Must be done   C. with special permission

23. _____ Required by the Code

24. _____ Up to the electrician

25. Find *NEC® Section 250-52(c)(1)* and record the first three words.

26. Luminaire style F is four feet long. The length in SI units is

27. The gross area of the drugstore basement is 1395 square feet. The area in SI units is

Determine the following dimensions. Write the dimensions using unit names, not symbols (e.g., 1 foot not 1') and indicate where the information was found.

28. What is the inside clear distance of the interior stairway to the drugstore basement?

29. What is the gross square footage of each floor of the building?

30. What is the distance in the drugstore from the exterior block wall to the party block wall separating the drugstore from the bakery?

31. What is the finished floor to finished ceiling height on the second floor?

Properly cite the *NEC*® locations for the following:

32. The standard ampere rating for fused and fixed trip circuit breakers. _____

_____

33. The minimum bending radius for metal clad cable with a smooth sheath with an external diameter of 1 inch. _____

_____

34. The permission to use splices in busbars as grounding electrodes. _____

_____

Perform the following.

35. In examining the layout of the boiler room you become concerned about the working clearance between the engine generator and the control panel. Write a memo to your superior supporting your concern. Cite the applicable section(s) of the *NEC*.® (Propose a solution, if you wish.) _____

_____

_____

36. Write a letter to one of the organizations listed in this chapter requesting information about the organization and the services they provide.

UNIT 2

Reading Electrical Working Drawings – Entry Level

OBJECTIVES

After studying this unit, the student will be able to

- read and interpret electrical symbols used in construction drawings.
- identify the electrical installation requirements for the drugstore.

Electrical and architectural working drawings are the maps that the electrician must read and understand. Having this skill is essential to being able to install a complete electrical system and to coordinating related activities with those of workers in the other crafts. This unit will provide a first step in developing the ability to read symbols appearing on the drawings as they apply to the electrical work.

The units in this text that address electrical working drawings will apply information presented in preceding units, and will introduce special features of the building area being discussed. The questions at the end of each of these units will require that the student use the specifications, the drawings, and the 1999 *National Electrical Code.*

The user of this text is encouraged to peruse not only the electrical working drawings but the architectural working drawings, the loading schedules, the panelboard worksheets, and the panelboard directories. (The first two items are in the Appendix, the third on the working drawings.) Most of the information given on and in these items is yet to be discussed in detail, but much is self evident. Early familiarization will enhance future learning. See Figure 2-1.

ELECTRICAL SYMBOLS

On Sheet E-4 of the Commercial Building working drawings is an "Electrical Symbol Schedule" that lists all of the electrical symbols that are used in this set of drawings. Knowing the special characteristics of these symbols will improve your ability to remember them and to interpret other symbols that are not used in these drawings.

Surface Raceway

 This is, as the name implies, a raceway that is installed on a surface. The requirements for construction and installation are given in *NEC® Article 352.* The symbol may be used to indicate a variety of raceway types and the electrician must always check on the specific installation requirements. This raceway use is common in remodeling applications and in situations where electrical power, communications, and electronic signals need to be available at a series of locations. The installation of surface raceways is discussed in unit 9.

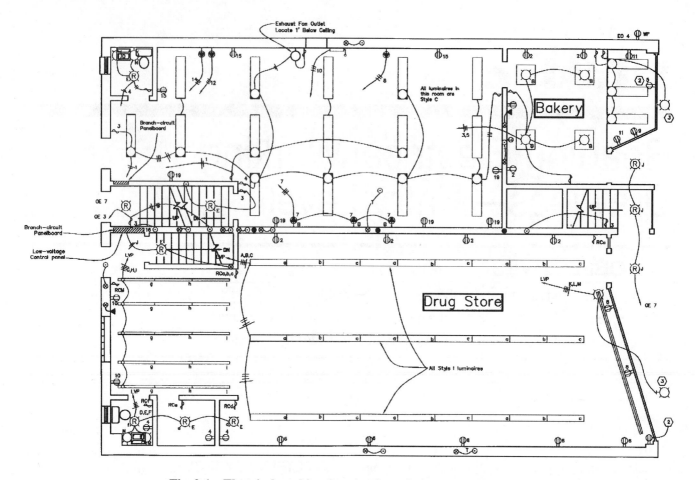

Fig. 2-1   Electrical working drawing for a drugstore and bakery.

Panelboard

Panelboards are distribution points for electrical circuits. They contain circuit protective devices. See *NEC® Article 100* for the official definition. There are two basic classes, power and lighting and appliance. A power panelboard is often installed as a distribution point serving lighting and appliance panelboards and other loads. The requirements for construction and application of panelboards are set forth in *NEC® Article 384*. In this building a panelboard is located in each occupancy so the tenant will have ready access to the overcurrent devices. The symbol is not drawn to scale and the installer must consult the shop drawings for specific dimensions.

Lighting Outlets

The next three symbols are for different applications of the same basic symbol, which is a circle with four ticks, at right angles, on the circumference as is shown for the ceiling outlet. In this case the luminaire would be surface mounted or stem hung with the outlet box installed so that the opening is flush with the ceiling surface. A box may not be recessed by more than one-quarter of an inch from the surface to be considered as being flush.

The second symbol has a stem protruding which, when drawn on the plans, will attach to a wall. This stem indicates that the outlet box is to be installed in the wall. The box must be installed so its opening is flush with the wall surface.

This third lighting outlet symbol is used when the luminaire is recessed. The outlet box will be attached to the luminaire when it is installed, often with a short section of flexible raceway. Usually a rough-in kit will be available as a part of the luminaire package and will be set in place before the ceiling material is installed.

Receptacle Outlets

The next nine symbols are for receptacle outlets. The first four are for duplex receptacle outlets, indicated by the two parallel lines drawn within the circle.

 For the first of these symbols the receptacle symbol is drawn within a box which indicates an outlet box is to be installed in the floor and a duplex receptacle is installed.

In the second of this set of four, the two parallel lines are extended beyond the box as stems. This indicates the outlet box is to be installed flush mounted in the wall. The receptacle that should be installed is a NEMA type 5, 15 ampere.

 The next symbol is identical to the previous one, except the space between the lines is darkened to indicate that a NEMA type 5, 20-ampere receptacle should be installed.

 This fourth symbol is the same as the previous one except that it is a very special hospital grade receptacle. It has transient voltage suppression and a grounding connection that is isolated from the raceway and the metal box should be installed.

The next five symbols are the same except for the uppercase letter. The stem indicates that the outlet box should be installed flush in the wall. The uppercase letter is used to indicate the type of receptacle that should be installed. This symbol is used when there is a limited number of that type to be installed.

Switches

The final three symbols in the left side of the schedule are for switches. The actual switch symbol is the S and the line drawn through the S and extended indicates that the switch box should be installed in

 the wall. For the first symbol a single pole switch should be installed.

 This symbol indicates a three-way switch.

 The last symbol on this side of the schedule indicates a four-way switch is to be installed.

Other Symbols

 A small number located adjacent to the symbol for an outlet, not a switch, indicates the branch-circuit number. This receptacle is connected to branch circuit number 2.

 A large rectangular or square box represents a luminaire. These are drawn to scale and may be individual or several in a row. A lowercase letter within the symbol indicates which switch is used to operate the luminaire(s). An uppercase letter in the symbol indicates the style of luminaire which is to be installed. The style designation is defined in the luminaire schedule. This luminaire is a style "A" and is operated by a switch marked with an "a."

 This symbol indicates the installation of a manual disconnecting means that may, or may not, be equipped with an overcurrent device. The construction specifications would need to be consulted to determine the exact requirements of this device.

 This symbol indicates that both a motor controller and a disconnect switch are to be installed. The symbol for a disconnect switch is not usually drawn to scale.

 When concealed raceways are drawn on the plans they are usually shown by curved or wavy lines. A straight line would indicate a surface-mounted raceway which is not usually specified for use in new construction. Hash marks drawn across these lines indicates the use of the installed conductors. A long line indicates a neutral (white) conductor, the short lines indicate phase "hot" conductors which may be any color but white or green. The bold dot indicates the presence of an insulated grounding (green) conductor. If switching conductors are being installed they are often designated by slanted lines.

 When an arrowhead is shown at the end of a branch-circuit symbol this indicates the raceway goes from this point to the panelboard but will no longer be drawn on the plans. This symbol is used to avoid the graphic congestion created if all the lines came into a single point on the plans. The small numbers indicate which branch circuits are to be installed in the raceway. As the overcurrent devices in a panelboard are usually numbered with odd numbers on the left and even on the right, it is common to see groups of odd or even numbers.

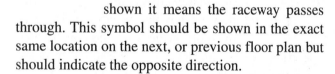 When raceways are installed vertically in the building from one floor to another the vertical direction may be represented by the arrow symbol inside a circle. A dot represents the head of the arrow and indicates that the raceway is headed upward. If a cross is shown this represents the tail of the arrow and the raceway is headed downwards. When both are shown it means the raceway passes through. This symbol should be shown in the exact same location on the next, or previous floor plan but should indicate the opposite direction.

 If the raceway is for use of the telephone system the line will be broken and an uppercase T inserted.

 The three parallel lines are used to indicate a lighting track which will accept special luminaires that the occupant can exchange or adjust.

 The last four symbols indicate different ways of labeling switches. The first indicates the switch is a part of a low-voltage wiring system.

 An RCM indicates that it is a master switch on a low-voltage control that will override the action of individual switches or perform the same action that several of the single switches would perform.

 The lowercase letter identifies this as switch "a." This letter will appear at all the luminaires that this switch controls.

 The final symbol indicates that this is a switch that can also be used to dim the connected lighting.

THE DRUGSTORE

A special feature of the drugstore wiring is the low-voltage remote-control system. See unit 20 for complete discussion. This system offers flexibility of control that is not available in the traditional control system. The switches used in this system operate on 24 volts, and the power wiring, at 120 volts, goes directly to the electrical load. This reduces branch-circuit length and voltage drop. A switching schedule gives details on the system operation, and a wiring diagram provides valuable information to the installer. One of the reasons for the low popularity of this system is the dearth of electricians who are prepared to install a low-voltage control system. The student is encouraged to request manufacturer's literature from any electrical distributor. Different types of illumination systems have been selected for many of the spaces in the building. The student should observe the differences in the wiring requirements.

In the merchandise area, nine luminaires are installed in a continuous row. It is necessary to install electrical power to only one point of a continuous row of luminaires. From this point the conductors are installed in the wiring channel of the luminaire. In the pharmacy area, a luminous ceiling is shown. This illumination system consists of rows of strip fluorescents and a ceiling that will transmit light. The installation of the ceiling, in many jurisdictions, is the work of the electrician. For this system to be efficient, the surfaces above the luminous ceiling must be highly reflective (white).

THE BAKERY

In the production area of the bakery a special luminaire is selected to prevent contamination of the bakery materials. These are totally enclosed individual units requiring a separate electrical connection to each luminaire. They maybe supplied by installing a conduit in the upper level slab or on the ceiling surface. In the sales area more attractive luminaires have been selected.

A conventional control system is to be installed for the lighting in the bakery. The goal of the system is to provide control at every entry point so a person is never required to walk through an unlighted space. Often this requires long switching circuits such as the three point control of the main lighting in the work area.

The electrician may be responsible from making changes in conductor size to compensate for excessive voltage drop. This requires the electrician to be alert for high loads on long circuits such as the control circuit on the bakery work area lighting.

REVIEW QUESTIONS

Refer to the *National Electrical Code®* or the working drawings when necessary. Where applicable, responses should be written in complete sentences. Write units using unit names, do not use abbreviations or symbols (1 foot, not 1' or 1 ft).

Answer questions 1–7 by identifying the symbol and the type of installation (wall, floor, or ceiling) for the boxes.

1. _____

2. _____

3. _____

4. _____

5. _____

6. _____

7. _____

Note: The drugstore basement does not include the service equipment area or the boiler room.

8. How many duplex receptacle outlets are to be installed in the drugstore basement?

_____

9. The duplex receptacle outlets in the drugstore basement are supplied from which panelboard? _____

10. The duplex receptacle outlets in the drugstore basement are to be connected to which branch circuit(s)? _____

11. How many lighting outlets are to be installed in the drugstore basement? _____

12. What style(s) of luminaires is/are to be installed in the drugstore basement?

_____

13. How may the luminaires for the drugstore basement be installed?

_____

14. Tabulate the luminaires required for installation in the drugstore. Give the style and the count for each style and the mounting method.

   Style _____   Count _____   Mounting method _____

   Style _____   Count _____   Mounting method _____

   Style _____   Count _____   Mounting method _____

   Style _____   Count _____   Mounting method _____

   Style _____   Count _____   Mounting method _____

   Style _____   Count _____   Mounting method _____

15. From the information you have studied about the various electrical symbols, identify the following symbols, some of which are not shown in the symbol schedule.

_____

_____

_____

_____

16. Where is the source of supply located for the disconnect switch installed in the drug-store basement? _____

UNIT 3

Computing the Electrical Load

OBJECTIVES

After studying this unit, the student will be able to

- determine the minimum lighting loading for a given area.
- determine the minimum receptacle loading for a given area.
- determine the minimum equipment loading.
- determine a reasonable connected load.
- tabulate the unbalanced load.

THE ELECTRICAL LOAD

To plan any electrical wiring project the first step is to determine the load that the electrical system is to serve. Only with this information can components for the branch circuit, the feeders, and the service be properly selected. The *NEC®* provides considerable guidance in determining the minimum loading that is appropriate for a given occupant. Often the electrician is asked to generate this information. This unit will provide a foundation for proper selection of electrical circuit components.

NEC® Article 220 establishes the procedure that is to be used to calculate electrical loads. Using the drugstore as an example case, the application of this procedure will be illustrated.

- The phrase "computed load" will be used to designate when the value is in compliance with the requirements of *NEC® Article 220*.

- The phrase "connected load" will be used to designate the value of the load as it actually exists.

The values discussed and calculated are shown in Table 3-1.

LIGHTING LOADING CALCULATIONS

The following guidelines should be followed when calculating the lighting load.

- Use *NEC® Table 220-3(a)* to find the value of volt-amperes per square foot for general lighting loads or use the actual volt-amperes if that value is higher.

- For show window lighting, allow 200 volt-amperes per linear foot or use the connected load if that value is higher. See *NEC® Section 220-12(a)*.

- Where lighting track is installed, 150 VA is to be allowed for every two feet of track or any fraction thereof. See *NEC® Section 220-12(b)*.

- At least one receptacle outlet shall be installed above each 12-foot (3.66 m) section of show window. See *NEC® Section 210-62*.

It not a requirement that the connected load be equal to or greater than these allowances. For example some energy codes may limit the connected lighting load to less than the values given in *NEC® Table 220-3(a)*. It is a requirement that the electrical system have sufficient capacity for these allowances.

General Lighting

The minimum lighting load to be included in the calculations for a given type of occupancy is determined from *NEC® Table 220-3(a)*. For a "store" occupancy, the Table indicates the unit load per square foot

Table 220-3(a). General Lighting Loads by Occupancies

Type of Occupancy	Unit Load per Square Foot (Volt-Amperes)
Armories and auditoriums	1
Banks	3½[b]
Barber shops and beauty parlors	3
Churches	1
Clubs	2
Court rooms	2
Dwelling units[a]	3
Garages — commercial (storage)	½
Hospitals	2
Hotels and motels, including apartment houses without provision for cooking by tenants[a]	2
Industrial commercial (loft) buildings	2
Lodge rooms	1½
Office buildings	3½[b]
Restaurants	2
Schools	3
Stores	3
Warehouses (storage)	¼
In any of the above occupancies except one-family dwellings and individual dwelling units of two-family and multifamily dwellings:	
Assembly halls and auditoriums	1
Halls, corridors, closets, stairways	½
Storage spaces	¼

Note: For SI units, 1 ft² = 0.093 m².

[a]See Section 220-3(b)(10).

[b]In addition, a unit load of 1 volt-ampere per square foot shall be included for general-purpose receptacle outlets where the actual number of general-purpose receptacle outlets is unknown.

Reprinted with permission from NFPA 70-1999.

is three volt-amperes. Therefore, for the first floor of the drug store the lighting load allowance is:

60 ft × 23.25 ft × 3 VA per sq ft = 4158 VA

Since this value is a minimum, it is also necessary to determine the connected load. The greater of the two values becomes the computed lighting load in accordance with the requirements of *NEC® Article 220*.

As presented in detail in a later unit, the illumination (lighting) in the sales area of the drugstore is provided by fluorescent luminaires (lighting fixtures) equipped with two lamps and a ballast. *This ballast can have a significant effect on the volt-ampere requirement for a luminaire.* According to the luminaire schedule shown on Sheet E4 of the working drawings, this lamp-ballast combination consumes 75 watts of power but has a load rating of 87 volt-amperes. (Watts can be used to determine the cost of operation; volt-amperes are used to deter-

mine the size of the conductors and overcurrent devices.)

The connected lighting load for the first floor of the drug store is:

27 Style I luminaires
@ 87 volt-ampere   = 2349 volt-ampere

4 Style E luminaires
@ 144 volt-ampere   = 576 volt-ampere

15 Style D luminaires
@ 74 volt-ampere   = 1110 volt-ampere

2 Style N luminaires
@ 60 volt-ampere   = 120 volt-ampere

Total Connected load   = 4155 volt-ampere

It is important that the connected load be tabulated as accurately as possible. See Table 3-1. The values will not only be used to select the proper electrical components but they may also be used to predict the cost of energy for operating the building.

Storage Area Lighting

According to *NEC® Table 220-3(a)*, the minimum load for basement storage space is 0.25 volt-ampere per square foot.

The load allowance for this space is:

1021 square feet × 0.25 VA per square foot
= 253 VA

The connected load for the storage space is:

9 Style L luminaires @ 87 VA = 783 VA

Show Window Lighting

The load allowance for a show window is given in *NEC® Section 220-12* as 200 volt-amperes per linear foot of the show window. The drugstore window is 16 feet long. The load allowance is:

16 ft × 200 VA per ft = 3200 VA

The allowance for lighting track is set forth in *NEC® Section 220-12*. That Section stipulates that 150 VA be allotted for each 2 feet, or fraction thereof of track. The track in the drugstore is 15 feet which is 8 units of track. The allowance is:

8 units @ 150 VA/unit = 1200 VA.

As the actual number of lighting fixtures supplied by the track is unknown and will vary during usage, the allowance is considered as the connected load. The total connected load is:

TABLE 3-1 Drugstore Loading Schedule							
	Count	**VA/Unit**	***NEC®***	**Actual**	**Computed**	**Balanced**	**Nonlinear**
General Lighting:							
NEC® Section 220-2	1395	3	4185				
Style I luminaire	27	87		2349			2349
Style E luminaire	4	144		576			576
Style D luminaire	15	74		1110			1110
Style N luminaire	2	60		120			
Totals:			4185	4155	4185		
Storage:							
NEC® Section 220-2	1012	0.25	253				
Style L luminaire	9	87		783			540
Totals:			253	783	783		
Show Window:							
NEC® Section 220-12	16	200	3200				
Receptacle outlets	3	500		1500			
Lighting Track	8	150		1200			
Totals:			3200	2700	3200		
Other Loads:							
Receptacle outlets	21	180		3780			
Roof receptacle	1	1500		1500			
Sign Outlet	1	1200		1200			
Totals:				6480	6480		
Motors & Appliances:	**Ampere**						
Air Conditioning:							
Compressor	20.2	360		7272		7272	
Condenser	3.2	208		666		666	
Evaporator	3.2	208		666		666	
Totals:				8604	8604		
TOTAL LOADS:					23252	8604	4575

3 receptacle outlets @ 500 VA per outlet = 1500 VA

8 units of lighting track @ 150 VA per unit
= 1200 VA

1500 VA + 1200 VA = 2700 VA

Although the connected load is less than the load allowance the two circuits supplying this track have capacity for:

20 A × 120 V × 0.8 × 2
= 3840 VA of continuous load.

The 0.8 multiplier is a factor applied to continuous loads. This will be discussed in detail later. This capacity is well above the minimum requirement.

OTHER LOADS

The remaining loads consist of miscellaneous receptacle outlets, a receptacle on the roof and a sign outlet.

Receptacle Outlets

- each single or multiple receptacle on one strap shall be considered at not less than 180 volt-amperes, *NEC® Section 220-3(b)(9)*. See Figure 3-1.

- allow actual rating for specific loads, *NEC® Section 220- 3(b)(1)*.

The allowance for the receptacle outlets is:

21 receptacle outlets @ 180 VA per outlet
= 3780 VA

Roof Receptacle

The roof receptacle outlet is a requirement of *NEC® Section 210-63*. This weatherproof receptacle outlet must be located within 25 feet of the equipment; *NEC® Section 210-8(b)(2)* requires that the receptacle be GFCI protected. No load allowance is stipulated. The arbitrary allowance is:

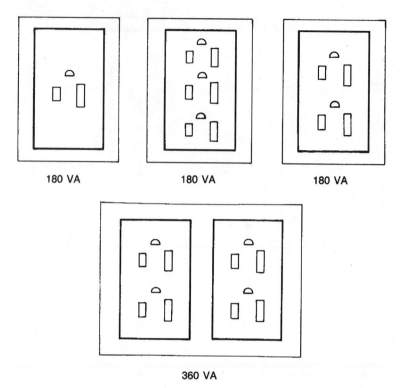

180 VA   180 VA   180 VA

360 VA

Fig. 3-1   Minimum receptacle outlet allowance.

1 roof receptacle outlet @ 1500 VA per outlet
= 1500 VA

Sign Outlet

The sign outlet is a requirement of *NEC®
Section 220-3(b)(6)* and is assigned a minimum
allowance of 1200 volt-amperes.

1 sign outlet @ 1200 VA per outlet = 1200 VA

MOTORS AND APPLIANCES

- *NEC® Article 422* sets forth the installation
  requirement for appliances.

- *NEC® Article 430*, unless "specifically
  amended" by *NEC® Article 422*, should be
  referred to for motor operated appliances.

- *NEC® Article 440* addresses air-conditioning
  and refrigeration equipment that incorporates
  hermetic refrigeration motor-compressor(s).

Hermetic Motor-Compressor

As a part of the specification, included in the
Appendix, are listed the electrical characteristics of
the air-conditioning equipment to be installed in the

commercial building. The load requirements for the
drugstore are copied here for convenience. The connection scheme is shown in Figure 3-2.

Supply Voltage:
   208-volt, three-phase, three-wire, 60 hertz
Hermetic refrigeration compressor-motor:
   Rated load current 20.2 amperes @ 208-volt,
   three-phase
Evaporator motor:
   Full load current 3.2 amperes @ 208-volt, single-
   phase

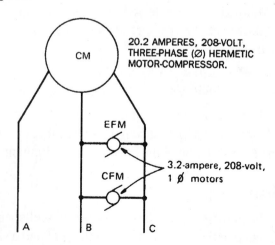

Fig. 3-2   Typical air conditioner load connections.

Condenser motor:

Full load current 3.2 amperes @ 208-volt, single-phase

It is necessary to convert these loads to volt-ampere values before they can be added to the other loads. For a 208-volt, three-phase supply the rated load current is to be multiplied by 360. For a 208-volt load, single-phase supply the rated load current is to be multiplied by 208.

The air-conditioning load is:

Compressor: 3 ph, 208 V, 20.2 A = 7272 VA
Evaporator: 1 ph, 208 V, 3.2 A = 666 VA
Condenser: 1 ph, 208 V, 3.2 A = 666 VA
Total Air Conditioning load   = 8604 VA

This is a load calculation and does not indicate the current that will occur in the conductors supplying the equipment. A typical connection scheme is shown in Figure 3.2. It can be seen that the current in the three phases would not be equal. This will be discussed in detail in later units.

Storage Area Lighting

According to *NEC® Table 220-3(a)*, the minimum load for storage spaces is 0.25 volt-amperes per square foot. The minimum load allowance for this storage space is:

1021 square feet × 0.25 VA/square foot
= 253 volt-amperes

The connected load for the storage area (basement) is:

9 Style L luminaires @ 87 volt-amperes
= 783 volt-amperes

Summary of Drugstore Loads

General Lighting: The actual load (4155 VA) is less than the load allowance (4185 VA) so the allowance is used.

Storage: The actual load (783 VA) is greater than the load allowance (253 VA) so the actual load is used.

Show Window: The actual load (2700 VA) is less than the load allowance (3200) so the allowance is used.

Other Loads: Specific allowances are used.

Motor loads: The actual motor loads are used.

REVIEW QUESTIONS

Refer to the *National Electrical Code®* or the working drawings when necessary. Where applicable, responses should be written in complete sentences. Write units using unit names, do not use abbreviations or symbols (1 foot, not 1' or 1 ft).

For questions 1 through 4, indicate the unit load that would be included in the branch circuit calculations as set forth in *NEC® Table 220-3(a)*.

1. A restaurant   _____

2. A school room   _____

3. A corridor in a school   _____

4. A corridor in a dwelling   _____

A different Style N luminaire is selected for use in the drugstore. The new luminaire is rated for 150 watts but only 100 watts will be installed.

5. The *NEC®* load for the 2 style N luminaires is _____.

6. After this revision the computed load for the general lighting will be _____.

Indicate the effect, if any, a change in loads can have on the feeders serving the area.

7. An increase in the computed load.

_____

8. An increase in the balanced load.

_____

9. An increase in the nonlinear load.

_____

Answer questions 10, 11, and 12 using the following information.

An addition is being planned to a rural school building. You have been asked to determine the load that will be added to the panelboard that will serve this addition.

The addition will be a building 80 feet by 50 feet. It will consist of four classrooms, each 40 feet by 20 feet and a corridor that is 10 feet wide.

The following loads will be installed.

Each classroom:
    12 fluorescent luminaires, 2 feet by 4 feet @ 85 VA each
    20 duplex receptacles
    1 AC unit 208 vole 1 phase @ 5000 VA

Corridor:
    5 fluorescent 1 foot by 8 feet @ 85 VA each
    8 duplex receptacles

Exterior:
    4 wall mounted luminaires @ 125 VA each
    4 duplex receptacles

Load	Count	VA/Unit	NEC®	Actual	Computed	Balanced	Nonlinear

10. The computed load is   _____ VA

11. The balanced load is   _____ VA

12. The nonlinear load is   _____ VA

UNIT 4

Branch Circuits

OBJECTIVES

After studying this Unit the student will be able to

- determine the required number of branch circuits for a set of loads.
- determine the correct rating for branch-circuit protective devices.
- determine the preferred type of wire for a branch circuit.
- determine the required minimum size conductor for a branch circuit.

CONDUCTOR SELECTION

When called upon to connect an electrical load such as lighting, motors, heating, or air-conditioning equipment, the electrician must have a working knowledge of how to select the proper type and size of conductors to be installed. Installing a conductor of the proper type and size will assure that the voltage at the terminals of the equipment is within the minimums as set forth by the *NEC®* and that the circuit will have a long uninterrupted life.

One of the first steps in understanding conductors is to refer to *NEC® Article 310*. This article contains such topics as insulation types, conductors in parallel, wet and dry locations, marking, maximum operating temperatures, permitted use, trade names, direct burial, ampacity tables, adjustment factors, and corrections factors.

Conductor Type Selection

An important step in selecting a conductor is the selection of an insulation type appropriate for the installation. An examination of *NEC® Table 310-16* will reveal that although there are several different types listed there are only three different temperature ratings, (140, 167, and 194°F) 60, 75 and 90°C. It should also be noted that the higher the temperature rating the higher the ampacity for a given size of wire.

Should the operational temperature of a conductor become excessive the insulation may soften, or melt, causing grounds or shorts and possible equipment damage and personal injury. This heat comes from two sources, the surrounding room temperature which is referred to as ambient heat and from the current in the wire. Heat generated in the wire is expressed by the formula:

$$H = I^2Rt$$

In this formula "H" is the heat in watt hours, "I" is the current in amperes, "R" the resistance in ohms and "t" is the time in hours. Many of the rules in the *NEC®* concerning the selection of conductors are related to this formula.

The resistance of a conductor is dependent on the size, material, and length. Resistance values are listed in *NEC® Chapter 9, Table 8*. Conductor resistance maybe lowered by selecting copper instead of aluminum or by increasing the size. (The circuit length is usually established by other factors.) The resistance of a conductor does not change appreciably in response to current or ambient conditions.

The current has a dramatic effect on the heat. A 25% load change, from 16 to 20 amperes, will result in 56% change in heat. The *NEC®* incorporates many features to address this issue. The most notable of these is in the selection of the conductor type. Certain types of conductor insulation can tolerate higher heat levels thus permitting a higher current with no change in wire size. The *NEC®* also restricts the connected load thus limiting the current. In addition it recognizes the heating effect of bundling conductors and requires the application of an adjustment factor.

There are many factors to consider when selecting a conductor *NEC® Table 310-13*.

- Column 1 provides the technical names for the conductors which are seldom used in the trade.

- Column 2 gives the type designation some of which are listed over the columns in *NEC® Table 310-16*. These are used extensively in the trade. There is some logic to their origin. In general R indicates a rubber-based covering, H indicates a high temperature rating, W is used when the conductor is usable in wet locations, T indicates a thermoplastic covering, and N is used when there is a nylon outer covering

- Column 3 lists the temperature ratings. Only (140, 167, and 194°F) 60, 75 and 90°C ratings are listed in *NEC® Table 310-16*.

- Column 4 is highly used as it gives the applications where the conductor is approved for use.

Turning to *NEC®* Article 100 and finding "Location" will provide complete definitions for dry, damp and wet. For example if a circuit is run underground in a conduit, conductors rated for wet locations must be used.

- Column 5 gives a brief, technical, description of the insulation.

- Column 6 gives the insulation thickness for various sizes. This value becomes very important when calculating conduit fill. For example in a 3/4 inch raceway: 16 THHN, or 11 TW, or 8 THW, size 12 AWG conductors maybe installed.

- Column 7 provides a description the outer coating. Popular Types THHN and THWN are listed as having a nylon jacket. This jacket tends to make installation easier but is not critical to the insulation value.

Conductor Size Selection

NEC® Table 310-16 is referred to regularly by electricians, engineers and electrical inspectors when information about wire sizing is needed. This table shows the allowable ampacities of insulated conductors. The temperature limitations, the types

TABLE 310-13

Trade Name	Type Letter	Maximum Operating Temperature	Application Provisions	Insulation	Thickness of Insulation		Outer Covering[1]
					AWG or kcmil	Mils	
Heat-resistant thermoplastic	THHN	90°C 194°F	Dry and damp locations	Flame-retardant, heat-resistant thermoplastic	14–12 10 8–6 4–2 1–4/0 250–500 501–1000	15 20 30 40 50 60 70	Nylon jacket or equivalent
Moisture- and heat-resistant thermoplastic	THHW	75°C 167°F 90°C 194°F	Wet location Dry location	Flame-retardant, moisture- and heat-resistant thermoplastic	14–10 8 6–2 1–4/0 213–500 501–1000	30 45 60 80 95 110	None
Moisture- and heat-resistant thermoplastic	THW[5]	75°C 167°F 90°C 194°F	Dry and wet locations Special applications within electric discharge lighting equipment. Limited to 1000 open-circuit volts or less (Size 14-8 only as permitted in Section 410-31)	Flame-retardant, moisture- and heat-resistant thermoplastic	14–10 8 6–2 1–4/0 213–500 501–1000 1001–2000	30 45 60 80 95 110 125	None

[1]Some insulations do not require an outer covering.

[5]Listed wire types designated with the suffix ''-2,'' such as RHW-2, shall be permitted to be used at a continuous 90°C (194°F) operating temperature, wet or dry.

Reprinted with permission from NFPA 70-1999.

Table 310-16. Allowable Ampacities of Insulated Conductors Rated 0 through 2000 Volts, 60°C through 90°C (140°F through 194°F) Not More than Three Current-Carrying Conductors in Raceway, Cable, or Earth (Directly Buried), Based on Ambient Temperature of 30°C (86°F)

Size	Temperature Rating of Conductor (See Table 310-13)						Size
	60°C (140°F)	75°C (167°F)	90°C (194°F)	60°C (140°F)	75°C (167°F)	90°C (194°F)	
AWG or kcmil	Types TW, UF	Types FEPW, RH, RHW, THHW, THW, THWN, XHHW, USE, ZW	Types TBS, SA, SIS, FEP, FEPB, MI, RHH, RHW-2, THHN, THHW, THW-2, THWN-2, USE-2, XHH, XHHW, XHHW-2, ZW-2	Types TW, UF	Types RH, RHW, THHW, THW, THWN, XHHW, USE	Types TBS, SA, SIS, THHN, THHW, THW-2, THWN-2, RHH, RHW-2, USE-2, XHH, XHHW, XHHW-2, ZW-2	AWG or kcmil
	COPPER			ALUMINUM OR COPPER-CLAD ALUMINUM			
18	—	—	14	—	—	—	—
16	—	—	18	—	—	—	—
14*	20	20	25	—	—	—	—
12*	25	25	30	20	20	25	12*
10*	30	35	40	25	30	35	10*
8	40	50	55	30	40	45	8
6	55	65	75	40	50	60	6
4	70	85	95	55	65	75	4
3	85	100	110	65	75	85	3
2	95	115	130	75	90	100	2
1	110	130	150	85	100	115	1
1/0	125	150	170	100	120	135	1/0
2/0	145	175	195	115	135	150	2/0
3/0	165	200	225	130	155	175	3/0
4/0	195	230	260	150	180	205	4/0
250	215	255	290	170	205	230	250
300	240	285	320	190	230	255	300
350	260	310	350	210	250	280	350
400	280	335	380	225	270	305	400
500	320	380	430	260	310	350	500
600	355	420	475	285	340	385	600
700	385	460	520	310	375	420	700
750	400	475	535	320	385	435	750
800	410	490	555	330	395	450	800
900	435	520	585	355	425	480	900
1000	455	545	615	375	445	500	1000
1250	495	590	665	405	485	545	1250
1500	520	625	705	435	520	585	1500
1750	545	650	735	455	545	615	1750
2000	560	665	750	470	560	630	2000

CORRECTION FACTORS

Ambient Temp. (°C)	For ambient temperatures other than 30°C (86°F), multiply the allowable ampacities shown above by the appropriate factor shown below.						Ambient Temp. (°F)
21–25	1.08	1.05	1.04	1.08	1.05	1.04	70–77
26–30	1.00	1.00	1.00	1.00	1.00	1.00	78–86
31–35	0.91	0.94	0.96	0.91	0.94	0.96	87–95
36–40	0.82	0.88	0.91	0.82	0.88	0.91	96–104
41–45	0.71	0.82	0.87	0.71	0.82	0.87	105–113
46–50	0.58	0.75	0.82	0.58	0.75	0.82	114–122
51–55	0.41	0.67	0.76	0.41	0.67	0.76	123–131
56–60	—	0.58	0.71	—	0.58	0.71	132–140
61–70	—	0.33	0.58	—	0.33	0.58	141–158
71–80	—	—	0.41	—	—	0.41	159–176

*See Section 240-3.

of insulation, the material the wire is made of, the conductor size in AWG (American Wire Gauge) or kcmil (thousand circular mils), the ampacity of the conductor and the correction factors for a range of ambient temperatures.

NEC® Table 310-16 is the most often used table since it is always referred to when conductors are to be installed in raceways. It is important to read and understand all the notes and footnotes to this table.

Correction Factors

If the environmental temperature anywhere along the length of a raceway is less than 26°C (78°F) or higher than 30°C (86°F) the ampacity of the conductors in the raceway, or cable, must be modified by a Correction Factor. These are provided in the extension to NEC® Table 310-16. A discussion of the conditions that require this action are set forth in NEC® Section 310-10. The selection is dependant on the conductor's temperature rating. As THHN is the conductor of choice for the commercial building the correction factor will be taken from the 90°C (194°F) column.

As an example, the air-conditioning equipment, circuit 16, is located on the roof where the temperature would reach 38°C (100°F). A correction factor of 0.91 is applied to the THHN conductors serving this equipment.

Adjustment Factors

The use of adjustment factors is another response to excessive heat affecting the current carrying capacity of a conductor. See Table 4-1. This is the collective heat generated by a group of conduc-

tors such as those placed in a raceway or cable or otherwise bundled in a way that prevents air flow from dissipating the heat. Whenever there is a grouping of more than three conductors a factor from NEC® Table 310-15(b)(2) must be applied. Drugstore circuits 16 and 17 are examples of the application of adjustment factors. Five current carrying conductors are installed in a raceway serving the air-conditioning equipment. The adjustment factor for 5 conductors is 80% or 0.8. See Table 4-1.

Conductors must be derated according to NEC® Section 310-15(b)

- If more than three current-carrying-conductors are installed in a raceway of cable. See Figure 4-1(A) and NEC® Section 310-15(b)(2).

- When single conductors or cable assemblies are stacked or bundled without spacing, as in a cable tray, for lengths greater than 24 inches. See Figure 4-1(B) and NEC® Section 310-15(b)(2).

- If the number of cable assemblies, (i.e., non-metallic sheathed cable) are run together for distances more than 24 inches. See Figure 4-1(C) and NEC® Section 310-15(b)(2).

It is important to note that the NEC® refers to current-carrying conductors for the purpose of derating when more than three conductors are installed in a raceway or cable. Following are the basic rules: (the word circuit includes branch circuits, feeders and services)

- DO count all current carrying wires.

TABLE 4-1 Adjustment Factors.	
Number of Current-Carrying Conductors	Percent of Values in Tables as Adjusted for Ambient Temperature if Necessary
4 through 6	80
7 through 9	70
10 through 20	50
21 through 30	45
31 through 40	40
41 and above	35

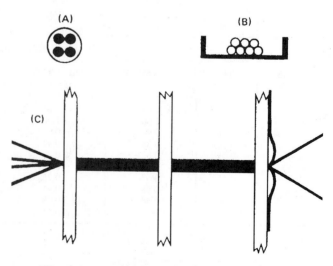

Fig. 4-1   Bundled and stacked conductors.

- DO count neutrals of a 3-wire, 1-phase circuit when the system is 4-wire, 3-phase, wye connected. See Figure 4-2 and *NEC® Section 310-15(b)(4)(b)*.

- DO count neutrals of a 4-wire 3-phase wye connected circuit when the major portion of the load is electric discharge lighting (fluorescent, mercury vapor, high pressure sodium, etc.), data processing, and other loads where the neutral carries third harmonic current. See Figure 4-3 and *NEC® Section 310-15(b)(4)(c)*.

- DO NOT count neutrals of a 3-wire, 1 phase circuit where the neutral carries only the unbalanced current of the phase conductors. See Figure 4-4 and *NEC® Section 310-15(b)(4)(a)*.

- DO NOT count equipment grounding conductors that are run in the same raceway with circuit conductors. See Figure 4-5. Grounding conductors must be included when calculating raceway fill.

- DO NOT derate conductors in sections of raceway 24 inches or less in length. See Figure 4-6.

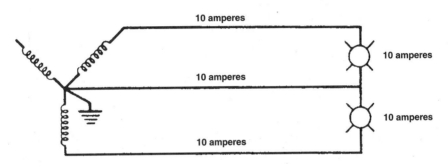

Fig. 4-2   Four-wire, wye system, three-wire circuit.

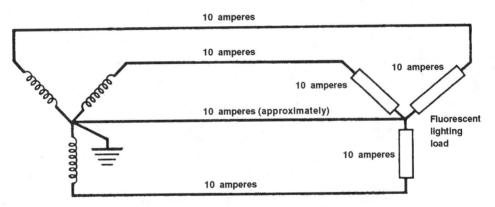

Fig. 4-3   Four-wire, wye system, nonlinear load.

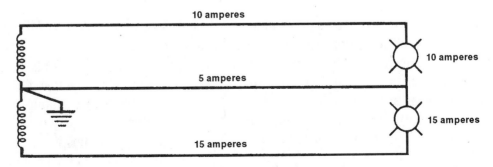

Fig. 4-4   Single phase, three-wire circuit.

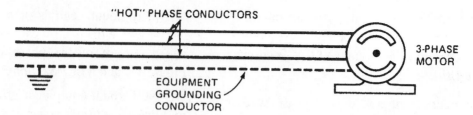

Fig. 4-5   Counting equipment grounding conductors.

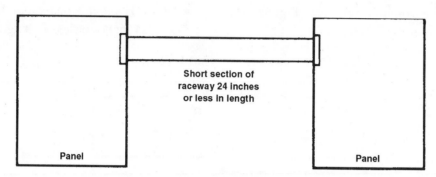

Fig. 4-6   Raceways 24 inches or less in length.

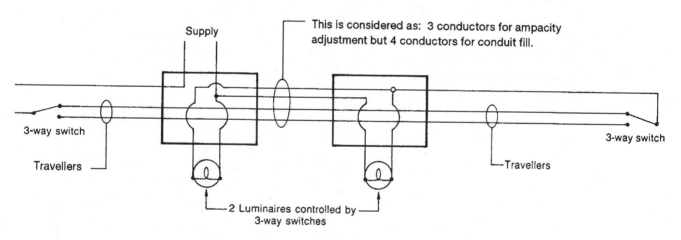

Fig. 4-7   Counting three-way switch travelers.

- DO NOT count the non-current-carrying "dummy" or "traveler" on three-way or four-way switching arrangements. Only one of the travelers is carrying current at anyone time, so only one conductor need be counted when determining ampacity adjustment factor. See Figure 4-7. Both travelers are counted when computing raceway fill.

DETERMINING CIRCUIT COMPONENTS
Conductors

- The conductor type must be selected to meet the criteria of *NEC® Section 110-14(c)*. If the circuit, i.e. the OCPD, rating is 100 amperes or less the conductor may be selected from either of the three columns in *NEC® Table 310-16*. If the circuit rating is greater than 100 amperes it must either be a 75°C (167°F) or 90°C (194°F) rated conductor.

- The conductor type must comply with the requirements of the location, e.g., dry, damp, wet.

- In compliance with *NEC® Section 210-19(a)*, the allowable ampacity (the value given in *NEC® Table 310-16*) of the conductor must be equal to or greater than the noncontinuous load plus 125% of the continuous load. In these statements and in the panelboard worksheet this value is referred to as the "OCPD selection amperes."

- For circuits rated 100 amperes of less, or if any of the terminations are marked for No. 1 AWG conductors or less, enter the 60°C (140°F) column of *NEC® Table 310-16* and identify the allowable ampacity that is equal to or next greater than the OCPD selection amperes. The size of this conductor is the minimum size allowed for the load.

- For circuits with a rating greater than 100 amperes, or if all the terminations are rated for 75°C or higher, enter the 75°C column *NEC® Table 310-16* and identify the allowable ampacity that is equal to or next greater than the OCPD selection amperes. The size of this conductor is the minimum size allowed for the load.

- The derated ampacity must be sufficient to allow the use of the required OCPD and equal to or higher than the computed load (continuous plus the noncontinuous load).

- When serving a motor or an air-conditioning unit the conductor must have an ampacity of not less than 125% of the full-load current rating. Ratings for single phase motors shall be selected from *NEC® Table 430-148*, for three-phase motors from *NEC® Table 430-150*.

Overcurrent Protection

It is the purpose of an overcurrent protective device (OCPD) to protect the circuit wiring and devices, and to some extent the equipment served by the circuit. The following *NEC®* references must be consulted when selecting an OCPD. The student should read these references with great care for only a project specific synopsis is presented. Further details on OCPD selection follow in a later units.

- A continuous load will operate at 100% of the maximum current for a period of three hours or more. Office and store lighting are common examples of continuous loads. See *NEC® Article 100*.

- *NEC® Section 210-20(a)* states that the OCPD shall have a rating equal to or greater than the noncontinuous load plus 125% of the continuous load.

- The rating of the overcurrent protective device becomes the rating of the circuit regardless of the conductor size or load type. See Figure 4-8 and *NEC® Section 210-3*.

- In general, branch circuit conductors shall be protected in accordance with their ampacities. See *NEC® Section 240-3*.

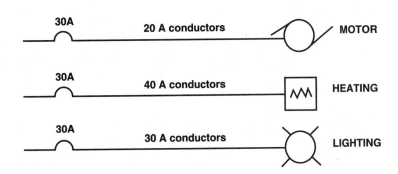

Fig. 4-8   Three circuits rated 30-ampere.

- A footnote to *NEC® Table 310-16* must be observed when selecting No. 14, 12, or 10 AWG conductors. In the table an obelisk (†), also called a dagger, is used to mark most of the ampacities of the above sizes. It will be noted that the allowable ampacity of these sizes is greater than the permitted OCPD rating.

- The derated ampacity of the circuit conductors maybe less than the circuit rating **provided** that the circuit is rated at 800 amperes or less and it is not a multioutlet branch circuit supplying cord- and plug-connected portable loads. See *NEC® Section 240-3(b).*

- Multioutlet branch circuits ratings are limited to 15, 20, 30, 40, and 50-ampere except on industrial premises where may they have a higher rating. See *NEC® Section 210-3.*

- Overcurrent protective devices are only available in the standard ampere ratings listed in *NEC® Section 240-6.*

- Branch circuits serving storage-type water heaters with a capacity of 120 gallon or less shall have a rating of 125% of the nameplate load. See *NEC® Section 422-13.*

- For motor loads of less than 100 amperes the rating of an inverse time circuit breaker (as used in the commercial building) shall not exceed 400% of the load. A rating of 250% is considered as appropriate. See *NEC® Section 430-52(c)* and *Table 430-152.*

- When a branch circuit supplies an air-conditioning system that consists of a hermetic motor-compressor and other loads (as occurs in the commercial building) the circuit rating shall not exceed 225% of the hermetic motor load plus the sum of the additional loads. See *NEC® Section 440-22(b)(1).*

Selection Criterion

An abbreviated circuit component selection criterion for motors, motor-compressor units and other circuits is shown in Table 4-2. For motors, the OCPD's are inverse time circuit-breakers.

DEFINING THE BRANCH CIRCUITS

In Unit 3 the electrical loads to be installed in the drugstore were identified and quantified. The next step usually would be to draw, or sketch, a layout of the space. For this exercise it will suffice to review Working Drawing E3 observing the location of the various electrical outlets.

The information in this unit concentrates on the branch circuits including the selection of the overcurrent devices and the conductor type and sizes. It will be useful to refer to the panelboard worksheet shown in this unit. The first five columns essentially define the branch circuit. Much of this information is taken for the loading schedule. The next four columns result in the determining of the circuit

TABLE 4-2 Circuit Component Selection			
Circuit size/ rating Load type	Minimum Allowable Ampacity	OCPD Rating	Maximum OCPD Rating
C – Continuous	125% load	not less than 125% load	Next higher over ampacity
H – Hermetic Motor Compressor	125% compressor load plus other loads	175% compressor load plus other loads	225% compressor load plus other loads[1]
N – Noncontinuous	Load	Load	Next higher over ampacity
M – Motor (100 Ampere Maximum)	125% load plus other loads	225% load plus other loads	400% load[2]
R – Cord & Plug Branch Circuit	Not less than OCPD rating	Not less than load	Not higher than ampacity

1. See *NEC®* Section 930-52 before applying this value
2. See *NEC®* Section 440-22(C)(1) before applying this value

rating. The remaining columns are related to the sizing of conductors.

Laying out an electrical system usually starts with a sketch or drawing of the space(s). Symbols, as discussed in Unit 2, are placed on the plans representing the location of the equipment. Then lines are drawn showing the circuiting of the equipment. Following this branch circuits can be assigned. The actual number of branch circuits to be installed in the drugstore is determined by referring to the plans or specifications. The following discussion will illustrate the process employed to arrive at the final number.

The specifications (see Appendix) for this commercial building direct the electrician to install copper conductors of a size not smaller than No. 12 AWG. The maximum load on a circuit is limited by the rating of the overcurrent protective device. For a No. 12 AWG copper, referring to the note at the end of *NEC® Table 310-16*, that is 20 amperes. On a 120-volt circuit this limits the load to:

$$120 \text{ V} \times 20 \text{ A} = 2400 \text{ VA}.$$

Minimum Number Branch Circuits

Referring to the loading schedule developed in Unit 3, the connected lighting load is 8138 volt-amperes. Add the 3780 volt-ampere allowance for the 21 receptacles for a total connected load of 11,918 volt-amperes. Divide this total load by the maximum load per branch circuit, rounding up to the next integer, to determine the minimum number of branch circuits for general lighting and power.

11,918 VA / 2400 VA per branch circuit
= 5 branch circuits.

Add the 3 special circuits, one for the roof receptacle, one for the sign and one for the cooling system to arrive at 8 circuits, the absolute minimum number of circuits for the drugstore.

Actual Number Branch Circuits

This simplified procedure does not account for the continuous loads that limit branch-circuit loads to 80% of the rating of the overcurrent device.

Other factors, such as switching arrangements and convenience of installation, are additional important considerations in determining the number of circuits to be used. This is illustrated in the drugstore, as 17 circuits are scheduled.

At this point it is necessary to refer to the loading schedule developed in the previous unit. An examination of the first five columns of the panelboard worksheet will reveal the first steps in arriving at a final number of branch circuits.

USING THE PANELBOARD WORKSHEET, COLUMNS A–E

The information studied in this unit is now applied to the drugstore and is illustrated in a panelboard worksheet. This worksheet is presented in three parts. In the first part the information necessary to do the computations is recorded in columns A through E, Table 4-3. In the second part, columns

		TABLE 4-3 Panelboard Worksheet – Columns A–E		
A	**B**	**C**	**D**	**E**
Phase	Circuit Number	Load/Area Served	Computed Volt-ampere	Computed Amperes
A	1	9 Style I,	783	6.5
A	2	4 Rec., Merchandise	720	6
B	3	18 Style I,	1566	13
B	4	3 Rec., Toilet area.	540	4.5
C	5	3 Style E, 2 N,	552	4.6
C	6	5 Rec., Merchandise	900	7.5
A	7	2 Rec., Pharmacy	360	3
A	8	3 Rec., Show Window	1500	13
B	9	9 Style L, 1 E,	684	5.7
B	10	15 Style D, Pharmacy	1110	9.3
C	11	Track Show Window	600	5
C	12	4 Rec., S. Basement	720	6
A	13	Track Show Window	600	5
A	14	3 Rec., N. Basement	540	4.5
B	15	Sign	1200	10
B		Evaporator		3.2
C	16	Compressor	8604	20
A		Condenser		3.2
C	17	1 Receptacle, Roof	1500	13

F through K, the rating of the overcurrent protective device is determined along with the minimum conductor size and derated ampacity. In the final part, columns L through S, the permitted wire size is selected and listed in column Q. The final two columns verify the selection by demonstrating that the derated ampacity of the selected conductor is equal to or greater than the minimum derated ampacity given in column K.

Following each part is a discussion of the action performed by each of the columns. If there is a question, students are encouraged to review the information presented at the beginning of this unit.

Phase Connection (Column A)

The letter in this column indicates which phase(s) is/are being used by the circuits. The letters A, B and C are used to designate the different phases.

Circuit Number (Column B)

If the viewer is facing the panelboard, the circuits connected on the left side are given odd numbers with the even numbers being assigned to the circuits on the right side. Thus circuits 1, 3 & 5 would be the top three circuits on the left side, 2, 4 & 6 on the right side. If it is a three-phase panelboard two or three circuits can be grouped together with a single neutral and the conductors installed in a raceway, or cable, to serve two or three single phase 120 volt loads.

A typical case is circuits 1 and 3. These are connected to phases A and B and installed in a conduit serving the luminaires (lighting fixtures) in the main sales area. The lighting load served by these three circuit 2349 VA. This load will continue for more than three hours thus is a continuous load and is limited to:

$$120 \text{ V} \times 20 \text{ A} \times 0.8 = 1920 \text{ VA.}$$

There are three switches for this lighting which could be served by 2 or 3 circuits. To place all on one circuit would overload the circuit:

$$783 \text{ VA} + 1566 \text{ VA} = 2349 \text{ VA.}$$

A circuit could have been provided for each switch but that strategy would resulted in an excessive use of circuits. They would have be loaded to:

$$2349 \text{ VA} / 3 = 783 \text{ VA}$$

The choice was to connect 2 switches to one circuit and the other switch to a second circuit thus one circuit has a load of 1566 VA the other 783 VA.

Load/Area Served (Column C)

In this column is a listing of the loads and the areas of the occupancy. The information this column comes from the drugstore loading schedule prepared in the previous unit.

Computed Volt-43Ampere (Column D)

The loads in VA (volt-ampere) were also taken from the information provided by the drugstore loading schedule.

Computed Amperes (Column E)

The value given in Column 3 is converted to amperes. For a single pole circuit (circuits 1-15 and 17) the VA is divided by 120, if the circuit were 2 pole the divisor is 208, and if the circuit is 3 pole (circuit 16) the divisor is 360.

USING THE PANELBOARD WORKSHEET, COLUMNS F–K

Columns F (Load Type), G (Load Modifier), H (OCPD Selection Amperes), I (OCPD Ampere Rating), J (Minimum Conductor Size) and K (Minimum Derated Ampacity) are used to record the critical first steps in the circuit component determination process ampacity, Table 4-4. The content of these columns is to a great extent dictated by compliance with *NEC® Section 110-14(c)*.

How these conditions impact the circuit values will be demonstrated in this and later examples. In the commercial building the lighting loads are in general considered continuous loads. Receptacles may be placed in C, N or R class depending on how they are to be used. Circuit 8 serves receptacles in the show window which will supply lighting thus the circuit is considered a continuous load. Circuit 17 is only used when the AC Unit is being repaired or maintained so is considered a noncontinuous

TABLE 4-4 Panelboard Worksheet – Columns F–K					
F	G	H	I	J	K
Load Type	Load Modifier	OCPD Selection Amperes	OCPD Rating	Minimum Conductor Size AWG	Minimum Derated Ampacity
C	1.25	9	20	12	16
R	1	6	20	12	21
C	1.25	17	20	12	17
R	1	5	20	12	21
C	1.25	6	20	12	16
R	1	8	20	12	21
R	1	3	20	12	21
C	1.25	16	20	12	16
C	1.25	8	20	12	16
C	1.25	12	20	12	16
C	1.25	7	20	12	16
R	1	6	20	12	21
C	1.25	7	20	12	16
R	1	5	20	12	21
C	1.25	13	20	12	16
H	1 2.25 1	52	50	10	32
N	1	13	20	12	16

load. Circuit 16 is a Hermetic unit consisting of three motors, the evaporator, the compressor and the condenser.

Load Types (Column F)

The load types are, Continuous (C), Non-continuous (N), Hermetic Units (H), Motor (M), and Receptacles on branch circuits that supply cord- and plug-connected portable loads (R). Following is a summary of the *NEC®* requirements for these load types as they apply to the commercial building.

Continuous Loads (Column C)

If the load is continuous:

- the OCPD ampere rating must not be less than 125% of the load or 20-ampere is as required by the contract specifications.

- the derated minimum ampacity must be not less than the load and must permit the use of the use of the minimum OCPD. This would be an ampacity greater that the next lower rating taken from *NEC® Section 240-6*. For example in this section 15 is the next lower rating to 20-ampere. Thus, an ampacity of 16 amperes would permit the use of a 20-ampere OCPD. Actually, the ampacity could be as low as 15.51 amperes for rounding up is permitted.

Noncontinuous Loads (Column N)

If the load is noncontinuous:

- the OCPD ampere rating must be not less than the load.

- the minimum derated ampacity must be not less than the load.

Hermetic Units (Column H)

If the circuit serves a hermetic refrigeration unit:

- the OCPD selection amperes is determined adding the condenser and evaporator loads to 175% of the motor-compressor load

- the OCPD rating must not be greater the sum of the condenser and evaporator loads plus 225% of the compressor load.

- the circuit conductors must have a derated ampacity not less than 125% of the motor-compressor load plus the sum of the other loads

- Motor leads and motor controllers have a 75°C (167°F) temperature rating thus that column of *NEC® Table 310-16* shall be used when determining the minimum size conductor regardless of the load or the conductor size. See *NEC® Section 110-14*.

Motors (Column M)

If the circuit supplies a motor:

- the OCPD ampere rating of an inverse time circuit breaker, for a load of 100 amperes or less the maximum rating is 250% not to exceed 400%, of the motor load. See *NEC® Section 450-52(c)(1)* and *Table 430-152*.

- the minimum derated ampacity shall be not less than 125% of the largest motor plus the sum of the other motors. See *NEC® Section 430-24.*

- The leads and controllers for hermetic units have a 75°C (167°F) temperature rating thus that column of *NEC® Table 310-16* shall be used when determining the minimum size conductor regardless of the load or the conductor size. See *NEC® Section 110-14.*

Receptacles Supplying Cord- and Plug-Connected Portable Loads (Column R)

If a branch circuit supplies cord- and plug-connected portable loads:

- the OCPD ampere rating shall be not less than the load.

- the minimum derated ampacity shall be not less than the OCPD rating.

Load Modifier (Column G)

- If the load type is "C" the load is continuous and is to be increased by 25%, in accordance with *NEC® Section 210-19.*

- If the load type is "M" the load is to be increased by 25% in accordance with *NEC® Section 430-22(a).*

- If the load is an "H" it shall comply with *NEC® Section 440-22(b)(1).*

- If the load is a water heater it shall be included as a continuous load.

- Other loads remain unchanged.

Selection Amperes (Column H)

The value in this column is the product of connected load in amperes and the load modifier. This value is used to select the overcurrent protective device.

OCPD Ampere Rating (Column I)

The OCPD ampere rating is selected from *NEC® Section 240-6.* It is the rating next larger than the OCPD selection amperes. On this project the minimum rating for a circuit is 20-ampere.

Minimum Conductor Size (Column J)

For loads 100-ampere or less, where one or more of the termination are rated at 60°C (140°F), using No. 1 AWG or smaller conductors, the minimum conductor size is determined by:

- locating the ampacity in the 60°C (140°F) column of *NEC® Table 310-16* that is equal to or next greater than the OCPD selection amperes and recording the conductor size of that ampacity

If the load is greater than 100 amperes, or if conductors are larger than No, 1 AWG or if all the terminations have a rating of 75°C (167°F), the minimum conductor size is determined by:

- locating the ampacity in the 75°C (167°F) column of *NEC® Table 310-16* that is equal to or greater than the OCPD selection amperes and recording the conductor size of that ampacity.

Minimum Derated Ampacity (Column K)

The values in the two of the previous columns are used to determine the minimum derated ampacity value.

1. If the load type is "C" or "N" the minimum derated ampacity must be greater than
   - the ampere value in column H and
   - the rating from *NEC® Section 240-6* that is next less than the ampere rating in column I.

2. If the load type is "R" the minimum derated ampacity must be greater than
   - the OCPD ampere rating.

3. If the load type is "H" the minimum derated ampacity must be not less than:
   - the sum of all other loads plus 125% of the motor-compressor load.

4. If the load type is "M" the minimum derated ampacity must be not less than:
   - 125% of the largest motor load plus the sum of other loads.

USING THE PANELBOARD WORKSHEET, COLUMNS L–S

Beginning with the column titled Ambient Temperature the remaining columns, Table 4-5, are used to develop the information necessary to select the minimum conductor size and applicable data.

Ambient Temperature (Column L)

The ambient temperature in degrees centigrade, as discussed above and in *NEC® Section 310-10*, is recorded in this column. A default value of 30 degrees Centigrade is prerecorded.

Correction Factor (Column M)

This factor is taken from the table located as an extension of *NEC® Table 310-16*. The factor is based on the ambient temperature of the environment where the circuit is installed. *NEC® Section 310-10* sets forth the requirements for the use of correction factors.

Current-Carrying Conductors (Column N)

The number of current-carrying conductors in a specific raceway is recorded in this column. The column is blank unless there are 4 or more.

Adjustment Factor (Column O)

This factor is taken from *NEC® Table 310-15(b)(2)* and is used to compensate for the increase in temperature caused by grouping current carrying conductors in a raceway or cable.

Minimum Allowable Ampacity (Column P)

The minimum allowable ampacity is the minimum derated ampacity divided by the product of the adjustment factor and correction factor.

Conductor Size (Column Q)

The conductor size is determined by entering *NEC® Table 310-16* with the minimum allowable ampacity for the required type (THHN) and material (copper). A size is selected that has an ampacity equal to or greater than the minimum allowable ampacity (column K) and a size that is as large or larger than the minimum conductor size (column J).

Allowable Ampacity (Column R)

The conductor ampacity taken for *NEC® Table 310-16* for the specified type (THHN) and the computed conductor size.

Derated Ampacity (Column S)

The derated ampacity is the ampacity from column R multiplied by the correction and adjustment factors. This value should be equal to or greater than the minimum derated ampacity of column K.

TABLE 4-5 Panelboard Worksheet – Columns L–S

L Ambient Temp. C	M Correction Factor	N Current Carrying Conductors	O Adjustment Factor	P Minimum Allowable Ampacity	Q Conductor Size AWG	R Allowable Ampacity	S Derated Ampacity
30	1		1	16	12	30	30
30	1		1	21	12	30	30
30	1		1	17	12	30	30
30	1		1	21	12	30	30
30	1		1	16	12	30	30
30	1		1	21	12	30	30
30	1		1	21	12	30	30
30	1		1	16	12	30	30
30	1		1	16	12	30	30
30	1		1	16	12	30	30
30	1		1	16	12	30	30
30	1		1	21	12	30	30
30	1		1	16	12	30	30
30	1		1	21	12	30	30
30	1		1	16	12	30	30
38	0.9	5	0.8	44	8	55	40
38	0.9	5	0.8	22	12	30	22

REVIEW QUESTIONS

Refer to the *National Electrical Code* ® or the working drawings when necessary. When applicable, responses should be written in complete sentences. Write units using unit names, do not use abbreviations or symbols (1 foot not 1' or 1 ft).

Complete the following table using the following circuits.

1. 11, 120 volt, duplex receptacles for general use.

2. 10, 120 volt, fluorescent luminaires,150 VA each

3. 8, 120 volt, incandescent luminaires, 200 watt each

4. 1, 240 volt, exhaust fan, 5000 watts,

5. 1, 120/240 1-phase, 3-wire feeder, 40 kVA continuous load, originating in room with ambient of 100°F

Circuits 1, 2, 3, and 4 are installed in same raceway for a distance of 10 feet, and the maximum ambient temperature is 86°F. Type THHN conductors will be used for all circuits.

Circuit number	1	2	3	4	5
Computed load volt-amperes					
Computed load amperes					
Load type					
Load modifier					
OCPD selection amperes					
OCPD rating					
Min. Conductor Size					
Min. Derated ampacity					
Ambient Temperature					
Correction factor					
Current carrying conductors					
Adjustment factor					
Min. Allowable Ampacity					
Conductor size					
Allowable ampacity					
Derated ampacity					

UNIT 5

Switches and Receptacles

OBJECTIVES

After studying this unit, the student will be able to

- select switches and receptacles with the proper rating for a particular application.
- install various types of receptacles correctly.
- connect single-pole, three-way, four-way, and double-pole switches into control circuits.

During the course of the work, an electrician selects and installs numerous receptacles and switches. Therefore, it is essential that the electrician know the important characteristics of these devices, and how they are to be connected into the electrical system.

RECEPTACLES

The National Electrical Manufacturers Association (NEMA) has developed standards for the physical appearance of locking and nonlocking plugs and receptacles. The differences in the plugs and receptacles are based on the ampacity and voltage rating of the device. For example, the two most commonly used receptacles are the NEMA 5-15R, Figure 5-1, and the NEMA 5-20R, Figure 5-2. The NEMA 5-15R receptacle has a 15-ampere, 125-volt

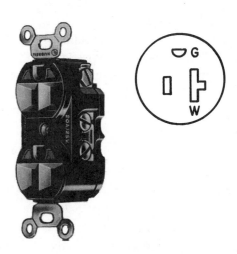

Fig. 5-1   NEMA 5-15R.

Fig. 5-2   NEMA 5-20R.

Fig. 5-3   NEMA 5-15P.

Fig. 5-4   NEMA 5-20P.

Fig. 5-6   NEMA 6-15P or 6-20P.

Fig. 5-5   NEMA 6-20R.

Fig. 5-7   NEMA 6-30R.

rating and has two parallel slots and a ground pin-hole. This receptacle will accept the NEMA 5-15P plug only, Figure 5-3. The NEMA 5-20R receptacle has two parallel slots and a "T" slot. This receptacle is rated at 20 amperes, 125 volts. The NEMA 5-20R will accept either a NEMA 5-15P or 5-20P plug, Figure 5-4. A NEMA 6-20R receptacle is shown in Figure 5-5. This receptacle has a rating of 20 amperes at 250 volts and will accept either the NEMA 6-15P or 6-20P plug, Figure 5-6. The acceptance by most of the receptacles of 15- and 20-ampere plugs complies with *NEC® Table 210-21(b)(3)* which permits the installation of either the NEMA 5-15R or 5-20R receptacle on a 20-ampere branch circuit. Another receptacle which is specified for the commercial building is the NEMA 6-30R, Figure 5-7. This receptacle is rated at 30 amperes, 250 volts.

The NEMA standards for general-purpose non-locking and locking plugs and receptacles are shown in Tables 5-1, 5-2, and 5-3. A special note should be made of the differences between the 125/250 (NEMA 14) devices and the 3φ, 250-volt (NEMA 15) devices. The connection of the 125/250-volt

receptacle requires a neutral, a grounding wire, and two-phase connections. For the 3φ, 250-volt receptacle, a grounding wire, and three-phase connections are required.

TABLE 5-1 NEMA Terminal Identification for Receptacles and Plugs.	
Green Colored Terminal (marked G, GR, GN, or GRND)	This terminal has a hexagon shape. Connect equipment grounding conductor ONLY to this terminal. (Green, bare, or green with yellow stripe.)
White (Silver) Colored Terminal (marked W)	Connect white or gray grounded circuit conductor ONLY to this terminal.
Brass Colored Terminal (marked "X," "Y," "Z")	Connect "HOT" conductor to this terminal (black, red, blue, etc.). There is no NEMA standard for a specific color conductor to be connected to a specific letter. It is good practice to establish a color code and stick to it throughout the installation.

TABLE 5-2   NEMA Configurations for General-Purpose Nonlocking Plugs and Receptacles.

		15 AMPERE		20 AMPERE		30 AMPERE		50 AMPERE		60 AMPERE	
		RECEPTACLE	PLUG	RECEPTACLE	PLUG	RECEPTACLE	PLUG	RECEPTACLE	PLUG	RECEPTACLE	PLUG
2-POLE 2-WIRE	125 V [1]	1-15R	1-15P		1-20P		1-30P				
	250 V [2]		2-15P	2-20R	2-20P	2-30R	2-30P				
	277 V AC [3]	(RESERVED FOR FUTURE CONFIGURATIONS)									
	600 V [4]	(RESERVED FOR FUTURE CONFIGURATIONS)									
2-POLE 3-WIRE GROUNDING	125 V [5]	5-15R	5-15P	5-20R	5-20P	5-30R	5-30P	5-50R	5-50P		
	250 V [6]	6-15R	6-15P	6-20R	6-20P	6-30R	6-30P	6-50R	6-50P		
	277 V AC [7]	7-15R	7-15P	7-20R	7-20P	7-30R	7-30P	7-50R	7-50P		
	347 V AC [24]	24-15R	24-15P	24-20R	24-20P	24-30R	24-30P	24-50R	24-50P		
	480 V AC [8]	(RESERVED FOR FUTURE CONFIGURATIONS)									
	600 V AC [9]	(RESERVED FOR FUTURE CONFIGURATIONS)									
3-POLE 3-WIRE	125/250 V [10]			10-20R	10-20P	10-30R	10-30P	10-50R	10-50P		
	3 Ø 250 V [11]	11-15R	11-15P	11-20R	11-20P	11-30R	11-30P	11-50R	11-50P		
	3 Ø 480 V [12]	(RESERVED FOR FUTURE CONFIGURATIONS)									
	3 Ø 600 V [13]	(RESERVED FOR FUTURE CONFIGURATIONS)									
3-POLE 4-WIRE GROUNDING	125/250 V [14]	14-15R	14-15P	14-20R	14-20P	14-30R	14-30P	14-50R	14-50P	14-60R	14-60P
	3 Ø 250 V [15]	15-15R	15-15P	15-20R	15-20P	15-30R	15-30P	15-50R	15-50P	15-60R	15-60P
	3 Ø 480 V [16]	(RESERVED FOR FUTURE CONFIGURATIONS)									
	3 Ø 600 V [17]	(RESERVED FOR FUTURE CONFIGURATIONS)									
4-POLE 4-WIRE	3 Ø 208Y/120 V [18]	18-15R	18-15P	18-20R	18-20P	18-30R	18-30P	18-50R	18-50P	18-60R	18-60P
	3 Ø 480Y/277 V [19]	(RESERVED FOR FUTURE CONFIGURATIONS)									
	3 Ø 600Y/347 V [20]	(RESERVED FOR FUTURE CONFIGURATIONS)									
4-POLE 5-WIRE GROUNDING	3 Ø 208Y/120 V [21]	(RESERVED FOR FUTURE CONFIGURATIONS)									
	3 Ø 480Y/277 V [22]	(RESERVED FOR FUTURE CONFIGURATIONS)									
	3 Ø 600Y/347 V [23]	(RESERVED FOR FUTURE CONFIGURATIONS)									

TABLE 5-3   NEMA Configurations for General-Purpose Locking Plugs and Receptacles.

			15 AMPERE		20 AMPERE		30 AMPERE		50 AMPERE		60 AMPERE	
			RECEPTACLE	PLUG	RECEPTACLE	PLUG	RECEPTACLE	PLUG	RECEPTACLE	PLUG	RECEPTACLE	PLUG
2-POLE 2-WIRE	125 V	L1	L1-15R	L1-15P								
	250 V	L2			L2-20R	L2-20P						
	277 V AC	3			(RESERVED FOR FUTURE CONFIGURATIONS)							
	600 V	4			(RESERVED FOR FUTURE CONFIGURATIONS)							
2-POLE 3-WIRE GROUNDING	125 V	5	L5-15R	L5-15P	L5-20R	L5-20P	L5-30R	L5-30P	L5-50R	L5-50P	L5-60R	L5-60P
	250 V	6	L6-15R	L6-15P	L6-20R	L6-20P	L6-30R	L6-30P	L6-50R	L6-50P	L6-60R	L6-60P
	277 V AC	7	L7-15R	L7-15P	L7-20R	L7-20P	L7-30R	L7-30P	L7-50R	L7-50P	L7-60R	L7-60P
	347 V AC	24			L24-20R	L24-20P						
	480 V AC	8			L8-20R	L8-20P	L8-30R	L8-30P	L8-50R	L8-50P	L8-60R	L8-60P
	600 V AC	9			L9-20R	L9-20P	L9-30R	L9-30P	L9-50R	L9-50P	L9-60R	L9-60P
3-POLE 3-WIRE	125/250 V	10			L10-20R	L10-20P	L10-30R	L10-30P				
	3 Ø 250 V	11	L11-15R	L11-15P	L11-20R	L11-20P	L11-30R	L11-30P				
	3 Ø 480 V	12			L12-20R	L12-20P	L12-30R	L12-30P				
	3 Ø 600 V	13					L13-30R	L13-30P				
3-POLE 4-WIRE GROUNDING	125/250 V	14			L14-20R	L14-20P	L14-30R	L14-30P	L14-50R	L14-50P	L14-60R	L14-60P
	3 Ø 250 V	15			L15-20R	L15-20P	L15-30R	L15-30P	L15-50R	L15-50P	L15-60R	L15-60P
	3 Ø 480 V	16			L16-20R	L16-20P	L16-30R	L16-30P	L16-50R	L16-50P	L16-60R	L16-60P
	3 Ø 600 V	17					L17-30R	L17-30P	L17-50R	L17-50P	L17-60R	L17-60P
4-POLE 4-WIRE	3 Ø 208Y/120 V	18			L18-20R	L18-20P	L18-30R	L18-30P				
	3 Ø 480Y/277 V	19			L19-20R	L19-20P	L19-30R	L19-30P				
	3 Ø 600Y/347 V	20			L20-20R	L20-20P	L20-30R	L20-30P				
4-POLE 5-WIRE GROUNDING	3 Ø 208Y/120 V	21			L21-20R	L21-20P	L21-30R	L21-30P	L21-50R	L21-50P	L21-60R	L21-60P
	3 Ø 480Y/277 V	22			L22-20R	L22-20P	L22-30R	L22-30P	L22-50R	L22-50P	L22-60R	L22-60P
	3 Ø 600Y/347 V	23			L23-20R	L23-20P	L23-30R	L23-30P	L23-50R	L23-50P	L23-60R	L23-60P

Hospital Grade Receptacles

In locations where severe abuse or heavy use is expected, hospital grade receptacles are recommended. These are a high-quality product. They meet UL requirements as Hospital Grade receptacles. These receptacles are marked with a small dot, Figure 5-8.

Electronic Equipment Receptacles

Circuits with receptacle outlets that serve microcomputers, solid-state cash registers, or other sensitive electronic equipment should be served by receptacles that are specially designed. These receptacles may be constructed for an isolated ground, Figure 5-9, they may have transient voltage surge protection, they may be hospital grade, or any combination of these three. Where the isolated grounding is desired, an insulated grounding conductor is to be run from the neutral terminal at the service entrance to the isolated terminal on the receptacle. ▶ "Isolated ground" receptacles are required by *NEC® Section 410-56(c)* to have an orange triangle on their face as illustrated in Figure 5-9. Prior to the 1996 edition of the *NEC*, the entire face of the receptacle was permitted to be orange in color, or identified by an orange triangle on the face of the receptacle. ◀

Electronic Equipment Grounding

Most commercial buildings today contain electronic equipment such as computers, copy machines, fax machines, data processing equipment, telephone systems, security systems, medical diagnostic instruments, HVAC and similar electronic controls, electronic cash registers, and similar types of loads.

Electronic equipment is extremely sensitive and susceptible to line disturbances caused by such things as:

voltage sags, spikes, surges
static charges
lightning strikes and surges on the system
electromagnetic interference (EMI)
radio frequency interference (RFI)
improper grounding

Metal raceways in a building act like a large

FRONT VIEW   BACK VIEW

Fig. 5-8   Hospital grade receptacle.

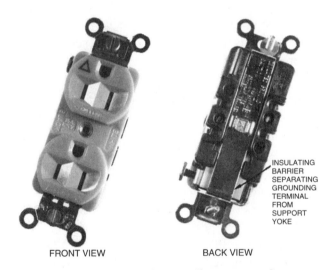

INSULATING BARRIER SEPARATING GROUNDING TERMINAL FROM SUPPORT YOKE

FRONT VIEW   BACK VIEW

Fig. 5-9   Isolated grounding receptacle.

antenna, and can pick up electromagnetic interference (electrical noise) that changes data in computer/ data processing equipment.

Electric utilities are responsible for providing electricity within certain voltage limits. Switching surges or lightning strikes on their lines can cause problems with electronic equipment. Static, lightning energy, radio interference, and electromagnetic interference can cause voltages that exceed the tolerance limits of the electronic equipment. These can cause equipment "computing" problems. Electronic data processing equipment will put out "garbage" that is unacceptable in the business world. Electronic equipment must be protected from these disturbances.

Quality voltage provides a "clean" sine wave that is free from distortion, Figure 5-10. Poor voltage (dirty power) might show up in the form of sags, surges, spikes or impulses, notches or dropouts, or total loss of power, Figure 5-11. Disturbances that can change the sine wave may be caused by other electronic equipment operating in the same building.

Proper grounding of the AC distribution system, and the means by which electronic equipment is grounded, is critical, as these can affect the operation of electronic equipment. It is important that the ungrounded (HOT) conductor, the grounded (NEUTRAL) conductor, and the equipment (GROUND) conductor be properly sized, tightly connected, and correctly terminated. The equipment grounding conductor serves a vital function for a computer in that the computer's dc logic has one side directly connected to the metal frame of the computer, thus using ground as a reference point for the processing of information. An acceptable impedance for the grounding path associated with normal equipment grounding for branch-circuit wiring is 1 to 2 ohms. Applying Ohm's Law:

$$\text{Amperes} = \frac{\text{Volts}}{\text{Ohms}} = \frac{120}{1} = 120 \text{ amperes}$$

A ground-fault value of 120 amperes will easily cause a 15- or 20-ampere branch-circuit breaker to trip "off."

This meets the requirements of *NEC® Section 250-2(d)* which states that the ground path have "sufficiently low impedance to facilitate the operation of the circuit protective devices under fault conditions." *NEC® Section 250-118* lists the different items that are considered acceptable equipment grounding conductors.

The acceptable impedance of the ground path for branch-circuit wiring is *not* acceptable for grounding electronic equipment such as computer/data processing equipment. The acceptable impedance of a ground path for grounding electronic equipment is much more critical than the safety equipment ground impedance for branch-circuit wiring. The maximum impedance for grounding electronic equipment is 0.25 ohms. That is why a separate insulated isolated equipment grounding conductor is needed for the grounding of computer/data processing equipment to assure a low impedance path to ground. Furthermore, stray currents on the equipment grounding conductor can damage the electronic equipment, or destroy or

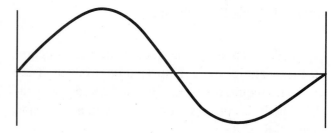

Fig. 5-10   Quality voltage.

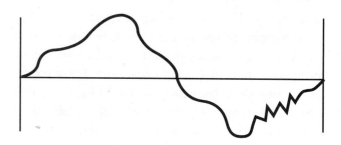

Fig. 5-11   Poor voltage.

alter the accuracy of the information being processed. Proper equipment grounding minimizes these potential problems. A separate insulated isolated equipment grounding conductor is sized according to *NEC® Table 250-122*. For longer runs, a larger size equipment grounding conductor is recommended to keep the impedance to 0.25 ohms or less. The electrical engineer designing the circuits and feeders for computer rooms and data processing equipment considers all of this. *NEC® Article 645* covers information technology equipment and systems. The National Fire Protection Association publication *NFPA-75* entitled Protection of Electronic Computer/Data Processing Equipment contains additional information.

For circuits supplying computer/data processing equipment, *never* use the grounded metal raceway as the *only* equipment ground for the electronic data processing equipment. Instead of allowing the grounded metal raceway system to serve as the equipment ground for sensitive electronic equipment, use a separate green *insulated* equipment grounding conductor that:

- connects to the isolated equipment grounding terminal of an isolated type receptacle.

- does not connect to the metal yoke of the receptacle.

- does not connect to the metal outlet or switch box.

- must be "isolated" from the metal raceway system.

- does not touch anywhere except at the connection at its source and at the grounding terminal of the isolated receptacle.

- must be run in the same raceway as the HOT and NEUTRAL wires.

- provides a ground path having an impedance of 0.25 ohms or less.

This separate insulated equipment grounding conductor is permitted by the *NEC®* to pass through a panelboard, and be carried all the way back to the main service, or to the transformer such as a 480-volt primary, 208/120-volt wye connected secondary step-down transformer. See *NEC® Sections 250-146(d) 384-20, Exception.*

To summarize grounding when electronic computer/data processing equipment is involved, generally there are two equipment grounds: (1) the metal raceway system that serves as the safety equipment grounding means for the metal outlet boxes, switch boxes, junction boxes, and similar metal equipment that is part of the premise wiring; and (2) a second insulated isolated equipment ground that serves the electronic equipment. This is sometimes referred to as a "dedicated" equipment ground.

NEC® Sections 250-146, FPN and *250-96, FPN* state that the "use of an isolated equipment grounding conductor does not relieve the requirement for grounding the raceway system and outlet box."

Figure 5-12 illustrates one way that receptacles serving computer type equipment can be connected.

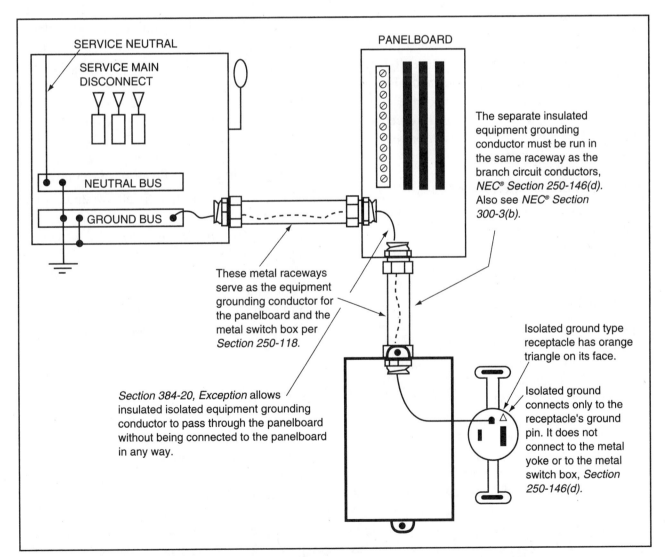

Fig. 5-12   Diagram showing an insulated isolated ground to reduce "noise" on circuits supplying computers.

The panel might have been provided with an equipment ground bus insulated from the panel. This would provide a means of splicing the one or more separate insulated isolated equipment grounding conductors running from the ground pin of the isolating type receptacles. One larger size insulated isolated equipment grounding conductor would then be run from the source (in this figure, the ground bus in the main switch) to the ground bus in the panel.

Another problem that occurs where there is a heavy concentration of electronic equipment is that of overheated neutral conductors. To minimize this problem on branch circuits that serve the electronic loads such as computers, it is highly recommended that separate neutrals be run for each phase conductor of the branch-circuit wiring rather than running a common neutral for multi-wire branch circuits. This is done to eliminate the problem of harmonic currents overheating the neutral conductor.

If surge protection is desired, this will absorb high voltage surges on the line and further protect the equipment, Figure 5-13. These are highly recommended in areas of the country where lightning strikes are common. When a comparison is made between the cost of the equipment, or the cost of recreating lost data, and the cost of the receptacle, probably under $50, this is inexpensive protection.

Ground-Fault Circuit Interrupter Receptacles

NEC® Section 210-8 sets forth the requirements for the installation of ground-fault circuit interrupters (GFCI). In commercial buildings only the 15 and 20-ampere receptacles installed on roof tops or in bathrooms are required to provide GFCI protection. Should a commercial building include for example a residential type kitchen, then serious consideration should be given to installing GFCI receptacles. For complete information on this type of installation it is suggested that *Electrical Wiring Residential*, Delmar Publishers be consulted.

Electrocutions and personal injury have resulted because of electrical shock from appliances such as radios, shavers, and electric heaters. This shock hazard exists whenever a person touches both the defective appliance and a conducting surface such as a water pipe, metal sink, or any conducting material that is grounded. To protect against this possibility of shock, 15- and 20-ampere branch circuits can be

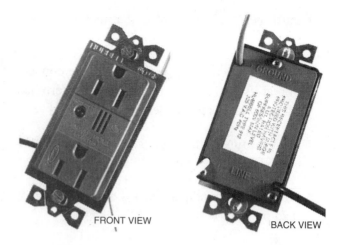

FRONT VIEW BACK VIEW

Fig. 5-13 Surge suppressor receptacle.

protected with GFCI receptacles or GFCI circuit breakers. The receptacles are the most commonly used.

The Underwriters Laboratories require that Class A GFCIs trip on ground-fault currents of 4 to 6 milliamperes (0.004 to 0.006 ampere). Figure 5-14 illustrates the principle of how a GFCI operates.

In bathrooms and on rooftops of commercial buildings all of the 15 and 20-ampere, 125-volt, receptacles shall provide GFCI protection.

Receptacles in Electric Baseboard Heaters

Electric baseboard heaters are available with or without receptacle outlets. Figure 5-15 shows the relationship of an electric baseboard heater to a receptacle outlet. *NEC® Article 424* of the Code states the requirement for fixed electric space heating equipment. This receptacle may be counted as one of the required outlets for the wall space utilized by the heater. See *NEC® Section 210-52*.

SNAP SWITCHES

The term "snap switch" is rarely used today except in the *National Electrical Code®* and the Underwriters Laboratories standards. Most electricians refer to snap switches as toggle switches or wall switches. Refer to Figure 5-16. These switches are divided into two categories.

Category 1 contains those ac/dc general-use snap switches that are used to control:

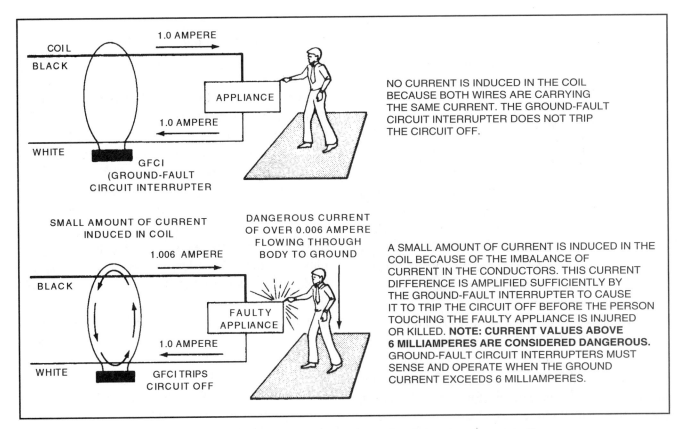

Fig. 5-14 Basic principle of ground-fault circuit interrupter operation.

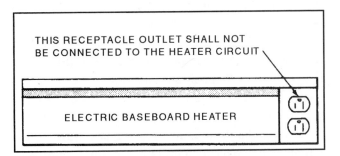

Fig. 5-15 Factory-mounted receptacle on permanently installed electric baseboard heater.

- alternating-current or direct-current circuits.
- resistive loads not to exceed the ampere rating of the switch at rated voltage.
- inductive loads not to exceed one-half the ampere rating of the switch at rated voltage.
- tungsten filament lamp loads not to exceed the ampere rating of the switch at 125 volts when marked with the letter "T." (A tungsten filament lamp draws a very high current at the instant the circuit is closed. As a result, the switch is subjected to a severe current surge.)

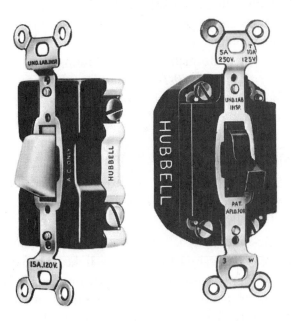

Fig. 5-16 General-use snap switches.

The ac/dc general-use snap switch normally is not marked *ac/dc*. However, it is always marked with the current and voltage rating, such as *10A-125V*, or *5A-250V-T*.

Category 2 contains those ac general-use snap switches that are used to control:

- alternating-current circuits only.

- resistive, inductive, and tungsten-filament lamp loads not to exceed the ampere rating of the switch at 120 volts.

- motor loads not to exceed 80% of the ampere rating of the switch at rated voltage.

Ac general-use snap switches may be marked *ac only*, or they may also be marked with the current and voltage rating markings. A typical switch marking is *15A, 120-277V ac*. The 277-rating is required on 277/480-volt systems. See *NEC® Section 210-6* for additional information pertaining to maximum voltage limitations.

Terminals of switches rated at 20 amperes or less, when marked *CO/ALR* are suitable for use with aluminum, copper, and copper-clad aluminum conductors, *NEC® Section 380-14(c)*. Switches not marked *CO/ALR* are suitable for use with copper and copper-clad aluminum conductors only.

Screwless pressure terminals of the conductor push-in type may be used with copper and copper-clad aluminum conductors only. These push-in type terminals are not suitable for use with ordinary aluminum conductors.

Further information on switch ratings is given in *NEC® Section 380-14* and in the Underwriters Laboratories *Electrical Construction Equipment Directory*.

Snap Switch Types and Connections

Snap switches are readily available in four basic types: single-pole, three-way, four-way, and double-pole.

Single-Pole Switch.   A single-pole switch is used where it is desired to control a light or group of lights, or other load, from one switching point. This type of switch is used in series with the ungrounded (hot) wire feeding the load. Figure 5-17 shows typical applications of a single-pole switch controlling a light from one switching point, either at the switch or at the light. Note that red has been used for the switch return (switch leg/switch loop) in the raceway (conduit) diagrams. Any color other than white, natural gray, or green could have been used for the switch return. When using cable for the

wiring of switches *NEC® Section 200-7(c)(2)* permits the use of a conductor with insulation that is white, natural gray, or having three continuous white stripes as a switch return *only* if the conductor is permanently re-identified by painting or other effective means at each location where the conductor is visible and accessible.

Three diagrams are shown for each of the switching connections discussed. The first diagram is a schematic drawing and is valuable when visualizing the current path. The second diagram represents the situation where a raceway will be available for installing the conductors. The third diagram illustrates the connection necessary when using a cable such as armored cable (AC), metal-clad cable (MC), or nonmetallic-sheathed cable (NMC). Grounding conductors and connections are not shown in order to keep the diagrams as simple as possible.

Three-Way Switch.   A three-way switch has a *common terminal* to which the switch blade is always connected. The other two terminals are called the *traveler terminals*, Figure 5-18. In one position, the switch blade is connected between the common terminal and one of the traveler terminals. In the other position, the switch blade is connected between the common terminal and the second traveler terminal. The three-way switch can be identified readily because it has no *On* or *Off* position. Note that *On* and *Off* positions are not marked on the switch handle in Figure 5-18. The three-way switch is also identified by its three terminals. The common terminal is darker in color than the two traveler terminals which have a natural brass color. Figure 5-19 shows the application of three-way switches to provide control at the light or at the switch. Travelers are also referred to as "kickers" or "dummies."

In Figure 5-19(A) and (B), red has been used for the switch return (switch leg/switch loop) in the raceway (conduit) diagrams. Any color other than white, natural gray, or green could have been used for the switch return. For the *travelers*, select a color that is different than the switch return, such as a pair of yellows, a pair of blues, a pair of browns, etc. When using cable for the wiring of three-way switches *NEC® Section 200-7(c)(2)* permits the use of a conductor with insulation that is white, natural gray, or having three continuous white stripes as a switch return *only* if the conductor is permanently

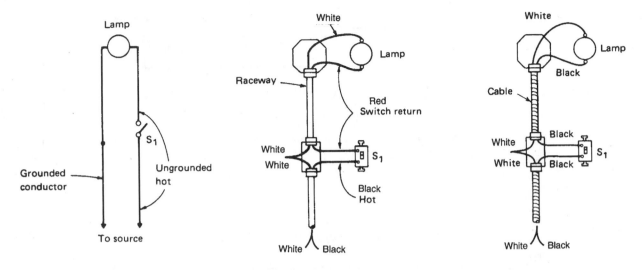

(A) Circuit with single-pole switch-feed at switch

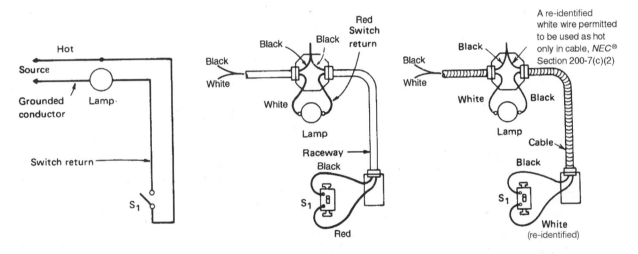

Fig. 5-17 Single-pole switch connection.

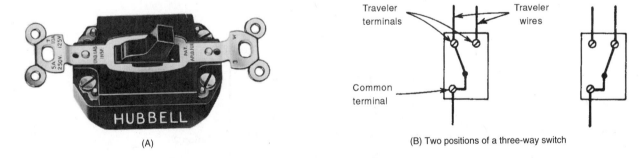

(A)   (B) Two positions of a three-way switch

Fig. 5-18 Three-way switch and positions.

re-identified by painting or other effective means at each location where the conductor is visible and accessible.

Four-Way Switch. A four-way switch is similar to the three-way switch in that it does not have *On*

and *Off* positions. However, the four-way switch has four terminals. Two of these terminals are connected to traveler wires from one three-way switch and the other two terminals are connected to traveler wires from another three-way switch, Figure 5-20. In

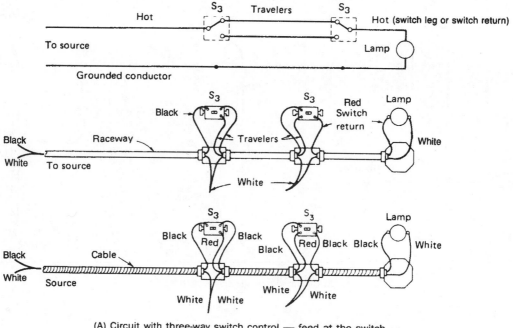

(A) Circuit with three-way switch control — feed at the switch

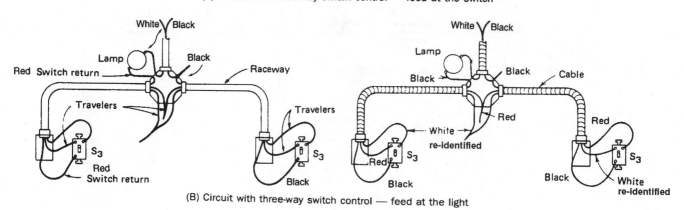

(B) Circuit with three-way switch control — feed at the light

Fig. 5-19   Three-way switch connections.

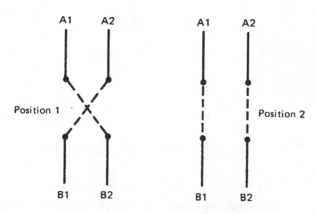

Two positions of four-way switch

Fig. 5-20   Four-way switch operation.

Figure 5-20, terminals A1 and A2 are connected to one three-way switch and terminals B1 and B2 are connected to the other three-way switch. In position 1, the switch connects A1 to B2 and A2 to B1. In position 2, the switch connects A1 to B1 and A2 to B2.

The four-way switch is used when a light or a group of lights, or other load, must be controlled from more than two switching points. The switches that are connected to the source and the load are three-way switches. At all other control points, however, four-way switches are used. Figure 5-21 illustrates a typical circuit in which a lamp is controlled from any one of three switching points. Note that red has been used for the switch return (switch leg/switch loop) in the raceway (conduit) diagrams. Any color other than white, natural gray, or green could have been used for the switch return. For the

travelers, select a color that is different than the switch return, such as a pair of yellows, a pair of blues, a pair of browns, etc. For cable wiring, the white conductor may not be used as the switch return, unless it has been permanently reidentified. In Figure 5-21, the arrangement is such that white conductors were spliced together. Care must be used to ensure that the traveler wires are connected to the proper terminals of the four-way switch. That is, the two traveler wires from one three-way switch must be connected to the two terminals on one end of the four-way switch. Similarly, the two traveler wires from the other three-way switch must be connected to the two terminals on the other end of the four-way switch.

Double-Pole Switch. A double-pole switch is rarely used on lighting circuits. As shown in Figure 5-22, a double-pole switch can be used for those

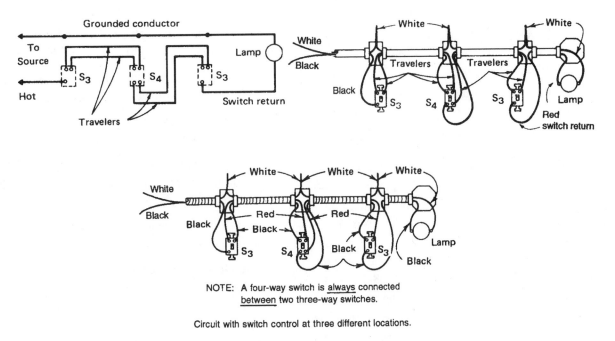

NOTE: A four-way switch is always connected between two three-way switches.

Circuit with switch control at three different locations.

Fig. 5-21   Four-way switch connections.

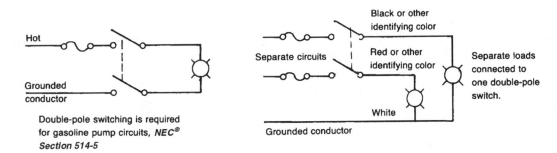

Fig. 5-22   Double-pole switch connections.

installations where two separate circuits are to be controlled with one switch. All conductors of circuits supplying gasoline dispensing pumps, or running through such pumps, must have a disconnecting means. Thus, the lighting mounted on gasoline dispensing islands may require two-pole switches, *NEC® Section 514-5.*

CONDUCTOR COLOR CODING

Some color coding is mandated by the *National Electrical Code®.* These are the *grounded conductors* and *equipment grounding conductors.* Other color coding choices are left up to the electrician. Dozens of colors are available. Reference to color coding is found in *NEC® Sections 200-6, -7, 210-5, 215-8,* and *310-12®.*

Grounded Conductor.   A grounded conductor is often referred to as a neutral conductor. A neutral conductor is always grounded under the requirements of *NEC® Section 250-20,* but a grounded conductor is not always a neutral conductor, such as in the case of the grounded phase conductor of a three-phase, grounded B phase system.

NEC® Section 200-6(a) requires that a grounded conductor, sizes No. 6 or smaller, be identified by a continuous white or natural gray color. *NEC® Section 200-6(b)* permits grounded conductors larger than No. 6 AWG to be identified at its terminations. For example, this would allow a black conductor to be identified in a disconnect switch or panel with white paint.

When more than one system is installed in the same raceway, such as combinations of 208/120 volt wye and 480/277 volt wye circuits, the grounded conductor of both systems must be white or gray in color, but one of the grounded conductors must be identified further by "an identifiable colored stripe (not green) running along the insulation . . . or other and different means of identification." You must be able to determine which white or gray grounded conductor belongs to which set of "hot" conductors.

A white or natural gray conductor must not be used as a "hot" phase conductor except when reidentified according to *NEC® Section 200-7,* which requires it be a part of a cable or flexible cord.

Equipment Grounding Conductor.   An equipment grounding conductor may be bare, green, or green with one or more yellow stripes, see *NEC® Section*

250-119. Never use a green insulated conductor for a "hot" phase conductor.

Typical Color Coding (For Conduit Wiring)

The following are some examples of color coding quite often used for the conductors of branch circuits, feeders, and services. These color coding recommendations are not code mandated, other than for the white (or natural gray) grounded conductor. Certainly other color combinations are permitted. Just be sure that you do not use white, natural gray, green, and bare conductors other than for their permitted uses discussed previously.

Two-wire, single phase: black, white

Three-wire, single phase: black, white, red

Three-wire, three-phase delta: brown, orange, yellow

Three-wire, three-phase delta grounded B phase: black, red, white for the "B" phase

Four-wire (three-phase, four-wire wye 208/120 volt): black, white, red, blue

Four-wire (three-phase, four-wire wye 480/277 volt): brown, white, yellow, orange

Four-wire (three-phase, four-wire delta with "high leg"): brown, white, orange (the high leg), yellow

Switch legs: Use a different color than the phase conductors.

Travelers: Travelers are used for three-way and four-way switch connections. Use conductors that have a color different than the phase conductors or the actual switch leg; for example, a pair of blue conductors, or a pair of brown conductors.

Typical Color Coding (For Cable Wiring)

When wiring with cable (NMC, AC, or MC), the choice of colors is limited to those that are furnished with the cable.

Two-wire cable: black, white, bare or green equipment grounding conductor

Three-wire cable: black, white, red, bare or green equipment grounding conductor

Four-wire cable: black, white, red, blue, bare or green equipment grounding conductor

Five-wire cable: black, white, red, blue, yellow, bare or green equipment grounding conductor

Several of the wiring diagrams in this unit illustrate how the white wire is correctly used in switching circuits.

SWITCH AND RECEPTACLE COVERS

The cover that is placed on a recessed box containing a receptacle or a switch is called a faceplate, and a cover placed on a surface mounted box, such as a 4-inch square, is called a raised cover.

Faceplates come in a variety of colors, shapes, and materials, but for our purposes they can be placed in two categories, insulating and metal. See *NEC® Section 410-56(d)*. Metal faceplates can become a hazard because they can conduct electric-ity; this problem is addressed in *NEC® Section 380-9*. Metal faceplates are considered to be effectively grounded through the No. 6-32 screws that fasten the faceplate to the grounded yoke of a receptacle or switch. If the box is nonmetallic, the yoke must be grounded and both switches and receptacles are available with a grounding screw for grounding the metal yoke.

In the past it was common practice for a receptacle to be fastened to a raised cover by a single No. 6-32 screw. *NEC® Section 410-56(f)(3)* now prohibits this practice and requires that two screws or another approved method be used to fasten a receptacle to a raised cover.

Receptacles that are installed outdoors in damp or wet locations, as stated in *NEC® Section 410-57(a) & (b)*, must be covered with weatherproof covers that maintain their integrity when the receptacle is in use. The cover would be deep enough to shelter the attachment plug.

REVIEW QUESTIONS

Refer to the *National Electrical Code®* or the working drawings when necessary. Where applicable, responses should be written in complete sentences. Write units using unit names, do not use abbreviations or symbols (1 foot, not 1' or 1 ft).

Indicate which of the following switches may be used to control the loads listed in questions 1–7.

A. Ac/dc   10 A-125 V   5 A-250 V
B. Ac only   10 A-120 V
C. Ac only   15 A-120/277 V
D. Ac/dc   20 A-125 VT   10 A-250 V

1. A 120-volt incandescent lamp load (tungsten filament) consisting of ten 150-watt lamps _____

2. A 120-volt fluorescent lamp load (inductive) of 1500 volt-amperes _____

3. A 277-volt fluorescent lamp load of 625 volt-amperes _____

4. A 120-volt motor drawing 10 amperes _____

5. A 120-volt resistive load of 1250 watts _____

6. A 120-volt incandescent lamp load of 2000 watts _____

7. A 230-volt motor drawing 3 amperes _____

Using the information in Tables 5-2 and 5-3, select by number the correct receptacle for each of the following loads.

8. 120/208-volt, single-phase, 20-ampere load _____

9. 230-volt, three-phase, 50-ampere load _____

10. 208-volt, single-phase, 30-ampere load _____

11. 120-volt, single-phase, 20-ampere, hospital grade _____

12. The metal yoke of an isolated grounding type receptacle (is) (is not) connected to the green equipment grounding terminal of the receptacle. (Circle the correct answer.)

13. Check the correct statement:

☐ The recommended method to ground computer/data processing equipment is to use the grounded metal raceway as the equipment grounding conductor.

☐ The recommended method to ground computer/data processing equipment is to install a separate insulated equipment grounding conductor in the same raceway that the branch-circuit conductors are installed in.

14. Check the correct statement:

When a separate insulated equipment grounding conductor is installed to minimize line disturbances on the circuit supplying an isolated grounding type receptacle, it:

☐ must terminate in the panelboard where the branch circuit originates.

☐ may run through the panelboard where the branch circuit originates, back to the main disconnect (the source) for the building.

15. Check the correct statement:

To eliminate the overheated neutral problem associated with branch-circuit wiring that supplies computer/data processing equipment:

☐ install a separate neutral for each phase conductor.

☐ install a common neutral on multi-wire branch circuits.

An electrical system has been installed for three-way control of a light. The power source is available at one of the three-way switches; the load is located between the three-way switches.

16. Draw a wiring diagram illustrating the connection of the conductors.

17. Draw in the conductors in the raceway diagram below. Draw in the connections to the switches and the light. Label the conductors for color and function.

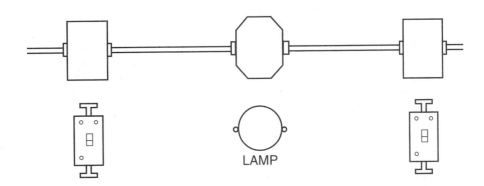

LAMP

18. Below is the same arrangement using nonmetallic sheathed cable. Draw in connections to the switches and the light. Label the conductors for color and function. A two-wire cable would contain black and white conductors, a three-wire cable would contain red, black, and white conductors:

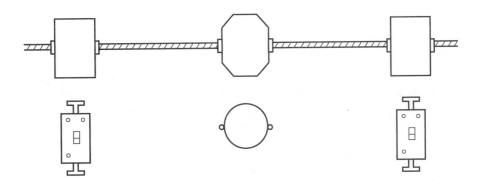

19. Test questions 20 and 21 show the installation of the three-way control of two lamps where the switches are installed between the two lamps. Draw a wiring diagram of the connection of the system.

20. Below is a diagram showing the installation of a metallic raceway system connecting two lamps and two three-way switches. Draw in the necessary conductors and show their connections.

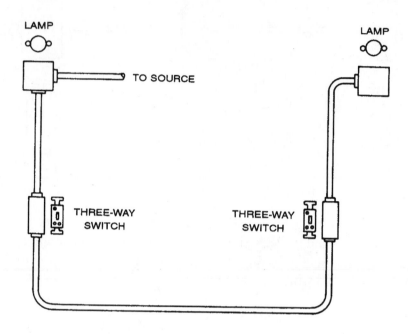

21. Below is a diagram showing the installation of a nonmetallic sheathed cable system connecting two lamps and two three-way switches. Draw in the necessary conductors and show their connections.

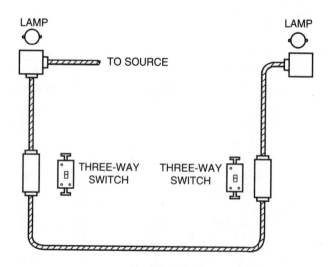

22. Test questions 23 and 24 show the installation of the three-way and four-way controls of a lamp where the supply is brought to the lamp outlet and the raceway from the lamp goes to the center of the three switches. Draw a wiring diagram of the connection of the system.

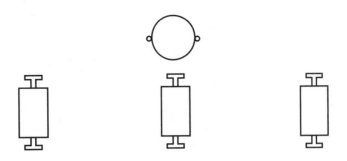

23. Below is a diagram showing the installation of a metallic raceway system connecting a lamp outlet to a four-way switch system. Draw in the necessary conductors and show their connections.

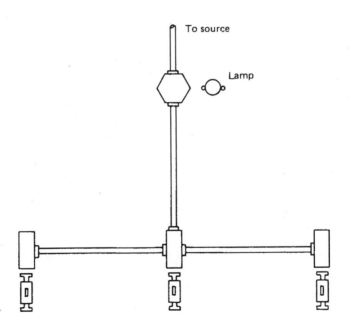

24. Below is a diagram showing the installation of a nonmetallic sheathed cable system connecting a lamp and a four-way switch system. Draw in the necessary conductors and show their connections.

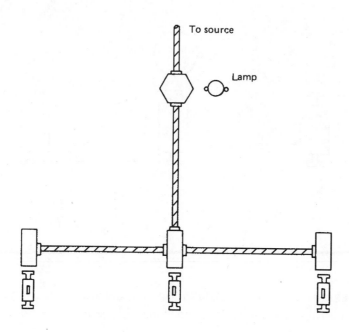

UNIT 6

Branch-Circuit Installation

OBJECTIVES

After studying this unit, the student will be able to

- select the proper raceway for the conditions.
- identify the installation requirements for a raceway.
- select the proper raceway size dependent upon the conductors to be installed.
- select the proper size of box dependent upon the fill.
- select the proper size of box dependent upon entering raceways.

In a commercial building, the major part of the electrical work is the installation of the branch-circuit wiring. The electrician must have the ability to select and install the correct materials to insure a successful job.

The term *raceway*, which is used in this unit as well as others, is defined by the *NEC®* as a channel for holding wires, cables, or bus bars that is designed and used expressly for this purpose.

The following paragraphs describe several types of materials that are classified as raceways, including rigid metal conduit, electrical metallic tubing, intermediate metal conduit, flexible conduit, rigid nonmetallic conduit, and electrical nonmetallic tubing.

It will be useful to review some of the terms used in the *NEC®* concerning raceways:

- A ferrous conduit is made of iron or steel.
- The common nonferrous raceway metal is aluminum.
- Metal or metallic would include both ferrous and nonferrous.
- Nonmetallic raceway is commonly referred to as plastic.
- Couplings are used to couple sections of raceway.

- Connectors are used to fasten raceways to boxes or fittings.
- Integral couplings/connectors are formed into the raceway.
- Associated couplings/connectors are separate items.
- A running thread is a double-length thread cut on one conduit. Connection to another conduit is achieved by screwing a coupling on the running thread, butting the conduits together, then backing the coupling on to the second conduit. These are not allowed.

RIGID METAL CONDUIT (RMC)

Rigid metal conduit, Figure 6-1, is of heavy-wall construction to provide a maximum degree of physical protection to the conductors run through it. Rigid conduit is available in either steel or aluminum. The

Fig. 6-1   Rigid steel conduit.

conduit can be threaded on the job, or nonthreaded fittings may be used where permitted by local codes, Figure 6-2. See *NEC® Article 346* and Figure 6-2.

RMC bends can be purchased or they can be made using special bending tools. Bends in ½-inch, ¾-inch, and 1-inch conduit can be made using hand benders or hickeys, Figure 6-3. Hydraulic benders must be used to make bends in larger sizes of conduit.

INTERMEDIATE METAL CONDUIT (IMC)

Intermediate metal conduit has a wall thickness that is between that of rigid metal conduit and EMT. This type of conduit can be installed using either

Fig. 6-2   Rigid metal conduit fittings.

Fig. 6-3   Tubing benders (without handles).

threaded or nonthreaded fittings. *NEC® Article 345* defines intermediate metal conduit, the uses permitted and not permitted, the installation requirements, and the construction specifications.

ELECTRICAL METALLIC TUBING (EMT)

EMT is a thin wall metal raceway that is not to be threaded, Figure 6-4. The specifications for the commercial building permits the use of EMT for all branch-circuit wiring. *NEC® Article 348* should be consulted for exact requirements for installation.

Fittings

Electrical metallic tubing is nonthreaded, thin-wall conduit. Since EMT is not to be threaded, conduit sections are joined together and connected to boxes, other fittings, or cabinets by fittings called *couplings* and *connectors*. Several styles of EMT fittings are available, including setscrew, compression, and indenter styles.

Setscrew.   When used with this type of fitting, the EMT is pushed into the coupling or connector and is secured in place by tightening the setscrews, Figure 6-5. This type of fitting is classified as concrete-tight.

Compression.   EMT is secured in these fittings by tightening the compressing nuts with a wrench or pliers, Figure 6-6. These fittings are classified as raintight and concrete-tight types.

Indenter.   A special tool is used to secure EMT in this style of fitting. The tool places an indentation in both the fitting and the conduit. It is a standard

Fig. 6-4   Electrical metallic tubing.

Fig. 6-5   EMT connector and coupling, setscrew type.

Fig. 6-6   EMT connector and coupling, compression type.

Fig. 6-7   EMT connector, indenter type.

wiring practice to make two sets of indentations at each connection. This type of fitting is classified as concrete-tight. Figure 6-7 shows a straight indenter connector.

Installing

The efficient installation of EMT requires the use of a bender, Figure 6-8. This tool is commonly available in hand-operated models for EMT in sizes from ½ inch to 1 inch, and in power-operated models for EMT in sizes greater than 1 inch.

Three kinds of bends can be made with the use of the bending tool. The stub bend, the back-to-back bend, and the angle bend are shown in Figure 6-9. The manufacturer's instructions that accompany each bender indicate the method of making each type of bend.

Installation of Metallic Raceway

RMC, IMC, and EMT are to be installed according to the requirements of *NEC® Articles 346, 345 and 348* respectively. The following points summarize the contents of these articles. All conduit runs should be level, straight, plumb, neat in a good workmanship manner. Don't do sloppy work.

The rigid types of metal conduit:

- may be installed in concealed and exposed work.

- may be installed in or under concrete when of the type approved for this purpose.

- must not be installed in or under cinder concrete or cinder fill that is subject to permanent moisture unless the conduit is encased in at least two inches (50.8 mm) of noncinder concrete, or is at least 18 inches (457 mm) under the fill, or is of corrosion-resistant material suitable for the purpose.

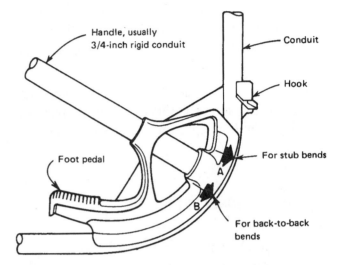

Fig. 6-8   EMT bender.

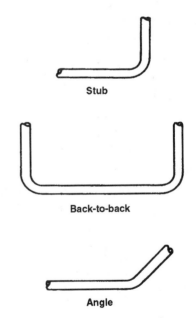

Fig. 6-9   Conduit bends.

- must not be installed where subject to severe mechanical damage (an exception to this is rigid metal conduit which may be installed in a location where it is subject to damaging conditions).

- may contain up to four quarter bends (for a total of 360 degrees) in any run.

- must be fastened within three feet (914 mm) of each outlet box, junction box, cabinet, or conduit body.

- must be securely supported at least every 10 feet (3.05 m).

- may be installed in wet or dry locations if the conduit is of the type approved for this use.

- must have the ends reamed to remove rough edges.

- conduit is considered adequately supported when run through drilled, bored, or punched holes in framing members such as studs or joists.

FLEXIBLE CONNECTIONS
(NEC® Articles 350 and 351)

The installation of certain equipment requires flexible connections, both to simplify the installation and to stop the transfer of vibrations.

The two basic types of material used for these connections are flexible metal conduit (FMC), Figure 6-10, and liquidtight flexible metal conduit (LFMC), Figure 6-11.

NEC® Article 350 regulates the use and installation of FMC and LFMC. FMC is similar to armored cable, except that the conductors are installed by the electrician. For armored cable, the cable armor is wrapped around the conductors at the factory to form a complete cable assembly.

Some of the more common installations using flexible metal conduit are shown in Figure 6-12.

Fig. 6-11   Liquidtight flexible conduit and fitting.

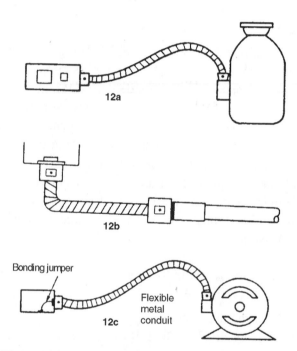

Fig. 6-12   Installations using flexible metal conduit. See *Article 350*. Also see Underwriters Laboratories White Book.

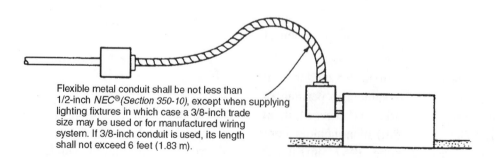

Flexible metal conduit shall be not less than 1/2-inch *NEC®(Section 350-10)*, except when supplying lighting fixtures in which case a 3/8-inch trade size may be used or for manufactured wiring system. If 3/8-inch conduit is used, its length shall not exceed 6 feet (1.83 m).

Fig. 6-10   Flexible metal conduit.

Note that the flexibility required to make the installation is provided by the flexible metal conduit. The figure calls attention to the *National Electrical Code®* and Underwriters Laboratories restrictions on the use of flexible metal conduit with regard to relying on the metal armor as a grounding means.

The use and installation of LFMC is described in *NEC® Article 351*. LFMC has a tighter fit of its spiral turns as compared to FMC. LFMC has a thermoplastic outer jacket that is liquidtight and is commonly used as a flexible connection to central air-conditioning units located outdoors, Figure 6-13. Flexible metal and nonmetallic conduit is commonly used to connect recessed lighting fixtures. When used in this manner, electricians refer to it as a "fixture whip."

FMC of a trade size of 3/4 inch or less may be used as a grounding means if

* it is listed as a grounding means.
* it is not over 6 feet in length.
* it is connected with fittings listed for grounding purposes.
* the circuit is rated at 20-ampere or less.

See Figure 6-12a.

FMC of a trade size larger than ¾ inch may be used as a grounding means if:

* it is listed as a grounding means.
* it is not over 6 feet in length.
* it is connected with fittings marked "GRND."

See Figure 6-12b.

FMC of any size may be installed with a bonding jumper

* outside the conduit if it is not over 6 feet in length.
* inside the conduit.

See Figure 6-12c

LFMC of a trade size of ½ inch or less may be used as a grounding means if:

* it is listed as a grounding means.
* it is not over 6 feet in length.
* it is connected with fittings listed for grounding purposes.
* the circuit is rated at 20-ampere or less.

See Figure 6-13

LFMC of a trade size of 3/4 inch, 1 or 1 1/4-inch may be used as a grounding means if:

* it is listed as a grounding means.
* it is not over 6 feet in length.
* it is connected with fittings listed for grounding purposes.
* the circuit is rated at 60-ampere or less.

See Figure 6-13

LFMC of any size or length may be installed with a required bonding jumper

See Figure 6-14.

Figures 6-13, 6-14 and 6-15 illustrate the limitations placed on the use of liquidtight flexible metal conduit as a grounding means. These limitations are given in *NEC® Section 351-9*.

Liquidtight flexible nonmetallic conduit (LFNC) may be used:

* in exposed or concealed locations.

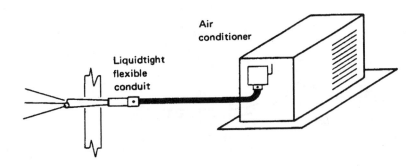

Fig. 6-13   Use of liquidtight flexible metal conduit.

Fig. 6-14   Limitations on the use of liquidtight flexible metal conduit.

- where flexibility is required.
- for direct burial, when so listed.
- where not subject to damage.
- where the combination of ambient and conductor temperature does not exceed that for which the flexible conduit is approved.
- in lengths not over 6 feet (1.83 m) unless certain provisions are adhered to.
- with fittings identified for use with the flexible conduit.
- in sizes ½ inch to 2 inches. For enclosing motor leads, ⅜ inch is suitable.
- for fixture whips not over 6 feet in length permitted or required by *NEC® Section 410-67(c)*, or for flexible connection to equipment, such as a motor.

When metal fittings are not marked *GRND*, it can be assumed that they are not approved for equipment grounding purposes. In this case, a separate equipment grounding conductor must be installed. The conductor is sized according to *NEC® Table 250-122*. Figure 6-15 illustrates the application of this table.

There are three types of liquidtight flexible nonmetallic conduit, which are marked:

LFNC-A for layered conduit

LFNC-B for integral conduit

LFNC-C for corrugated conduit

Both metal and nonmetallic fittings listed for us with the various types of LFNC will be marked with the "A," "B," or "C" designations.

Liquidtight flexible nonmetallic conduit is restricted to a six foot length unless flexibility needs require a longer length; see *NEC® Section 351-23(b)(3)*.

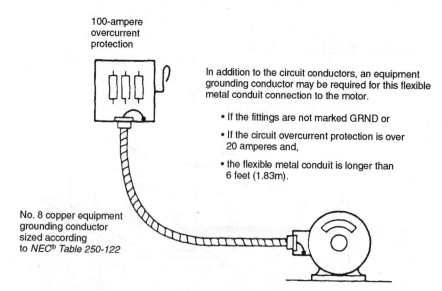

100-ampere
overcurrent
protection

In addition to the circuit conductors, an equipment grounding conductor may be required for this flexible metal conduit connection to the motor.

- If the fittings are not marked GRND or
- If the circuit overcurrent protection is over 20 amperes and,
- the flexible metal conduit is longer than 6 feet (1.83m).

No. 8 copper equipment grounding conductor sized according to *NEC® Table 250-122*

Fig. 6-15   Equipment grounding conductor in flexible metal conduit. See *NEC® Section 350-14*.

RIGID NONMETALLIC CONDUIT (RNC)

RNC is made of polyvinyl chloride, which is a plastic material. Solvent welded fittings may be used for RNC connections and terminations.

RNC may be used:

- concealed in walls, floors, and ceilings.

- in cinder fill.

- in wet and damp locations.

- for exposed work where not subject to physical damage unless identified for such use. (Schedule 80 is so listed.)

- for underground installations.

RNC may not be used:

- in hazardous (classified) locations except for limited use as set forth in *NEC® Sections 501-4(b), 504-20, 514-8,* and *515-5.*

- for support of lighting fixtures.

- where exposed to physical damage unless so identified (Schedule 80 is so listed).

- where subject to ambient temperatures exceeding 50°C (122°F).

- for conductors whose insulation temperature limitations would exceed those for which the conduit is listed. (Check the marking or listing label to obtain this temperature rating.)

- for patient care areas in hospitals and health care facilities.

- in places of assembly and theaters except as provided in *NEC® Articles 518* and *520.*

On an outdoor exposed run of RNC, it is important that expansion joints be installed to comply with the requirements set forth in *NEC® Section 347-9,* where expansion characteristics of RNC are given. An equipment grounding conductor must be installed in RNC, and, if metal boxes are used, the grounding conductor must be connected to the metal box. See *NEC® Article 347* for additional information for the installation of RNC.

RNC Fittings and Boxes

A complete line of fittings, boxes, and accessories are available for RNC. Where RNC is connected to a box or enclosure, a box adapter is recommended over a male connector because the box adapter is stronger and provides a gentle radius for wire pulling.

ELECTRICAL NONMETALLIC TUBING (ENT)

ENT is a pliable, corrugated raceway that is made of polyvinyl chloride, a plastic material that is also used to make RNC. ENT is hand bendable and is available in coils and reels to facilitate a single length installation between pull points. ENT is designed for use within a building or encased in concrete and is not intended for outdoor use. The specification for the commercial building allows the use of ENT for feeders, branch circuits, and telephone wiring where a raceway size of 2 inches or less is required. ENT is available in a variety of colors, which allows the color coding of the various systems; electrical circuits could be blue, telephone, red, etc.

ENT may be used:

- exposed, where not subject to physical damage, in buildings not more than three floors above grade.

- concealed, without regard to building height, in walls, floors, and ceilings where the walls, floors, or ceilings are of a material with a 15-minute finish rating. (½" UL classified gypsum board has a 15-minute finish rating.) Check with manufacturer for further information.

- subject to corrosive influences (check with manufacturer's recommendations).

- in damp and wet locations, with fittings identified for the use.

- above a suspended ceiling where the ceiling title has a 15-minute finish rating. The ceiling title does not need a finish rating if the building is not more than three floors above grade.

- embedded in poured concrete, including slab-on-grade or below grade, where fittings identified for the purpose are used.

ENT may not be used:

- in hazardous (classified) locations except as permitted by *NEC® Section 504-20.*

- for lighting fixture or equipment support.

- subject to ambient temperature in excess of 50°C (122°F).

- for conductors whose insulation temperature would exceed those for which ENT is listed. (Check the UL label for the actual temperature rating.)

- for direct earth burial.

- where the voltage is over 600 volts.

- in places of assembly and theaters except as provided in *NEC® Articles 518* and *520*.

- See *NEC® Article 331* for other ENT installation requirements.

ENT Fittings

Two styles of listed mechanical fittings are available for ENT; one piece snap-on and a clamshell variety. Solvent cement PVC fittings for RNC are also listed for use with ENT. The package label will indicate which of these fittings are concrete tight without tape.

ENT Boxes and Accessories

Wall boxes and ceiling boxes, with knockouts for ½"-1" ENT, are available. ENT can also be connected to any RNC box of access fitting. ENT mud boxes listed for fixture support are available, with knockouts for ½"-1" sizes.

RACEWAY SIZING

▶ The conduit size required for an installation depends upon three factors: (1) the number of conductors to be installed, (2) the cross-sectional area of the conductor, and (3) the permissible raceway fill. The relationship of these factors is defined in *NEC® Chapter 9, Notes 1* through *9* and *Tables 1, 4,* and *5*. After examining the working drawings and determining the number of conductors to be installed to a certain point, either of the following procedures can be used to find the conduit size. Refer to Tables 6-1 and 6-2. ◀

▶ If all of the conductors have the same insulation, the raceway size can be determined directly by referring to *NEC® Chapter 9, Appendix C*.

For example, assume that 3 No. 8 AWG, Type THWN conductors are to be installed in electrical metallic tubing to an air-conditioning unit. Referring to *NEC® Chapter 9, Table C1* it will be determined that a ½-inch raceway is acceptable.

Next, assume that a short section of the EMT is replaced by liquidtight flexible metal conduit. To comply with *NEC® Section 351-9*, this change requires the installation of a grounding conductor. Since the rating of the circuit would be greater than 20-ampere and less than 60-ampere *NEC® Table 250-122* specifies the use of a No. 10 AWG. The grounding must be included in the conductor fill calculations in compliance with *NEC® Chapter 9, Tables, Note 3*.

TABLE 6-1 Area of EMT for Permitted Combinations of Conductors					
(excerpted from *NEC® Chapter 9, Table 4*.)					
ELECTRICAL METALLIC TUBING					
Trade Size in Inches	Internal Diameter in Inches	Total Area 100% in Sq. In.	2 Wires 31% in Sq. In.	Over 2 Wires 40% in Sq. In.	1 Wire 53% in Sq. In.
1/2	0.622	0.304	0.094	0.122	0.161
3/4	0.824	0.533	0.165	0.213	0.283
1	1.049	0.864	0.268	0.346	0.458
1 1/4	1.380	1.496	0.464	0.598	0.793
1 1/2	1.610	2.036	0.631	0.814	1.079
2	2.067	3.356	1.040	1.342	1.778
2 1/2	2.731	5.858	1.816	2.343	3.105
3	3.356	8.846	2.742	3.538	4.688
3 1/2	3.834	11.545	3.579	4.618	6.119
4	4.334	14.753	4.573	5.901	7.819

Reprinted with permission from NFPA 70-1999.

TABLE 6-2	Cross-Sectional Area of Commonly Used Sizes and Types of Conductors	
	(excerpted from *NEC® Chapter 9, Table 5*)	
Conductor Type	**Size**	**Approximate Cross-Sectional Area in Square Inches**
TW	12	0.0181
	10	0.0243
	8	0.0437
	6	0.0726
THW	12	0.0260
	10	0.0333
	8	0.0556
	6	0.0726
THHN	12	0.0133
	10	0.0211
	8	0.0366
	6	0.0507
THWN	12	0.0133
	10	0.0211
	8	0.0366
	6	0.0507

Reprinted with permission from NFPA 70-1999.

1 No. 10 AWG, Type
  THWN @ 0.0133 sq. in.   0.0133 sq. in.
3 No. 8 AWG, Type
  THWN @ 0.0366 sq. in.   0.1098 sq. in.
Total cross-sectional area   0.1231 sq. in.

Referring to *NEC® Chapter 9, Table 4*: for electrical metallic tubing the maximum allowable fill for a ½ inch raceway is 0.122 square inch, for liquidtight flexible metal conduit the maximum allowable fill for a ½ inch raceway is 0.125 square inch. As both fill values must be 0.1231 sq. in. or greater, the use of a ½ raceway is not permitted.

TABLE 6-3	Chapter 9, Tables and Examples	
A. Tables		
Table 1. Percent of Cross Section of Conduit and Tubing for Conductors		
Number of Conductors	1	2	Over 2
All Conductor Types	53	31	40

(FPN): Table 1 is based on common conditions of proper cabling and alignment of conductors where the length of the pull and the number of bends are within reasonable limits. It should be recognized that, for certain conditions, a larger size conduit or a lesser conduit fill should be considered.

(continued)

SUMMARY OF CONDUIT AND TUBING TYPES						
	Electrical Nonmetallic Tubing	**Intermediate Metal Conduit**	**Rigid Metal Conduit**	**Rigid Nonmetallic Conduit**	**Electrical Metallic Tubing**	**Flexible Metallic Tubing**
NEC®	Article 331	Article 345	Article 346	Article 347	Article 348	Article 349
Min.-Max. Size.	½"–2"	½"–4"	½"–6"	½"–6"	½"–4"	⅜"–¾"
Min. Radii of Bends.	Table 346-10	Table 346-10	Table 346-10	Table 346-10	Table 346-10	Table 349-20(a)
Min. Spacing of Supports.	3' outlet. 3' apart.	3' box/outlet. 10' apart.	3' box/outlet. 10' apart or see Table 346-12.	3' box/outlet. See Table 347-8.	3' box/outlet. 10' apart.	No stipulation.
Uses Allowed.	In buildings not over 3 floors. Concealed if 15 min. fire rating. Dry & damp.	All conditions. Under cinder fill if down 18" or in 2" concrete.	All conditions. Under cinder fill if down 18" or in 2" concrete.	All conditions.	Under cinder fill if down 18" or in 2" concrete.	Dry, accessible locations.
Uses Not Allowed.	Hazardous. Support of equipment. Direct burial. In theaters.[1]	See *Section 300-6* for protection.	See *Section 300-6* for protection.	In hazardous. In theaters.[1]	Where subject to damage. Hazardous. See *Section 300-6* for protection.	Hazardous. Underground. In concrete. Lengths over 6'.
Miscellaneous.	Pliable. Corrugated. Non-grounding.	Grounding. Threaded with taper die.	Grounding. Threaded with taper die.	Expansion joints required. Non-grounding.	Not threaded.	Liquidtight without a nonmetallic jacket. See *Table 350-12* for fill of ⅜".

[1]O.K. if encased in 2" or more of concrete.

TABLE 6-3   (continued)
Notes to Tables

Note 1:   See Appendix C for the maximum number of conductors and fixture wires, all of the same size (total cross-sectional area including insulation), permitted in trade sizes of the applicable conduit or tubing.

Note 2:   Table 1 applies only to complete conduit or tubing systems and is not intended to apply to sections of conduit or tubing used to protect exposed wiring from physical damage.

Note 3:   Equipment grounding or bonding conductors, where installed, shall be included when calculating conduit or tubing fill. The actual dimensions of the equipment grounding or bonding conductor (insulated or bare) shall be used in the calculation.

Note 4:   Where conduit or tubing nipples having a maximum length not to exceed 24 in. (610 mm) are installed between boxes, cabinets, and similar enclosures, the nipples shall be permitted to be filled to 60 percent of their total cross-sectional area, and Article 310, Note 8(a) of Notes to Ampacity Tables of 0 to 2000 Volts, need not apply to this condition.

Note 5:   For conductors not included in Chapter 9, such as multiconductor cables, the actual dimensions shall be used.

Note 6:   For combinations of conductors of different sizes, use Tables 5 and 5A in Chapter 9 for dimensions of conductors and Table 4 in Chapter 9 for the applicable conduit or tubing dimensions.

Note 7:   When calculating the maximum number of conductors permitted in a conduit or tubing, all of the same size (total cross-sectional area including insulation), the next higher whole number shall be used to determine the maximum number of conductors permitted when the calculation results in a decimal of 0.8 or larger.

Note 8:   Where bare conductors are permitted by other sections of this *Code*, the dimensions for bare conductors in Table 8 of Chapter 9 shall be permitted.

Note 9:   A multiconductor cable of two or more conductors shall be treated as a single conductor for calculating percentage conduit fill area. For cables that have elliptical cross sections, the cross-sectional area calculation shall be based on using the major diameter of the ellipse as a circle diameter.

Note 10:   When pulling three conductors or cables into a raceway, if the ratio of the raceway (inside diameter) to the conductor or cable (outside diameter) is between 2.8 and 3.2, jamming can occur and the next larger size raceway should be used. While jamming can occur when pulling four or more conductors or cables into a raceway, the probability is very low.

Reprinted with permission from NFPA 70-1999.

SPECIAL CONSIDERATIONS

Following the selection of the circuit conductors and the branch-circuit protection, there are a number of special considerations that generally are factors for each installation. The electrician is usually given the responsibility of planning the routing of the conduit to insure that the outlets are connected properly. As a result, the electrician must determine the length and the number of conductors in each raceway. In addition, the electrician must derate the conductors' ampacity as required, make allowances for voltage drops, recognize the various receptacle types, and be able to install these receptacles correctly on the system.

BOX STYLES AND SIZING

The style of box required on a building project is usually established in the specifications. However, the sizing of the boxes is usually one of the decisions made by the electrician.

Switchboxes (Device Boxes)

Switchboxes are 2" × 3" in size and are available with depths ranging from 1½ inches to 3½ inches, Figure 6-16. These boxes can be purchased for either ½-inch or ¾-inch conduit, or with cable clamps. Each side of the switchbox has holes through which a 20-penny nail can be inserted for nailing to wood studs. The boxes can be ganged by removing the common sides of two or more boxes and connecting the boxes together. Plaster ears may be provided for use on plasterboard or for work on old installations.

Masonry Boxes

Masonry boxes are designed for use in masonry block or brick. These boxes do not require an extension cover, but will accommodate devices directly, Figure 6-17. Masonry boxes are available with depths of 2½ inches and 3½ inches and in through-the-wall depths of 3½ inches, 5½ inches, and 7½ inches. Boxes of this type are available with knockouts up to one inch in diameter.

Handy Box

Handy boxes, Figure 6-18, are generally used in exposed installations and are available with ½-, ¾-,

3 × 2 inch conduit switch boxes in assorted depths are available with or without plaster ears.

Fig. 6-16   Switch box.

1, 2, 3, 4, or 5-gang masonry boxes

Fig. 6-17   Four-gang masonry box.

Fig. 6-18   Handy box and receptacle cover.

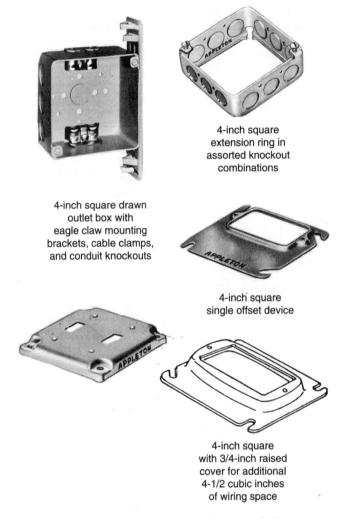

4-inch square
extension ring in
assorted knockout
combinations

4-inch square drawn
outlet box with
eagle claw mounting
brackets, cable clamps,
and conduit knockouts

4-inch square
single offset device

4-inch square
with 3/4-inch raised
cover for additional
4-1/2 cubic inches
of wiring space

Fig. 6-19   Four-inch square boxes and covers.

or 1-inch knockouts. Since these boxes range in depth from 1¼ inches to 2½ inches, they can accommodate a device without the use of an extension cover.

4-Inch Square Boxes

A 4-inch square box, Figure 6-19, is used commonly for surface or concealed installations. Extension covers of various depths are available to accommodate devices in those situations where the box is surface mounted. This type of box is available with knockouts up to one inch in diameter.

Octagonal Boxes

Boxes of this type are used primarily to install ceiling outlets. Octagonal boxes are available either for mounting in concrete or for surface or concealed mounting, Figure 6-20. Extension covers are available, but are not always required. Octagonal boxes are commonly used in depths of 1½ inches and 2⅛ inches. These boxes are available with knockouts up to an inch in trade diameter.

Fig. 6-20   Octagonal box on telescopic hanger and box extension.

Table 370-16(a). Metal Boxes

Box Dimension in Inches, Trade Size, or Type	Minimum Capacity (in.³)	Maximum Number of Conductors*						
		No. 18	No. 16	No. 14	No. 12	No. 10	No. 8	No. 6
4 × 1¼ round or octagonal	12.5	8	7	6	5	5	4	2
4 × 1½ round or octagonal	15.5	10	8	7	6	6	5	3
4 × 2⅛ round or octagonal	21.5	14	12	10	9	8	7	4
4 × 1¼ square	18.0	12	10	9	8	7	6	3
4 × 1½ square	21.0	14	12	10	9	8	7	4
4 × 2⅛ square	30.3	20	17	15	13	12	10	6
4¹¹⁄₁₆ × 1¼ square	25.5	17	14	12	11	10	8	5
4¹¹⁄₁₆ × 1½ square	29.5	19	16	14	13	11	9	5
4¹¹⁄₁₆ × 2⅛ square	42.0	28	24	21	18	16	14	8
3 × 2 × 1½ device	7.5	5	4	3	3	3	2	1
3 × 2 × 2 device	10.0	6	5	5	4	4	3	2
3 × 2 × 2¼ device	10.5	7	6	5	4	4	3	2
3 × 2 × 2½ device	12.5	8	7	6	5	5	4	2
3 × 2 × 2¾ device	14.0	9	8	7	6	5	4	2
3 × 2 × 3½ device	18.0	12	10	9	8	7	6	3
4 × 2⅛ × 1½ device	10.3	6	5	5	4	4	3	2
4 × 2⅛ × 1⅞ device	13.0	8	7	6	5	5	4	2
4 × 2⅛ × 2⅛ device	14.5	9	8	7	6	5	4	2
3¾ × 2 × 2½ masonry box/gang	14.0	9	8	7	6	5	4	2
3¾ × 2 × 3½ masonry box/gang	21.0	14	12	10	9	8	7	4
FS — Minimum internal depth 1¾ single cover/gang	13.5	9	7	6	6	5	4	2
FD — Minimum internal depth 2⅜ single cover/gang	18.0	12	10	9	8	7	6	3
FS — Minimum internal depth 1¾ multiple cover/gang	18.0	12	10	9	8	7	6	3
FD — Minimum internal depth 2⅜ multiple cover/gang	24.0	16	13	12	10	9	8	4

Note: For SI units, 1 in.³ = 16.4 cm³.

*Where no volume allowances are required by Sections 370-16(b)(2) through 370-16(b)(5).

Table 370-16(b). Volume Allowance Required per Conductor

Size of Conductor (AWG)	Free Space Within Box for Each Conductor (in.³)
18	1.50
16	1.75
14	2.00
12	2.25
10	2.50
8	3.00
6	5.00

Note: For SI units, 1 in.³ = 16.4 cm³.

4¹¹/₁₆-Inch Boxes

These more spacious boxes are used where the larger size is required. Available with ½-, ¾-, 1-, and 1¼-inch knockouts, these boxes require an extension cover or a raised cover to permit the attachment of devices, Figure 6-21.

Box Sizing

NEC® Section 370-16 states that outlet boxes, switch boxes, and device boxes must be large enough to provide ample room for the wires in that box without having to jam or crowd the wires in the box.

When conductors are the same size, the proper box size can be determined by referring to *NEC® Table 370-16(a)*. When conductors are of different sizes, refer to *NEC® Table 370-16(b)*.

NEC® Table 370-16(a) does not consider fittings or devices such as fixture studs, cable clamps, hickeys, switches, pilot lights, or receptacles which may be in the box.

Table 6-4 provides a list of conditions that may arise when evaluating box fill and the appropriate response to each of those conditions. Figure 6-22

Fig. 6-21   4¹¹/₁₆-inch box.

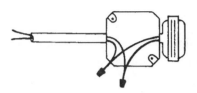

Fig. 6-22   Box containing four conductors, transform leads No. 18 AWG or larger are to be included.

covers the special case presented when transformer leads enter a box. Figure 6-23 addresses the common situation of fixture wires in a box. If there are less than four fixtures wires, and they are No. 14

TABLE 6-4   Checklist for Use When Determining Proper Box Size

• If box contains no fittings, devices, fixture studs, cable clamps, hickeys, switches, receptacles, or equipment grounding conductors . . .	⇨ • refer directly to *NEC® Table 370-16(a)* or *-16(b)*.
• **Clamps.** If box contains one or more internal cable clamps . . .	⇨ • add a single-volume based on the largest conductor in the box.
• **Support Fittings.** If box contains one or more fixture studs or hickeys . . .	⇨ • add a single-volume for each type based on the largest conductor in the box.
• **Device or Equipment.** If box contains one or more wiring devices on a yoke . . .	⇨ • add a double-volume for each yoke based on the largest conductor connected to a device on that yoke.
• **Equipment Grounding Conductors.** If a box contains one or more equipment grounding conductors . . .	⇨ • add a single-volume based on the largest equipment grounding conductor in the box.
• **Isolated Equipment Grounding Conductor.** If a box contains one or more additional "isolated" (insulated) equipment grounding conductors as permitted by *NEC® Section 250-146(d)*, for "noise" reduction . . .	⇨ • add a single-volume based on the largest equipment grounding conductor in the box.
• For conductors running through the box without being spliced . . .	⇨ • add a single-volume for each conductor that runs through the box.
• For conductors that originated outside of box and terminate inside the box . . .	⇨ • add a single-volume for each conductor that originates outside the box and terminates inside the box.
• If no part of the conductor leaves the box – for example, a "jumper" wire used to connect three wiring devices on one yoke, or pigtails as illustrated in figure 6-23 . . .	⇨ • don't count this (these). No additional volume required.
• For small equipment grounding conductors or not more than 4 fixture wires smaller than No. 14 that originate from a fixture canopy or similar canopy (like a fan) and terminate in the box . . .	⇨ • don't count this (these). No additional volume required.
• For small fittings, such as lock-nuts and bushings . . .	⇨ • don't count this (these). No additional volume required.

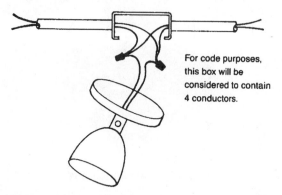

For code purposes, this box will be considered to contain 4 conductors.

Fig. 6-23   Box containing four conductors, fixture wires No. 14 AWG or smaller are not included.

AWG or smaller, they need not be counted. For further information on these cases see *NEC® Section 370-16(b)(1)*.

▶ **SELECTING THE CORRECT SIZE BOX**
Example:

When all conductors are the same size. A box contains one fixture stud and two internal cable clamps. Four No. 12 conductors enter the box.

four No. 12 conductors	4
one fixture stud	1
two cable clamps (only count one)	1
Total	6

Referring to *NEC® Table 370-16(a)*, we find that a 4" × 1½" octagonal box is suitable for this example. ◀

When the conductors are different sizes. When the box contains different size wires, refer to *NEC® Section 370-16*, and do the following:

- Size the box based upon the total cubic inch volume required for the conductors according to *NEC® Table 370-16(b)*.

- Then make the adjustment necessary due to devices, cable clamps, fixture studs, etc. This volume adjustment must be based on the cubic inch volume of the largest conductor in the box as determined by *NEC® Table 370-16(b)*.

- When conductors of different sizes are connected to a device(s) on one yoke or strap, the cross-sectional area of the largest conductor connected to the device(s) shall be used in computing the box fill.

- When more than one equipment grounding conductor is in a box the cross-sectional area of the largest grounding conductor shall be used in computing the box fill.

Example:

A box is to contain two devices, a duplex receptacle and a toggle switch. Two No. 12 conductors are connected to the receptacle, and two No. 14 conductors are connected to the toggle switch. There are two grounding conductors in the box, a No. 12 and a No. 14. There are two cable clamps in the box. The minimum box size would be computed as follows:

2 No. 12 conductors @ 2.25 cubic inches = 4.50 cubic inches

2 No. 14 conductors @ 2.00 cubic inches = 4.00 cubic inches

1 cable clamp @ 2.25 cubic inches = 2.25 cubic inches (based on size of largest conductor in box)

1 equipment grounding conductor @ 2.25 cubic inches = 2.25 cubic inches (based on size of largest grounding conductor)

1 duplex receptacle @ 2 × 2.25 cubic inches = 4.50 cubic inches (based on size of connecting conductors)

1 toggle switch @ 2 × 2.00 cubic inches = 4.00 cubic inches (based on size of connecting conductors)

Minimum box volume = 21.50 cubic inches

Therefore, select a box having a *minimum* volume of 21.50 cubic inches of space. The cubic inch volume may be marked on the box; otherwise, refer to the second column of *NEC® Table 370-16(a)*, entitled "Min. Cu. In. Cap."

The Code requires that all boxes OTHER than those listed in *NEC® Table 370-16(a)* be durable and legibly marked by the manufacturer with their cubic inch capacity. When sectional boxes are ganged together, the volume to be filled is the total cubic inch volume of the assembled boxes. Fittings may be used with the sectional boxes, such as plaster rings, raised covers, and extension rings.

When these fittings are marked with their volume and cubic inches, or have dimensions comparable to those boxes shown in *NEC® Table*

370-16(a), then their volume may be considered in determining the total cubic inch volume to be filled (see *NEC® Section 370-16(a)(1)*. The following example illustrates how the cubic inch volume of a plaster ring or of a raised cover, figure 6-19, is added to the volume of the box to increase the total cubic inch volume.

Example:

How many No. 12 conductors are permitted in this box and raised plaster ring (Figure 6-24)?

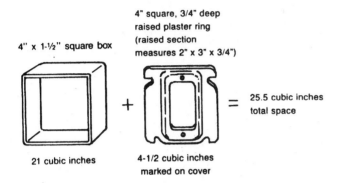

Figure 6-24   Box and raised plaster ring.

Answer:

See *NEC® Section 370-16* and *NEC® Tables 370-16(a)* and *370-16(b)* for the volume required per conductor. This box and cover will take:

$$\frac{25.5 \text{ cubic inch space}}{2.25 \text{ cubic inches per No. 12 conductor}} = \text{Eleven No. 12 conductors maximum, less the deductions for devices, clamps, and so on, per } NEC® \text{ Section 370-16(b)(2)}$$

Many devices such as GFCI receptacles, dimmers, and timers are much larger than the conventional receptacle or switch. The Code has recognized this problem by requiring that one or more devices on one strap requires a conductor reduction of two when determining maximum box capacity. It is always good practice to install a box that has ample room for the conductors, devices, and fittings rather than forcibly crowding the conductors into the box.

The sizes of equipment grounding conductors are shown in *NEC® Table 250-122*. The grounding

conductors are the same size as the circuit conductors in cables having No. 14 AWG, 12 AWG, or No. 10 AWG circuit conductors. Thus, box sizes can be calculated using *NEC® Table 370-16(a)*.

Another example of the procedure for selecting the box size is the installation of a four-way switch in the bakery. Four No. 12 AWG conductors and one device are to be installed in the box. Therefore, the box selected must be able to accommodate six conductors; a 3" × 2" × 2¾" switch box is adequate.

When the box contains fittings such as fixture studs, clamps, or hickeys, the number of conductors permitted in the box is one less for EACH type of device. For example, if a box contains two clamps, deduct one conductor from the allowable number shown in *NEC® Table 370-16(a)* (Figure 6-25).

An example of this situation occurs in the bakery at the box where the conduit leading to the four-way switch is connected.

- There are two sets of travelers (one from each of the three-way switches) passing through the box on the way to the four-way switch (4 conductors).

- There is a neutral, from the panelboard, that is spliced to a neutral that serves the luminaire attached to the box and to neutrals entering the two conduits leaving the box that go to other luminaires (4 conductors).

- There is a switch return that is spliced in the box to a conductor that serves the luminaire that is attached to the box and to switch returns that enter the conduits leaving the box to serve other luminaires (4 conductors).

- A fixture stud-stem assembly is used to connect the luminaire directly to the box (1 conductor).

- This is a total of 13 conductors.

From *NEC® Table 370-16(a)*, a 4 × 2⅛ square or a 4¹¹/₁₆ × 1½ square box would qualify. Since these boxes require plaster rings, a smaller size box could be used if the plaster ring volume added to the box volume were sufficient for the 13 conductors.

When necessary, conduit fittings, Figure 6-26, or pull boxes, Figure 6-27, are used. If pull boxes are used, they must be sized according to *NEC® Section 370-28*, as shown in figure 6-28. The following

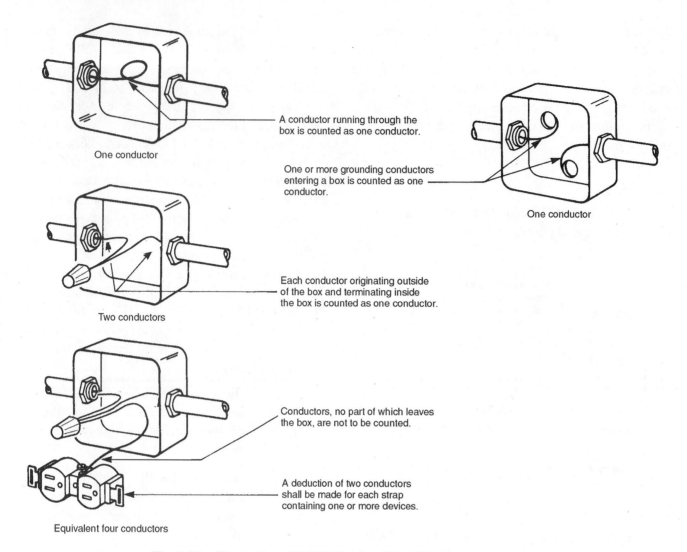

One conductor

Two conductors

Equivalent four conductors

A conductor running through the box is counted as one conductor.

One conductor

One or more grounding conductors entering a box is counted as one conductor.

Each conductor originating outside of the box and terminating inside the box is counted as one conductor.

Conductors, no part of which leaves the box, are not to be counted.

A deduction of two conductors shall be made for each strap containing one or more devices.

Fig. 6-25   Illustration of *NEC® Section 370-16(b)(1)* requirements.

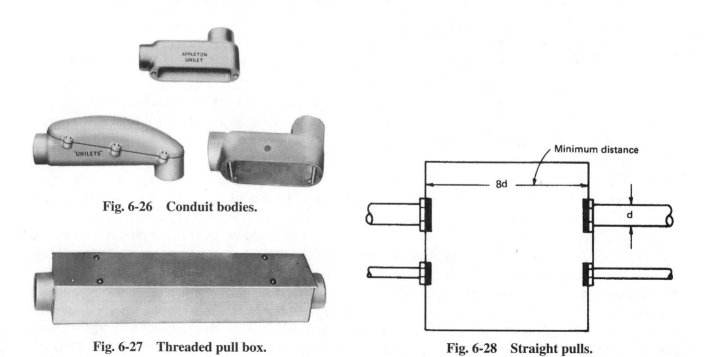

Fig. 6-26   Conduit bodies.

Fig. 6-27   Threaded pull box.

Minimum distance

8d

d

Fig. 6-28   Straight pulls.

NEC® requirements apply to the installation of pull boxes and junction boxes.

- For straight pulls, the box must be at least 8 times the diameter of the largest raceway when the raceways contain No. 4 AWG or larger conductors, *NEC® Section 370-28(a)*.

- For angle pulls, Figure 6-29, the distance *between* the raceways and the opposite wall of the box must be at least 6 times the diameter of the largest raceway. To this value, it is necessary to add the diameters of all other raceways entering in any one row the same wall of the box, Figure 6-30. When there is more than one row, the row with the maximum diameters must be used in sizing the box. This requirement applies when the raceways contain No. 4 AWG or larger conductors. See *NEC® Section 370-28(a)(2)*.

- Boxes may be smaller than the preceding requirements when approved and marked with the size and number of conductors permitted.
- Conductors must be *racked* or *cabled* if any dimension of the box exceeds 6 feet.
- Boxes must have an approved cover.
- Boxes must be accessible.

RACEWAY SUPPORT

There are many devices available that are used to support conduit. The more popular devices are shown in Figures 6-31 and 6-32.

The *NEC®* article that deals with a particular type of raceway also gives the requirements for supporting that raceway. Figure 6-33 shows the support requirements for rigid metal conduit, intermediate metal conduit, electrical metallic tubing, flexible metal conduit, and liquidtight flexible metal conduit.

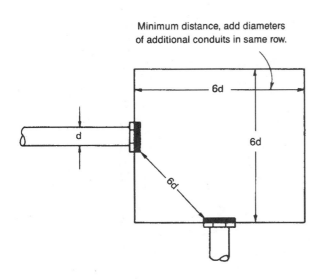

Fig. 6-29   Angle pull.

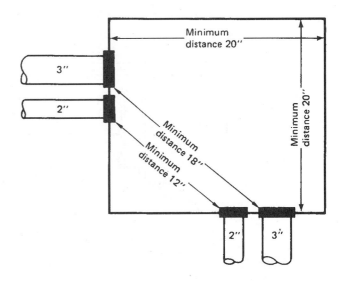

Fig. 6-30   Angle pulls with side by side conduits.

Fig. 6-31   Raceway support devices.

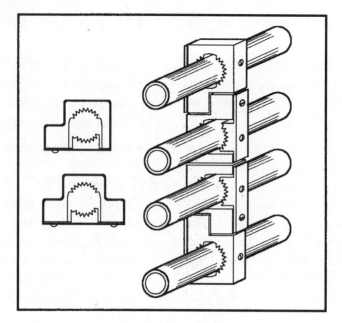

Fig. 6-32   Nonmetallic conduit and supports.

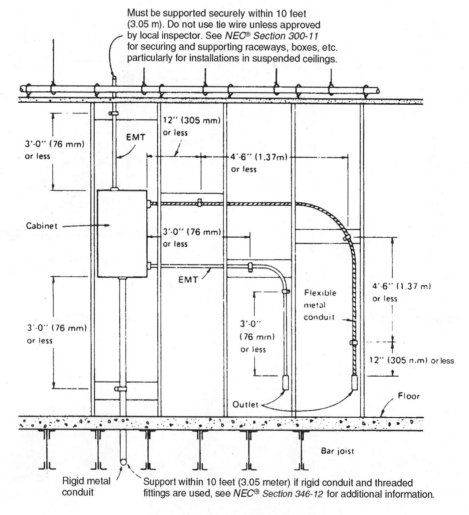

Must be supported securely within 10 feet (3.05 m). Do not use tie wire unless approved by local inspector. See *NEC® Section 300-11* for securing and supporting raceways, boxes, etc. particularly for installations in suspended ceilings.

3'-0" (76 mm) or less

EMT

12" (305 mm) or less

4'-6" (1.37m) or less

Cabinet

3'-0" (76 mm) or less

EMT

3'-0" (76 mm) or less

3'-0" (76 mm) or less

Flexible metal conduit

4'-6" (1.37 m) or less

12" (305 n.m) or less

Outlet

Floor

Bar joist

Rigid metal conduit

Support within 10 feet (3.05 meter) if rigid conduit and threaded fittings are used, see *NEC® Section 346-12* for additional information.

Fig. 6-33   Supporting raceway.

REVIEW QUESTIONS

Refer to the *National Electrical Code®* or the working drawings when necessary. Where applicable, responses should be written in complete sentences. Write units using unit names, do not use abbreviations or symbols (1 foot, not 1' or 1 ft).

When responding to questions 1 through 4, select from the following list of raceway types: (check letter[s] representing the correct response[s])

a. electrical metallic tubing
b. electrical nonmetallic tubing
c. flexible metallic tubing
d. rigid metal conduit
e. intermediate metal conduit
f. rigid nonmetallic conduit

1. Which raceway(s) may be used where cinder fill is present?
   ☐ a. ☐ b. ☐ c. ☐ d. ☐ e. ☐ f.

2. Which raceway(s) may be used in hazardous locations?
   ☐ a. ☐ b. ☐ c. ☐ d. ☐ e. ☐ f.

3. Which raceway(s) may be used in wet locations?
   ☐ a. ☐ b. ☐ c. ☐ d. ☐ e. ☐ f.

4. Which raceway(s) may be used for service entrances?
   ☐ a. ☐ b. ☐ c. ☐ d. ☐ e. ☐ f.

Give the raceway size required for each of the conductor combinations.

5. Five, 10 AWG, type THW   _____

6. Three, 1/0 AWG, type THHN   _____

7. Three, 1/0 AWG and one, 1 AWG, type THW   _____

Compute the correct box size for each of the following conditions.

8. Two nonmetallic sheathed cables with two 12 AWG conductors, a grounding conductor and a switch in a metal box.

9. A conduit run is serving a series of luminaires, connected to a total of three circuits. The luminaires are supplied by 120 volts from a 3-phase 4-wire system. Each box will contain two circuits running through the box and the third circuit connected to a luminaire which is hung from a fixture stud.

10. What is the minimum cubic inch volume that is permissible for the outlet box shown below? Show your calculations.

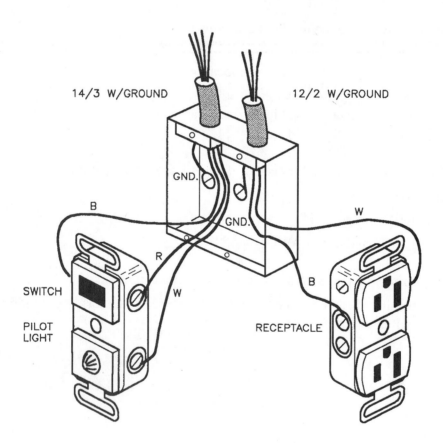

Two three-inch raceways enter a box directly across from each other. No other raceways enter the box. What are the minimum dimensions of the box?

11. Length _____

12. Width _____

13. Depth _____

Two three-inch raceways enter a box at right angles to each other. No other raceways enter the box. What are the minimum dimensions of the box?

14. Length _____

15. Width _____

16. Depth _____

17. If you had the freedom of choice, what type of raceway would you select for installation under the sidewalks and driveways of the commercial building? Support your selection. _____

_____

_____

_____

Compute the required raceway size for the following combination of conductors.

18. Four No. 8 AWG and four No. 12 AWG type THW

19. Three, 350 kcmil, and one, 250 kcmil Type TW conductors and a No. 4 AWG bare conductor

Determine the required box size for the following situations.

20. In a nonmetallic sheathed cable installation a 10/3 with ground is installed in a metal box to supply two 12/2 with ground branch-circuit cables. What is the minimum size box?

Two three-inch raceways enter a box, one through a side and the other in the back. No other raceways enter the box. What are the minimum dimensions of the box?

21. Length _____

22. Width _____

23. Depth _____

Compute the answers using information shown in the drawing.

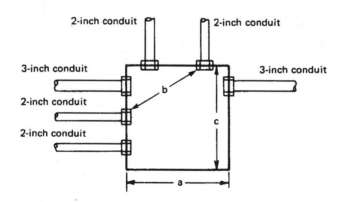

24. Dimension a must be at least _____ inches.

25. Dimension b must be at least _____ inches.

26. Dimension c must be at least _____ inches.

27. Do you foresee any difficulties in installing the conductors in these raceways? Explain.

UNIT 7

Motor and Appliance Circuits

OBJECTIVES

After studying this unit, the student will be able to

- use and interpret the word "appliance."
- use and interpret the term "utilization equipment."
- choose an appropriate method for installing electrical circuits to appliances.
- identify appropriate grounding for appliances.
- determine branch-circuit ratings, conductor sizes, and overcurrent protection for appliances and motors.
- understand the meaning of the terms "Type 1 and Type 2 Protection."

APPLIANCES

Most of the Code rules that relate to the installation and connection of electrical appliances are found in *NEC® Article 422*. For motor-operated appliances, reference will be made to *NEC® Article 430* (Motors, Motor Circuits, and Controllers). Where the appliance is equipped with hermetic refrigerant motorcompressor(s), such as refrigeration and air-conditioning equipment, *NEC® Article 440* (Air-Conditioning Equipment) applies.

The *National Electrical Code®* defines an appliance as utilization equipment, generally other than industrial, of standardized sizes or types, that is installed or connected as a unit to perform such functions as air-conditioning, washing, food mixing, and cooking, among others.

The Code defines utilization equipment as equipment that utilizes electric energy for mechanical, chemical, heating, lighting, or similar purposes.

In commercial buildings, one of the tasks performed by the electrician is to make the electrical connections to appliances. The branch-circuit sizing may be specified by the consulting engineer in the specifications or on the electrical plans. In other cases, it is up to the electrician to do all of the calcu-

lations, and/or make decisions based upon the nameplate data of the appliance.

The bakery in the commercial building provides examples of several different types of electrical connection methods that can be used to connect electrical appliances.

Appliance Branch-Circuit Overcurrent Protection

- For those appliances that are marked with a maximum size and type of overcurrent protective device, this rating shall not be exceeded, see *NEC® Section 422-11(a)*.

- If the appliance is not marked with a maximum size overcurrent protective device, *NEC® Section 422-11(a)* stipulates that *NEC® Section 240-3* applies. This section states that the fundamental rule is to protect conductors at their ampacities, as specified in *NEC® Table 310-16*.

There are several variations allowed to this fundamental rule, three of which relate closely to motor-operated appliances, air-conditioning, and refrigeration equipment.

NEC® Section 240-3(b) allows the use of the next standard "higher" ampere rating overcurrent device when the ampacity of the conductor does not match a standard ampere rating overcurrent device. There are two exceptions to this, if the conductors are part of a multioutlet branch-circuit supplying cord- and plug-connected portable loads or if the next standard rating exceeds 800 amperes then the next lower standard rating overcurrent device must be chosen. Standard ratings of overcurrent devices are listed in *NEC® Section 240-6*.

NEC® Section 422-11 lists overcurrent protection requirements for several types of appliances. Commercial appliances include:

- infrared lamp heating,
- surface heating elements,
- single nonmotor operated,
- resistance type heating elements,
- sheathed type heating elements,
- water heaters and steam boilers and
- motor operated appliances.

- For a single nonmotor-operated appliance, *NEC® Section 422-11(e)* states that the overcurrent protection shall not exceed:
   a. the overcurrent device rating if so marked on the appliance.
   b. 150% of the appliance's rated current, if the appliance is rated at more than 13.3 amperes. But if the 150% results in a nonstandard ampere overcurrent device value, then it is permitted to use the next standard higher-ampere-rating overcurrent device.

   For example, an appliance is rated at 4500 watts, 240 volts. What is the maximum size fuse permitted?

   $$\frac{4500}{240} = 18.75 \text{ amperes}$$

   $$18.75 \times 1.5 = 28.125 \text{ amperes}$$

   Therefore, the Code permits the use of a 30-ampere fuse for the overcurrent protection.

Appliance Grounding

- When appliances are to be grounded, the requirements are given in *NEC® Article 250*. Grounding is discussed in several places in this text. It is suggested that the reader consult the Code Index, in the Appendix, where all *NEC®* references are listed.

Appliance-Disconnecting Means

- Permanently connected appliances, *NEC® Section 422-31*:
   a. If the appliance rating does not exceed ⅛ horsepower or does not exceed 300 volt-amperes, the branch-circuit overcurrent device may serve as the appliance's disconnecting means, *NEC® Section 422-31(a)*.
   b. If the appliance rating exceeds 300 volt-amperes or exceeds ⅛ horsepower, the branch-circuit disconnect switch or breaker can serve as the appliance's disconnecting means, but only if they are within sight of the appliance, or are capable of being locked in the "off" position, *NEC® Section 422-31(b)*.

- For motor-driven, permanently connected appliances rated at more than ⅛ horsepower, the disconnecting means must be within sight of the motor controller, *NEC® Section 422-31(b)*.

- Many appliances have an integral "unit switch." Where this unit switch has a marked "off" position, and disconnects all of the ungrounded conductors, this unit switch can be considered to be the appliance's disconnecting means, but only if the branch-circuit switch or circuit breaker supplying the appliance is readily accessible, *NEC® Section 422-33(d)*. The term "readily accessible" is covered in the Definitions in the *NEC®*. Simply stated, this means that there must be easy access to the disconnect switch without the need to "climb over or remove obstacles or to resort to portable ladders, etc."

- The term "in sight" is covered in the Definitions in the *NEC®*. Further discussion is found in *NEC® Sections 430-102*.

- *NEC® Section 422-32* permits the use of a cord- and plug-connected arrangement to serve as the appliance's disconnecting means. Such cord and plug assemblies may be furnished and attached to the appliance by the manufacturer, to meet the requirements of specific Underwriters Laboratories standards.

THE BASICS OF MOTOR CIRCUITS

Disconnecting Means

The Code requires that all motors be provided with a means to disconnect the motor from its electrical supply. These requirements are found in *NEC® Article 430, Part J.*

For most applications, the disconnect switch for a motor must be horsepower rated, *NEC® Section 430-109.* There are some exceptions. The most common exceptions are:

- For motors ⅛ horsepower or less where the branch-circuit overcurrent device can serve as the disconnecting means.

- For stationary motors 2 horsepower or less, rated at 300 volts or less, in which case a general use switch is permitted.

- For stationary motors over 100 horsepower where a general-use disconnect switch or isolating switch is permitted when the switch is marked "Do Not Open Under Load."

- An attachment plug is permitted to serve as the disconnecting means if the attachment plug cap is horsepower rated. The horsepower rating of an attachment plug cap is considered to be the corresponding horsepower rating for 80% of the ampere rating of the attachment plug cap.

Example:

A two-wire, 125-volt, 20-ampere attachment plug cap:

$$20 \times 0.80 = 16 \text{ amperes}$$

Checking *NEC® Table 430-148* for single-phase, 115-volt, motor, we determine that a 1-horsepower motor has a full-load ampere rating of 16-ampere.

Thus, this attachment plug cap is permitted to serve as the disconnecting means for this motor.

NEC® Section 430-110(a) requires that the disconnecting means for motors rated at 600 volts or less must have an ampere rating of not less than 115% of the motor's full-load ampere rating.

The nameplate on all disconnect switches, as well as the manufacturer's technical data, furnishes the horsepower rating, the voltage rating, and the ampere rating of the disconnect switch.

For those motors that are turned "off" and "on" by a controller, often referred to as a motor starter, the disconnecting means must be in sight of the controller and shall disconnect the controller, *NEC® Section 430-102(a)*, Figure 7-1. This explains why combination starters are so popular. Combination starters have the controller and the disconnecting means in one enclosure as permitted in *NEC® Section 430-103.* The electrician does not have to mount a separate disconnect switch and a separate controller. This results in labor savings.

According to *NEC® Section 430-102(b)*, Figure 7-1, the disconnecting means must also be in sight of the motor location and the driven machinery location. However, there is an exception that says that the disconnecting means does not have to be in sight of the motor location and driven machinery if the disconnecting means that is required to be within sight of the controller is able to be locked in the "off" position.

For a motor and the motor-disconnecting means to meet the Code definition of being "in sight" of each other, both of the following conditions must be met:

- the disconnecting means must be visible from the motor location, and

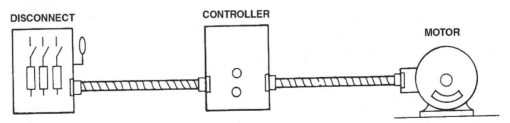

NEC® Section 430-102(a): The disconnect must be within sight of the controller and shall disconnect the controller.

NEC® Section 430-102(b): The disconnect must be within sight of the motor unless the disconnect that is within sight of the controller can be locked in the "off" position.

Fig. 7-1 Location of motor disconnect means.

- the disconnecting means must not be more than 50 feet (15.24 m) from the motor location. See *NEC® Article 100* for the definition of "In Sight From."

The reason the Code uses the term "motor location" instead of "motor" is that, in many instances, the motor is inside an enclosure, and is out of sight until an access panel is removed. If the term "motor" were to be used, then it would be mandatory to install the disconnect inside of the enclosure. This is not always practical. The rooftop air-conditioning units in the commercial building are good examples of this.

NEC® Section 440-14 does permit the disconnecting means to be installed on or within the air-conditioning unit, as might be the case for large air-conditioning units.

To turn the exhaust fan on and off, an ac general-use single-pole snap (toggle) switch is installed on the wall below the fan. The electrician will install an outlet box approximately one foot from the ceiling (see the electrical plan) from which he will run a short length of flexible metal conduit to the junction box on the fan. The fan exhausts the air from the bakery. The fan is through-the-wall installation, approximately 18" below the ceiling, Figure 7-2.

NEC® Section 430-111 permits this switch to serve as both the controller and the required disconnecting means.

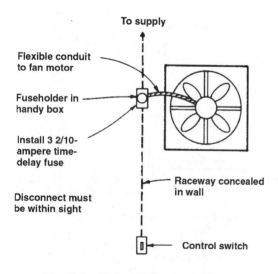

Fig. 7-2   **Exhaust fan installation.**

To supply

Flexible conduit to fan motor

Fuseholder in handy box

Install 3 2/10-ampere time-delay fuse

Disconnect must be within sight

Raceway concealed in wall

Control switch

Motor Circuit Conductors

The conductors that supply a single motor shall not be less than 125% of the full-load ampere rating of the motor, *NEC® Sections 430-6(a) and 22(a)*. The branch-circuit conductors supplying the exhaust fan are No. 12 THHN conductors, well within the minimum conductor size requirements.

Motor Overload Protection

NEC® Article 430, Part C addresses motor overload protection. This part discusses continuous and intermittent duty motors, motors that are larger than one horsepower, motors that are one horsepower or less automatically started, motors that are non-automatically started, motors that are automatically started, and motors that are marked with a service factor that have integral (built-in) overload protection.

The exhaust fan in the bakery draws 2.9 amperes, is less than one horsepower, and is non-automatically started. According to *NEC® Section 430-32(b)(2)*, second sentence, *"Any motor rated at 1 or less that is permanently installed shall be protected in accordance with NEC® Section 430-32(c)."* which in part states that the overload protection shall be as follows in Table 7-1.

The same values indicated Table 7-1 are also valid for continuous duty motors of more than one horsepower, see *NEC® Section 430-32(a)*.

Unit 17 covers in detail the subject of overcurrent protection using fuses and circuit breakers. Dual-element, time-delay fuses are an excellent choice for overload protection of motors. They can be sized close to the ampere rating of the motor. When the motor has integral overload protection, or where the motor controller has thermal overloads, then dual-element, time-delay fuses are selected to provide backup overload protection to the thermal overloads.

TABLE 7-1 Overload Protection	
Motor Nameplate Rating	**Overload Protection As a Percentage of Motor Nameplate Full-load Current Rating**
Service factor not less than 1.15	125%
Temperature rise not over 40°C	125%
All other motors	115%

The starting current of an electrical motor is very high for a short period of time as is illustrated in Figure 7-3. The current decreases to its rated value as the motor reaches full speed. Dual-element, time-delay fuses will carry five times their ampere

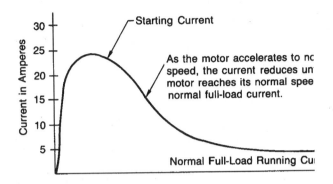

Fig. 7-3   Typical electric motor starting current.

Fig. 7-4   Type S plug fuse and adapter.

rating for at least 10 seconds. This allows the motor to start then provides good overload protection while it is running. For example, a 4-ampere, full-load rated motor could have a starting current as high as 24 amperes. The fuse must not open needlessly when it "sees" this inrush.

In this situation, a 15-ampere ordinary fuse or breaker might be required to allow the motor to start. This would provide the branch-circuit overcurrent protection, but it would not provide overload protection for the appliance.

If the motor does not have built-in overload protection, sizing the branch-circuit overcurrent protection as stated above would allow the motor to destroy itself if the motor were to be subjected to continuous overloaded or stalled conditions.

Abuse, lack of oil, bad bearings, and jammed V-belts are a few of the conditions that can cause a motor to draw more than normal current. Too much current results in the generation of too much heat within the windings of the motor. For every 10°C above the maximum temperature rating of the motor, the expected life of the motor is reduced by 50%. This is sometimes referred to as the "half-life rule."

The solution is to install a dual-element, time-delay fuse that permits the motor to start, yet opens on an overload before the motor is damaged. To provide this overload protection for single-phase, 120- or 240-volt equipment, fuses and fuseholders as illustrated in Figure 7-4 are often used.

Type S fuses and the corresponding Type S adaptor are inserted into the fuseholder, Figure 7-5.

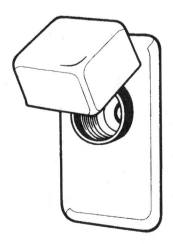

Fig. 7-5   Fuseholders.

Once the adaptor has been installed, it is extremely difficult to remove it. This makes it virtually impossible to replace a properly sized Type S fuse with a fuse having an ampere rating too large for the specific load.

Table 7-2 compares the load responses of ordinary fuses and dual-element, time-delay fuses.

Table 7-3(A) shows how to size dual-element, time-delay fuses for motors that are marked with a service factor of not less than 1.15 and for motors marked with a temperature rise of not over 40°C.

These charts show one manufacture's recommendations for the selection of dual-element time-delay fuses for motor circuits. The Design "E" premium efficient motors have a higher starting current than the older style motors. The nameplate should be checked to determine the design designation then refer to *NEC® Table 430-152* to determine proper branch-circuit, short-circuit, and ground-fault protection. Finally check the protective device time-current curve to determine is it is capable of carrying the starting current until the motor reaches full speed.

Table 7-3(B) shows how to size dual-element, time-delay fuses for motors that have a service factor of less than 1.15 and for motors with a temperature rise of more than 40°C.

For abnormal installations, dual-element, time-delay fuses larger than those shown in Table 7-3(A) and Table 7-3(B) may be required. These larger-rated fuses (or circuit breakers) will provide short-circuit protection only for the motor branch circuit. *NEC® Section 430-52* and *Table 430-152* show the maximum size fuses and breakers permitted. These percentages have already been discussed in this unit.

Abnormal conditions might include:

- Situations where the motor is started, stopped, jogged, inched, plugged, or reversed frequently.

- High inertia loads such as large fans and centrifugal machines, extractors, separators, pulverizers, etc. Machines having large flywheels fall into this category.

- Motors having a high Code letter and full-voltage start as well as some older motors without Code letters. Remember, the higher the Code letter, the higher is the starting current. For example, two motors have exactly the same full-load ampere rating. One motor is marked Code letter J. The other is marked Code letter D. The Code letter J motor will draw more momentary starting inrush current than the Code letter D motor. The Code letter of a motor is found on the motor nameplate. See *NEC® Section 430-7* and *Table 430-7(b)*.

Refer to the *National Electrical Code®* and to Bussmann's *Electrical Protection Handbook* for more data relating to motor overload and motor branch-circuit protection, disconnecting means, controller size, conductor size, conduit size, and voltage drop.

Dual-element, time-delay fuses are designed to withstand motor-starting inrush currents and will not open needlessly. However, they will open if a sustained overload occurs. Selecting dual-element, time-delay fuses of the Type S plug type, Figure 7-4, or of the cartridge type, as illustrated in unit 18 for motors, should be sized in the range of 115%, but not to exceed 125% of the motor full-load ampere rating.

TABLE 7-2 Load Response Time for Four Types of Fuses (Approximate Time in Seconds for Fuse to Blow)				
Load (amperes)	4-ampere Dual-element Fuse (time-delay)	4-ampere Ordinary Fuse (non time-delay)	8-ampere Ordinary fuse (non time-delay)	15-ampere Ordinary Fuse (non time-delay)
5	Over 300 sec.	Over 300 sec.	Won't blow	Won't blow
6	250 sec.	5 sec.	Won't blow	Won't blow
8	60 sec.	1 sec.	Won't blow	Won't blow
10	38 sec.	Less than 1 sec.	Over 300 sec.	Won't blow
15	17 sec.	Less than 1 sec.	5 sec.	Won't blow
20	9 sec.	Less than 1 sec.	Less than 1 sec.	300 sec.
25	5 sec.	Less than 1 sec.	Less than 1 sec.	10 sec.
30	2 sec.	Less than 1 sec.	Less than 1 sec.	4 sec.

TABLES 7-3 A & B   Selection of Dual-Element Fuses for Motor Overload Protection

A. MOTORS MARKED WITH NOT LESS THAN 1.15 SERVICE FACTOR OR TEMP. RISE NOT OVER 40°C

Motor (40°C or 1.15 S.F.) Ampere Rating	Dual-element Fuse Amp Rating (Max. 125%; 430-32)	Motor (40°C or 1.15 S.F.) Ampere Rating	Dual-element Fuse Amp Rating (Max. 125%; 430-32)
1.00 to 1.11	1 1/4	20.0 to 23.9	25
1.12 to 1.27	1 4/10	24.0 to 27.9*	30
1.28 to 1.43	1 6/10	28.0 to 31.9	35
1.44 to 1.59	1 8/10	32.0 to 35.9	40
1.60 to 1.79	2	36.0 to 39.9	45
1.80 to 1.99	2 1/4	40.0 to 47.9	50
2.00 to 2.23	2 1/2	48.0 to 55.9	60
2.24 to 2.55	2 8/10	56.0 to 63.9	70
2.56 to 2.79	3 2/10	64.0 to 71.9	80
2.80 to 3.19	3 1/2	72.0 to 79.9	90
3.20 to 3.59	4	80.0 to 87.9*	100
3.60 to 3.99	4 1/2	88.0 to 99.9	110
4.00 to 4.47	5	100 to 119	125
4.48 to 4.99	5 6/10	120 to 139	150
5.00 to 5.59	6 1/4	140 to 159	175
5.60 to 6.39	7	160 to 179*	200
6.40 to 7.19	8	180 to 199	225
7.20 to 7.99	9	200 to 239	250
8.00 to 9.59	10	240 to 279	300
9.60 to 11.9	12	280 to 319	350
12.0 to 13.9	15	320 to 359*	400
14.0 to 15.9	17 1/2	360 to 399	450
16.0 to 19.9	20	400 to 480	500

B. ALL OTHER MOTORS (i.e. LESS THAN 1.15 SERVICE FACTOR OR GREATER THAN 40° C RISE)

All other Motors–Ampere Rating	Dual-element Fuse Amp Rating (Max. 115%; 430-32)	All other Motors–Ampere Rating	Dual-element Fuse Amp Rating (Max. 115%; 430-32)
1.00 to 1.08	1 1/8	17.4 to 20.0	20
1.09 to 1.21	1 1/4	21.8 to 25.0	25
1.22 to 1.39	1 4/10	26.1 to 30.0*	30
1.40 to 1.56	1 6/10	30.5 to 34.7	35
1.57 to 1.73	1 8/10	34.8 to 39.1	40
1.74 to 1.95	2	39.2 to 43.4	45
1.96 to 2.17	2 1/4	43.5 to 50.0	50
2.18 to 2.43	2 1/2	52.2 to 60.0	60
2.44 to 2.78	2 8/10	60.9 to 69.5	70
2.79 to 3.04	3 2/10	69.6 to 78.2	80
3.05 to 3.47	3 1/2	78.3 to 86.9	90
3.48 to 3.91	4	87.0 to 95.6*	100
3.92 to 4.34	4 1/2	95.7 to 108	110
4.35 to 4.86	5	109 to 125	125
4.87 to 5.43	5 6/10	131 to 150	150
5.44 to 6.08	6 1/4	153 to 173	175
6.09 to 6.95	7	174 to 195*	200
6.96 to 7.82	8	196 to 217	225
7.83 to 8.69	9	218 to 250	250
8.70 to 10.0	10	261 to 300	300
10.5 to 12.0	12	305 to 347	350
13.1 to 15.0	15	348 to 391*	400
15.3 to 17.3	17 1/2	392 to 434	450
		435 to 480	500

*Note: Disconnect switch must have an ampere rating at least 115% of motor ampere rating. *NEC® Section 430-110(a)*. Next larger size switch with fuse reducers may be required.

For the exhaust fan motor that has a 2.9-ampere full-load current draw, select a dual-element, time-delay fuse in the range of:

$$2.9 \times 1.15 = 3.34 \text{ amperes}$$
$$2.9 \times 1.25 = 3.62 \text{ amperes}$$

Referring to Table 7-3(A) and Table 7-3(B), find dual-element, time-delay fuses rated at 3½ amperes. If the motor has built-in or other overload protection, the 3½-ampere size will provide backup overload protection for the motor.

Motor Branch-Circuit, Short-Circuit, and Ground-Fault Protection

It is beyond the scope of this text to cover every aspect of motor circuit design, as an entire text could be devoted to this subject. The basics of the typical motor installations will be presented, but for the "not so common" installations, refer to *NEC® Article 430*.

- Motor branch-circuit conductors, controllers, and the motor must be provided with short-circuit and ground-fault overcurrent protection, *NEC® Section 430-51*.

- The short-circuit and ground-fault overcurrent protection must have sufficient time-delay to permit the motor to be started, *NEC® Section 430-52*.

- *NEC® Section 430-52* and *Table 430-152* provide the maximum percentages permitted for fuses and circuit breakers, for different types of motors.

▶• When applying the percentages listed in *NEC® Table 430-152*, and the resulting ampere rating or setting does not correspond to the standard ampere ratings found in *NEC® Section 240-6*, we are permitted to "round up" to the next higher standard rating. See *NEC® Section 430-52(c)(1), Exception No. 1.* ◀

▶• If the size selected by "rounding up" to the next higher standard size will not allow the motor to start, such as might be the case on Design E (energy efficient motors) motor circuits, then we are permitted to size the motor branch-circuit fuse or breaker even larger, *NEC® Section 430-52(c)(1), Exception No. 2.* See Table 7-4. ◀

- Always check the manufacturer's overload relay table in a motor controller to see if the manufacturer has indicated a maximum size or type of overcurrent device. For example, "Maximum Size Fuse 25-ampere." Do not exceed this ampere rating, even though the percentage values listed in *NEC® Table 430-152* result in a higher ampere rating. Do not use a circuit breaker. To do so would be a violation of *NEC® Section 110-3(b)*.

TABLE 7-4 Sizing of Motor Branch-Circuit, Short-Circuit, and Ground-Fault.		
	Maximum Rating	**Absolute Maximum**
Non-time-delay fuses not over 600 amperes	300%	400%
Non-time-delay fuses over 600 amperes	300%	300%
Time-delay dual-element fuses not over 600 amperes	175%	225%
Fuses over 600 amperes	300%	300%
Inverse time circuit breakers Motor FLA 100 amperes or less	250%	400%
Inverse time circuit breakers Motor FLA over 100 amperes	250%	300%
Instant trip circuit breakers for *other* than Design E motors* or Design B energy efficient motors*	800%	1300%**
Instant trip circuit breakers for Design E motors* or Design B energy efficient motors*	1100%	1700%**
*Only permitted when part of a listed combination motor controller. **Only permitted where necessary after doing an engineering evaluation to prove that the 800% and 1100% sizing will not perform satisfactorily.		

Example:

▶ Determine the rating of dual-element, time-delay fuses for a Design E, premium efficiency, 7½-horsepower, 208-volt, 3-phase motor. First refer to *NEC® Table 430-150* to obtain the motor's full-load current rating, then refer to *NEC® Table 430-152*. The first column shows the type of motor, and the remaining columns show the maximum rating or setting for the branch-circuit, short-circuit, and ground-fault protection using different types of fuses and circuit breakers. Also refer to *NEC® Section 430-52*.

$$25.7 \times 1.75 = 44.975 \text{ amperes}$$

NEC® Section 240-6 lists 45-ampere as the next higher standard size.

If it is determined that 45-ampere dual-element, time-delay fuses will not hold the starting current of the motor, we are permitted to increase the ampere rating of the fuses, but not to exceed 225% of the motor's full-load current draw.

$$25.7 \times 2.25 = 57.825$$

(The next lower standard size is 50 amperes. 55-ampere fuses are available, but may not be stocked by most electrical distributors.) ◀

Motor Starting Currents/Code Letters

Motor branch-circuit short-circuit and ground-fault protection is based upon the type of motor and the Code Letter found on the motor's nameplate. The higher the Code Letter, the higher the starting current.

NEC® Table 430-7(b) lists Code Letters for electric motors. Because this table is based upon kVA per horsepower, it is easy to calculate the range of the motor's starting current in amperes.

Example:

Calculate the starting current for a 240-volt, 3-phase, 7½-horsepower motor.

$$I = \frac{kVA \times 1000}{E \times 1.73} = \frac{7.1 \times 7.5 \times 1000}{240 \times 1.73} = 128 \text{ amperes minimum}$$

$$I = \frac{kVA \times 1000}{E \times 1.73} = \frac{7.99 \times 7.5 \times 1000}{240 \times 1.73} = 144 \text{ amperes maximum}$$

Motor Design Designations

▶ *NEC® Section 430-52, Tables 430-151B*, and *152* reference *design* types of motors. The *design* types refer to the National Electrical Manufacturers Association (NEMA) designation for alternating current, polyphase induction electric motors. Different design types of motors have different applications as shown in Table 7-5.

Type 1 and Type 2 Coordination

Because these terms might be found on the nameplate of a motor controller, familiarity with their meaning is necessary.

These terms are used by manufacturers of motor controllers based on the IEC (International Electro-

TABLE 7-5 NEMA Design Applications			
NEMA Design	Starting Current	Torque	Application
A	medium	medium	A variation of Design B, having a higher starting current.
B	medium	medium	For normal starting torque for fans, blowers, rotary pumps, unloaded compressors, some conveyors, metal cutting machine tools, misc. machinery. Very common for general purpose across-the-line starting. Slight change of speed as load varies.
C	low to medium	high	For high inertia starts such as large centrifugal blowers, fly wheels, and crusher drums. Loaded starts, such as piston pumps, compressors, and conveyors, where rapid acceleration is needed. Slight change of speed as load varies.
D	medium	extra high	For very high inertia and loaded starts such as punch presses, shears and forming machine tools, cranes, hoists, elevators, oil well pumping jacks. Considerable change of speed as load varies.
E	high		These are the newer *premium efficiency* motors having very high starting currents, as high as 1800% of full-load current. Can cause nuisance opening of fuses or tripping of circuit breakers if not properly applied. Refer to *NEC® Section 430-52* and *Tables 430-150, -151B, and -152.* ◀

technical Commission) Standard No. 947-4-1. This is a European standard. Electrical equipment manufactured overseas to European standards is being shipped into the United States, and electrical equipment manufactured in this country to US standards (NEMA and UL Standard 508) is being shipped overseas. Conformance to one standard does not necessarily mean that the equipment conforms to the other standard.

In the United States, we prefer to use the term *protection* instead of *coordination* because coordination has a different meaning as discussed in unit 18.

Type 1 short-circuit protection means that under short-circuit conditions, the contactor and starter shall not cause danger for persons working on or near the equipment. However, a certain amount of damage is acceptable to the controller, such as welding of the contacts. Replacing the contacts of a controller does not constitute a hazard to personnel. Type 1 protection and UL 508 are similar in this respect. The steps necessary to get Type 1 protected controllers back on line after a short circuit or ground-fault occurs are: (1) disconnect the power, (2) locate and repair the fault, (3) replace the contacts in the controller if necessary, (4) replace the branch-circuit fuses or reset the circuit breaker, (5) restore power.

Type 2 short-circuit protection means that under short-circuit conditions, the contactor and starter shall not cause danger for persons working on or near the equipment, and in addition, the contactor or starter shall be suitable for further use. No damage to the controller is acceptable. The steps necessary to get Type 2 protected controllers back on line after a short circuit or ground-fault occurs are: (1) disconnect the power, (2) locate and repair the fault, (3) replace the branch-circuit fuses, (4) restore power.

At the time of this writing, there are no listed Type 2 protected motor controllers using circuit breakers as the overcurrent protective device.

Although Type 2 protected controllers are marked by the manufacturer with the maximum size and type of fuse, the maximum size might be larger than that permitted for motor branch-circuit protection when applying the rules of *NEC® Section 430-52* and *Table 430-152* for a particular motor installation. For example, the controller might be

marked "Maximum Ampere Rating Class J Fuse: 8 amperes." Yet, the proper size Class J time-delay fuse for a particular motor circuit, applying the maximum values according to *NEC® Section 430-52* and *Table 430-152* might be 4½ amperes. In this example, the proper maximum size fuse is 4½ amperes.

EQUIPMENT INSTALLATION

Equipment and Lighting on a Branch Circuit

The exhaust fan in the bakery is an example of an equipment load being connected on a branch circuit with lighting. Circuit No. 1 serves 1305 VA of lighting and the exhaust fan. This fan is a permanently connected motor-operated appliance. Permanently connected is often called "hard wired" differentiating from "cord- and plug-connected".

NEC® Section 210-23(a) permits, on 15- and 20-ampere branch-circuits, the supply of lighting and equipment provided the hard wired equipment does not exceed 50% of the circuit rating.

This particular exhaust fan has a, ¹⁄₁₀ horsepower, 120 volt motor with a full-load rating of 2.9 amperes. The load would be:

$$1305 \text{ VA} + (120 \text{ V} \times 2.9 \text{ A} \times 1.25) = 1740 \text{ VA}$$

$$1740 \text{ VA} / 120 \text{ V} = 14.5 \text{ A}.$$

In selecting the proper size conductor first check the 60°C column of *NEC® Table 310-16* to determine the minimum size. A No. 12 AWG has an ampacity of 25 ampere, thus qualifying as the minimum size conductor. No correction or adjustment factors are required so the No. 12 AWG becomes the circuit conductor. The standard OCPD rating of 20-ampere is the next higher than 14.5 amperes so that becomes the OCPD rating. The fan motor is under the 50% rating limitation.

Combination Load on Individual Branch Circuit

When a cord- and plug-connected appliance is supplied by an individual branch circuit the load is limited to 80% of the circuit rating by *NEC® Section 210-23(a)*. An example of this is circuit 9 serving the dough machine.

Motor load 792 VA

Heater load 2000 VA

Total 2792 VA

Conversion to Amperes 2792 VA /360 = 7.8 A

A 20-ampere rated circuit, the minimum allowed, is assigned to supply this load.

Supplying a Specified Load

The bake oven installed in the bakery is an electrically heated commercial-type bake oven, Figure 7-10. The oven has a marked nameplate indicating a load of 16000 VA, at 3 phase 208-volt. This load includes all electrical heating elements, drive motor, timers, transformers, controls, operating coils, lights, and all other electrical powered apparatus. The nameplate also indicates that all necessary protective devices are installed.

The instructions furnished with the bake oven, and on the nameplate, specify that the supply conductors have a minimum ampacity of 56 amperes and that the terminations are rated for 75°C (167°F).

If the panelboard were to have terminations rated at 60°C (167°F), the conductor selection would be made in the 60°C column of *NEC® Table 310-16*. A minimum size conductor of No. 4 AWG, which has the lowest ampacity higher than 56, would be required. If the panelboard terminations are rated at 75°C then the selection is made from the 75°C column and a No. 6 AWG having an ampacity of 65 amperes would qualify. See *NEC® Section 110-14(c)*. These sizes are required even though 90°C (194°F), Type THHN conductors are being used. A disconnect means, without overcurrent protection, is installed at the site to allow maintenance and adjustments to be made to the internal electrical system. This disconnect is not required by the *NEC®* if the branch circuit panelboard is in sight of the appliance and can be locked. See *NEC® Section 422-31(b)*.

Conductors Supplying Several Motors

NEC® Section 430-24 requires that the conductors supplying several motors on one circuit shall have an ampacity of not less than 125% of the largest motor plus the full-load current ratings of all other motors on that circuit.

For example, consider the bakery's multimixers and dough divider, which are supplied by a single branch-circuit and have full load ratings of 7.48, 2.2, and 3.96 amperes as determined by referring to the equipment's nameplate and installation instructions.

Therefore:

$$1.25 \times 7.48 = 9.35 \text{ amperes}$$
$$\text{plus} \quad 2.20 \text{ amperes}$$
$$\text{plus} \quad \underline{3.96 \text{ amperes}}$$
$$\text{Total} \quad 15.51 \text{ amperes}$$

The branch-circuit conductors must have an ampacity of 15.51 amperes minimum. Specifications for this commercial building call for No. 12 AWG minimum. Checking *NEC® Table 310-16* a No. 12 Type THHN copper conductor has an ampacity of 25 amperes, more than adequate to serve the three appliances. The ampacity of 25 amperes is from the 75°C (167°F) column. When adjusting ampacities for more than three conductors in one raceway, or when correcting ampacities for high temperatures, we would begin the derating by using the 90°C (194°F) column ampacity to determine the 90°C (194°F) Type THHN conductor's ampacity. This was covered in unit 4.

Several Motors or Loads on One Branch-Circuit

This topic is covered in *NEC® Section 430-53*. This kind of "group installation" does not have individual motor branch-circuit, short-circuit, and ground-fault protection. The individual motors do have overload protection. The individual motor controllers, circuit breakers of the inverse time type, and overload devices must be listed for "group installation." The label on the controller indicates the maximum rating of fuse or circuit breaker suitable for use with the particular overload devices in the controller(s).

The branch-circuit fuses or circuit breakers would be sized by:

1. determining the fuse or circuit breaker ampere rating or setting per *NEC® Section 430-52* for the highest rated motor in the group, plus

2. the sum of the full-load current ratings for all other motors in the group, plus

3. the current ratings for the "other loads."

Example:

To determine the maximum overcurrent protection using dual-element, time-delay fuses for a branch circuit serving three motors plus an additional load. The FLA rating of the motors are: 27, 14, and 11 amperes. There is 20 amperes of "other load" connected to the branch-circuit. First determine the maximum time-delay fuse for the largest motor:

$$27 \times 1.75 = 47.15 \text{ amperes}$$

The next higher standard ampere rating is 50 amperes; next add the remaining ampere ratings to this value.

$$50 + 14 + 11 + 20 = 95 \text{ amperes}$$

For these four loads the maximum ampere rating of dual-element, time-delay fuses is 95 amperes, or "rounded" up to the next higher standard rating . . . 100 amperes. The disconnect switch would be a 100-ampere rating.

Several Motors on One Feeder

NEC® *Sections 430-24* and *430-62* set forth the requirements for this situation. The individual motors will have branch-circuit, short-circuit, and ground-fault protection sized according to *NEC*® *Section 430-52*. The individual motors do have overload protection.

The feeder fuses or circuit breakers will be sized by:

1. determining the fuse or circuit breaker rating or setting from *NEC*® *Table 430-152* for the largest motor in the group,

2. adding the value determined in step 1 to the full-load current ratings for all other motors in the group.

Example:

To determine the maximum overcurrent protection using dual-element, time-delay fuses for a branch circuit serving three motors. The FLA ratings of the motors are 27, 14, and 11 amperes.

First determine the maximum time-delay fuse for the largest motor:

$$27 \times 1.75 = 47.15 \text{ amperes}$$

The next higher standard ampere rating is 50 amperes; next add the remaining ampere ratings to this value.

$$50 + 14 + 11 = 75 \text{ amperes}$$

For these three motors the maximum ampere rating of dual-element, time-delay fuses is 75 amperes.

Because 75-ampere dual-element, time-delay fuses are not a standard size as listed in *NEC*® *Section 240-6*, it would be permissible to install 80-ampere fuses. You could also install 70-ampere dual-element, time-delay fuses if you are confident that they would have sufficient time-delay to allow the motors to start. In either case, the size of the disconnect switch would be 100 amperes.

Sometimes there is confusion when one or more of the motors are protected with instant trip breakers, sized at 800% to 1700% of the motor's full-load ampere rating. This would result in a very large (and possibly unsafe) feeder overcurrent device. *NEC*® *Section 430-62(a)*, Exception states that in such cases, make the feeder calculation as though the motors protected by the instant trip breakers were protected by dual-element, time-delay fuses.

DISCONNECTING MEANS

Each of the three appliances is furnished with a four-wire cord-and-plug (three phases plus equipment ground). This meets the requirements for the disconnecting of cord- and plug-connected appliances, *NEC*® *Section 422-22*.

The disconnecting-means requirement for permanently connected appliances is covered in *NEC*® *Section 422-31*. *NEC*® *Section 422-35* requires that the disconnect means for motor driven appliances be located within sight of the appliance, or be capable of being locked in the "off" position.

The disconnecting means shall have an ampere rating not less than 115% of the motor full-load current rating, *NEC*® *Section 430-110*.

The disconnecting means for more than one motor shall not be less than 115% of the sum of the full-load current ratings of all of the motors supplied by the disconnecting means, *NEC*® *Section 430-110(c)(2)*.

GROUNDING

WHY?   *NEC*® *Section 250-2(a)* explains *why* equipment must be grounded.

WHERE?   *NEC*® *Section 250-110* explains locations *where* equipment fastened in place or connected by permanent wiring methods (fixed) must be grounded.

WHAT?   *NEC® Section 250-112* explains *what* equipment that is fastened in place or connected by permanent wiring methods (fixed) must be grounded. *Section 250-114* explains *what* equipment that is cord-and-plug connected must be grounded.

HOW?   *NEC® Section 250-134* explains *how* to ground equipment that is fastened in place and is connected by permanent wiring methods. *NEC® Section 250-138* explains *how* to ground equipment that is cord-and-plug connected.

OVERCURRENT PROTECTION

Overcurrent protection for appliances is covered in *NEC® Section 422-11*. If the appliance is motor-driven, then *NEC® Article 430, Part C* applies. In most cases, the overload protection is built into the appliance. The appliance meets Underwriters Laboratories standards.

Complete discussion for individual motors and appliances was given in the exhaust fan section earlier in this unit.

THE BAKERY EQUIPMENT

A bakery can have many types and sizes of food-preparing equipment, such as blenders, choppers, cutters, disposers, mixers, dough dividers, grinders, molding and patting machines, peelers, slicers, wrappers, dishwashers, rack conveyors, water heaters, blower dryers, refrigerators, freezers, hot-food cabinets, and others, Figures 7-6 and 7-7. All of this equipment is designed and manufactured by companies that specialize in food preparation equipment. These manufacturers furnish specifications that clearly state such things as voltage, current, wattage, phase (single or three phase), minimum required branch-circuit rating, minimum supply circuit conductor ampacity, and maximum overcurrent protective device rating. If the appliance is furnished with a cord-and-plug arrangement, it will specify the NEMA size and type, plus any

other technical data that is required in order to connect the appliance in a safe manner as required by the Code.

NEC® Article 422, Part E sets forth the requirements for the marking of appliances.

Fig. 7-6   Cake mixer. (*Courtesy* of Rondo.)

Fig. 7-7   Dough divider. (*Courtesy* of Rollequip International, Inc.)

The Mixers and Dough Dividers (Three Appliances Connected to One Circuit)

When more than one appliance is to be supplied by one circuit, a load calculation must be made. Note on the bakery electrical plans that three receptacle outlets are supplied by one three-phase, 208-volt branch circuit, Table 7-6.

The data for these food preparation appliances has been taken from the nameplate and installation instruction manuals of the appliances.

A NEMA 15-20R receptacle outlet, Figure 7-8, is provided at each appliance location. These are part of the electrical contract.

Each of these appliances is purchased as a complete unit. After the electrician provides the proper receptacle outlets, the appliances are ready for use as soon as they are moved into place and plugged in.

TABLE 7-6 Load Calculation for Appliances					
Type of Appliance	Voltage	Amperes	Phase	HP	How Connected
Multimixer	208	3.96	3	¾	NEMA 15-20P 4-wire cord and plug on appliance
Multimixer	208	7.48	3	1½	Same as above
Dough divider	208	2.2	3	½	Same as above

Fig. 7-8   NEMA 15-20R receptacle and plate.

THE DOUGHNUT MACHINE

An individual branch circuit provides power to the doughnut machine. This machine consists of (1) a 2000-watt heating element that heats the liquid used in frying and (2) a driving motor that has a full-load rating of 2.2 amperes.

As with most food preparation equipment, the appliance is purchased as a complete, prewired unit. This particular appliance is equipped with a four-wire cord to be plugged into a receptacle outlet of the proper configuration.

Figure 7-9 is the control circuit diagram for the doughnut machine. The following components are listed on the following page.

S   A manual switch used to start and stop the machine

T1   A thermostat with its sensing element in the frying tank; this thermostat keeps the oil at the correct temperature.

T2   Another thermostat with its sensing element in the drying tank; this thermostat controls the driving motor.

A   A three-pole contactor controlling the heating element

B   A three-pole motor controller operating the drive motor

M   A three-phase motor

OL   Overload units that provide overload protection for the motor; note one thermal overload unit in each phase.

P   Pilot light to indicate when power is "on."

Since this appliance is supplied by an individual circuit, its current is limited to 80% of the branch-circuit rating according to *NEC® Section 384-16(d)*. The branch-circuit supplying the doughnut machine must have sufficient ampacity to meet the minimum load requirements as indicated on the appliance nameplate, or in accordance with *NEC® Section 430-24*.

Heater load	= 2000 watts (VA)
Motor load	
2.2 amperes × 208 volts × 1.73	= 792 VA
Plus 25% of 792 VA	= 198 VA
Total	= 2990 VA

The maximum continuous load permitted on a 20-ampere, three-phase branch circuit is:

16 amperes × 208 volts × 1.73 = 5760 VA

The load of 2990 volt-amperes is well within the 5760 VA permitted loading of the 20-ampere branch circuit.

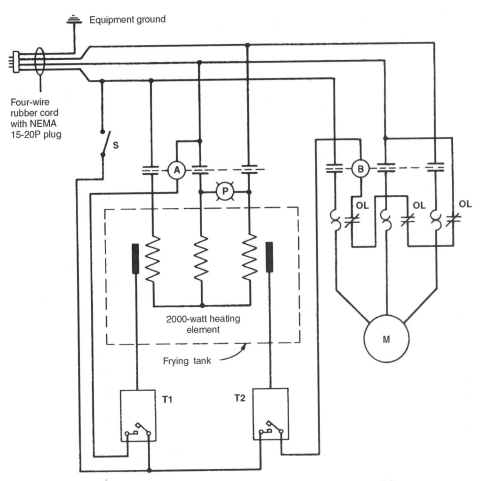

Fig. 7-9 Control wiring diagram for doughnut machine.

THE BAKE OVEN

The bake oven installed in the bakery is an electrically heated commercial-type bake oven. See Figure 7-10. The oven has a marked nameplate rating of 16,000 watts, 208 volts, three phase. This rated load includes all electrical heating elements, drive motor, timers, transformers, controls, operating coils, buzzers, lights, alarms, etc.

The instructions furnished with the bake oven, as well as the nameplate on the oven, specify that the supply conductors shall have a minimum ampacity of 60 amperes. The line-connecting terminals in the control panel and in the branch-circuit panelboard are suitable for 75°C (167°F). This would match the temperature rating of Type THW wire. Type THHN wire has a temperature rating of 90°C (194°F), but its ampacity is determined by referring to the 75°C (167°F) column of *NEC® Table 310-16*.

According to the Panel Schedule, the bake oven is fed with a 60-ampere, three-phase, three-wire feeder consisting of three No. 6 THHN copper con-

Fig. 7-10 Bake oven. (*Courtesy* of Cutler Industries, Inc.)

ductors. The metal raceway is considered acceptable for the equipment ground.

A three-pole, 60-ampere, 250-volt disconnect switch is mounted on the wall near the oven. From this disconnect switch, a conduit (flexible, rigid, intermediate metal conduit, or electrical metallic tubing as recommended by the oven manufacturer) is run to the control panel on the oven, Figure 7-11.

The bake oven is furnished as a complete unit. All overcurrent protection is an integral part of the circuitry of the oven. This meets the requirements of *NEC® Section 422-35*. The internal control circuit of the oven is 120 volts, which is supplied by an integral control transformer.

All of the previous discussion in this unit relating to circuit ampacity, conductor sizing, grounding, etc., also applies to the bake oven, dishwasher, and food-waste disposer. To avoid repetition, these Code requirements will not be repeated. Additional information relating to appliances is found in units 3 and 21.

For the bake oven:

1. Ampere rating   = 44.5 amperes
2. Minimum conductor ampacity =
   44.5×1.25 = 55.6 amperes
3. Branch-circuit overcurrent protection =
   44.5×1.25 = 55.6 amperes

It is permitted to install 55- or 60-ampere fuses in a 60-ampere disconnect switch.

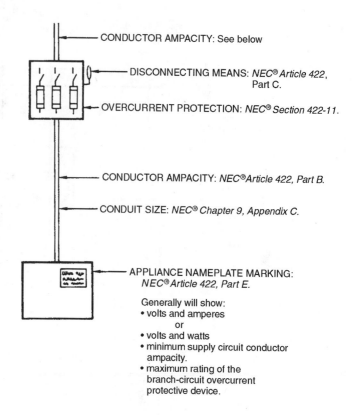

Fig. 7-11   Branch circuit for appliance.

REVIEW QUESTIONS

Refer to the *National Electrical Code®* or the working drawings when necessary. Where applicable, responses should be written in complete sentences. Write units using unit names, do not use abbreviations or symbols (1 foot, not 1' or 1 ft).

Cite the appropriate *NEC®* Article, Section or Table for each of the following.

1. Which *NEC®* Article refers specifically to air-conditioning equipment?

   _____

2. Which *NEC®* Article refers specifically to appliances? _____

3. Which *NEC®* Article refers specifically to disconnecting means for a motor?

   _____

4. Which *NEC®* Article refers specifically to grounding motor circuits?

   _____

The following questions refer to cord- and plug-connected appliances.

5. In your own words, define an *appliance* and give examples.

   _____

   _____

6. A 20-ampere branch circuit shall not serve a single appliance with a rating greater than _____.

7. An appliance is fastened in place and other cord- and plug-connected appliances are connected to a 20-ampere branch circuit. What is the maximum allowable load for the appliance that is fastened in place? _____

The following questions refer to motors and motor circuits.

8. Where shall the disconnect means for a 208-volt, three-phase, motor controller be located? _____

9. Where shall the disconnect means for a motor be located? _____

10. When selecting conductors, where is the motor full-load ampere rating to be taken from? _____

11. When selecting motor overload protective devices, where is the full-load ampere rating taken from? _____

12. A motor with a service factor of 1.25 should be protected by an overload device set to trip at not more than _____ of the motor full-load current rating.

13. The full-load current rating of a 5-horsepower, 208-volt, three-phase motor is 16.7 amperes. The minimum ampacity of the branch-circuit conductors supplying this motor would be not less than _____.

A 5-horsepower, three-phase motor has a nameplate current rating of 16 amperes. Dual-element, time-delay fuses are to be used for the motor branch-circuit protection. Show your calculations.

14. The preferred rating of the fuses is _____ amperes.

15. If the fuse with the preferred rating will not permit starting of the motor, the fuse rating may be increased to _____ amperes.

16. If the motor is to be started and stopped repeatedly, possibly resulting in nuisance opening of the circuit, the fuse rating may be increased to a maximum of _____ amperes.

The following questions apply to alternating-current motors.

17. Define *in sight* as it applies to a motor and its disconnect switch.

_____

_____

18. A motor has a full-load current rating of 74.8 amperes. What is the minimum allowable ampacity of the branch-circuit conductors?

_____

_____

19. A motor has a full-load current rating of 74.8 amperes and the terminations are rated for 75°C. What is the minimum conductor size and type?

_____

_____

20. Three, three-phase motors are supplied by a single set of conductors. The motors are 5, 10, and 15 horsepower. The minimum allowable ampacity of the conductors is _____ amperes.

21. A ½-horsepower, 115-volt motor is to be installed and a type S fuseholder and switch will be provided to control and protect the motor. What size fuse should be installed? _____ amperes.

22. Indicate the minimum and maximum starting currents for a 25-horsepower, 480-volt, Code Letter G motor.

23. Type 1 and Type 2 Protection are terms that might be found on the nameplate of a motor controller. Explain in your own words what these terms mean.

_____

_____

_____

_____

UNIT 8

Feeders

OBJECTIVES

After studying this unit, the student will be able to

- compute the feeder loading.
- determine the minimum overcurrent protective device rating.
- determine the minimum conductor size.
- determine the appropriate correction and adjustment factors.
- compute voltage drops.
- reduce the neutral size as appropriate.
- determine the minimum raceway size.

Feeders are that part of the electrical system that connects the branch circuit panelboards to the electrical service equipment. In the Commercial building a feeder is installed to each of the five occupancies, and one to the panelboard for the owner's circuit. (The supply to the boiler is a branch circuit.)

The feeder layout is shown in a riser diagram on working drawing E4. Specific requirements for feeders are given in *NEC® Article 215*, concerning installation requirements, and in *NEC® Article 220 Part B*, on computed loads and demand factors.

Some of the information presented in this unit has been introduced previously. This redundancy is intentional because of its application to both types of circuits, feeders and branch circuits, and because of its importance.

FEEDER REQUIREMENTS

Feeder Ampacity

The ampacity of a feeder must comply with the following requirements:

- A feeder must have an ampacity no less than the sum of the volt-amperes of the coincident loads of the branch circuits supplied by the feeder as reduced by demand factors. See *NEC® Article 220 Part B*.

- Demand factors are allowed in selected cases where it is unlikely that all the loads would be energized at the same time. See *NEC® Section 220-10*.

- Only the larger of any noncoincident loads (loads unlikely to be operated simultaneously such as the air-conditioning and the heating) need be included.

- Referring to *NEC® Section 220-22* a feeder neutral maybe reduced in size in certain situations. It may not be reduced in feeders consisting of 2-phase wires and a neutral of a 3-phase, 4-wire, wye connected system. The feeder to the doctor's office is an example of this condition.

The neutral of a 3-phase 4-wire, wye connected system maybe reduced in size provided it remains of a size sufficient to:

- carry the maximum unbalanced load
- carry the nonlinear load.

Overcurrent Protection

The details of selecting overcurrent protective devices are covered in Units 17, 18, and 19, and elsewhere in the text as indicated by the code index included in the appendix.

As previously stated, the rating of a circuits, is the rating of the circuit overcurrent protective device. See *NEC® Section 210-3*. The rating of a feeder shall not be less than the noncontinuous loading plus 125% of the continuous loading, see *NEC® Section 215-3*.

The basic requirement for overcurrent protection is that a conductor shall be protected in accordance with its ampacity, but several conditions exist where the rating of the overcurrent protective device (OCPD) can exceed the ampacity of the conductor being protected. For example, If the ampacity of a feeder does not match a standard OCPD rating then the next higher standard rating may be used, see *NEC® Section 240-3(b)*, provided the rating is 800-ampere or less. Feeder conductors with an ampacity of 130 amperes may be protected with a 150-ampere protective device.

Temperature Limitations

When selecting the components for a circuit, it is necessary to remember that no circuit is better than its terminations. *NEC® Section 110-14(c)* ensures that the consideration of terminations is included in the selection of the circuit components.

Terminations, like other components in a circuit, have temperature ratings. The terminations on breakers, switches or panelboards, with a rating of 100-ampere or less, may be rated for 60°C (140°F). *NEC® Section 110-14(c)* stipulates that unless equipment is marked with a 75°C (167°F) temperature rating (for example 60/75°C or 75°C) the circuit ampacity shall not exceed the value given in the 60°C (140°F) column of *NEC® Table 310-16*. The reason for this is that the heat generated by the current (amperes) in a conductor is dependent on the conductor size. The consequence of *NEC® Section 110-14(c)* is that it establishes a conductor size that is compatible with the temperature rating of the terminations. For example, the minimum conductor size for a noncontinuous load of 60 amperes would be a No. 4 AWG conductor unless the terminations are marked 75°C (167°F) in which case a No. 6 AWG would satisfy the requirements. The resistance of the No. 4 AWG is less than the No. 6 AWG thus the heat generated would be less and subsequently the termination rating may be less.

FEEDER CONDUCTORS

Wire Selection

The process of selecting a conductor begins with identifying the wire material. An examination of *NEC® Table 310-16* reveals that only two classifications of wire are listed. They are:

- copper
- aluminum or copper-clad aluminum.

The table indicates that for the same size and type a copper wire will have the higher ampacity. *NEC® Section 110-5* stipulates that the wire will be assumed to be copper unless stated otherwise.

Aluminum wire is seldom used in branch circuits but because of weight and cost factors it is frequently selected for feeders and services. When installing aluminum conductors the electrician must follow the directions and be especially careful to tighten all terminals to the recommended torque. It is also important to remember that listed connectors are required whenever it is necessary to join copper and aluminum conductors.

Some common problems associated with aluminum conductors when not properly connected maybe summarized as follows:

- A corrosive action is set up when copper and aluminum wires come in contact with one another if moisture is present.

- The surface of an aluminum conductor oxidizes as soon as it is exposed to air. If this oxidized surface is not broken through, a poor connection results. When installing aluminum conductors, particularly in larger sizes, an inhibitor is brushed onto the aluminum conductor, then the conductor is scraped with a stiff brush where the connection is to be made. The process of scraping the conductor breaks through the oxidation, and the inhibitor keeps the air from coming in contact with the conductor. Thus, further oxidation is prevented. Aluminum compression type connectors usually have an inhibitor paste installed inside the connector.

- Aluminum wire expands and contracts to a greater degree than does copper for an equal load. This characteristic is another possible cause of a poor connection. Crimp connections for aluminum conductors are usually longer that those for comparable copper conductors, resulting in greater contact surface of the conductor in the connector.

FEEDER COMPONENT SELECTION

Previously in this Unit the information necessary for selecting feeder and branch circuit components has been discussed in detail. Here, that information is first summarized then presented in sequential steps. While studying these steps it may be useful to review the detailed discussions and peruse the examples presented following the selection procedure.

Before the selection process can begin it is necessary that the following circuit parameters be known:

- The continuous and noncontinuous loadings, in amperes, calculated in accordance with *NEC® Article 220.*

- For the type of conductors that are to be installed see *NEC® Table 310-13.*

- The ambient temperature of the environment where the conductors are to be installed and the number of current carrying conductors that will be in the cable or raceway. See *NEC® Section 310-10.*

FEEDER CONDUCTOR SELECTION

Step 1: OCPD selection.

- If the sum of the noncontinuous load plus 125% of the continuous load is 800 amperes or less an OCPD shall be selected that has a rating equal to or next greater than that sum, see *NEC® Sections 240-3(b) and -6.*

- If the sum of the noncontinuous load plus 125% of the continuous load is greater than 800 amperes an OCPD shall be selected that has a rating equal to or less than that sum, see *NEC® Sections 240-3(c) and -6.*

Step 2: Minimum conductor size determination

- If the selected OCPD has a rating of 100-ampere or less or, if any of the terminations are marked for No. 1 AWG or smaller, or are rated at 60°C, the minimum size conductor is determined by entering the 60°C column of *NEC® Table 310-16,* and selecting the conductor size that has an allowable ampacity equal to or greater than the sum of the noncontinuous load plus 125% of the continuous load.

- If the selected OCPD has a rating greater than 100-ampere or, all the terminations are rated for 75°C, the minimum size conductor is determined by entering the 75°C column of *NEC® Table 310-16,* and selecting the conductor size that has an allowable ampacity equal to or greater than the sum of the noncontinuous load plus 125% of the continuous load.

Step 3: Conductor size determination

- If the selected OCPD has a rating of 100-ampere or less, the conductors to be installed may be selected from the 60°C, 75°C or 90°C columns of *NEC® Table 310-16.*

- If the selected OCPD has a rating greater than 100-ampere the conductors to be installed shall be selected from either the 75°C or 90°C columns of *NEC® Table 310-16.*

- To make the feeder conductor size determination enter *NEC® Table 310-16* and select a conductor size that is equal to or greater than the minimum size previously determined and has an allowable ampacity, in the column of choice, that after application of correction and adjustment factors has a derated ampacity that permits the use of the selected OCPD and is equal to or greater than the computed load (the continuous plus the noncontinuous loads).

Neutral Size Determination

NEC® Section 220-22 sets forth conditions for reducing the size of the neutral conductor. This can result in a reduction in wire and raceway cost. The procedure set forth is straight forward except when the supply is a three-phase four-wire wye-connected system such as is commonly specified for commercial type buildings. To determine a reasonable minimum size two types of loads must be considered, linear and nonlinear. Nonlinear loads are defined in *NEC® Article 100*. In the FPN following the definition the following items are listed: "Electronic equipment, electronic/electric discharge lighting and adjustable speed drives."

Linear loads

Linear loads may be calculated by the following formula:

$$N = \sqrt{A^2 + B^2 + C^2 - AB - BC - AC}$$

Where: N = Neutral current
         A = Phase A current = 30 Amperes
         B = Phase B current = 40 Amperes
         C = Phase C current = 50 Amperes

$$N = \sqrt{30^2 + 40^2 + 50^2 - (30 \times 40) - (40 \times 50) - (30 \times 50)}$$

$$N = \sqrt{900 + 1600 + 2500 - 1200 - 2000 - 1500}$$

$$N = \sqrt{300} = 17.32 \text{ Amperes}$$

Assuming phase loads of 30, 40, and 50 amperes the neutral load would be 17.32 amperes.

Additional calculation would indicate that if the 30 ampere load was disconnected the neutral load would be 43.6 amperes and if all but the 50-ampere load were disconnected the neutral load would be 50 amperes. For the neutral to comply with *NEC® Section 220-22* it must be sized for this maximum unbalanced condition.

Nonlinear Loads

Beginning in 1947, and continuing in each succeeding edition, the *NEC®* has been addressing the various problems of neutral currents from nonlinear/harmonic sources.

The following quotes are from the 1999 *National Electrical Code®*:

NEC® Article 100, Nonlinear load: *"A load where the wave shape of a steady state current does not follow the wave shape of the applied voltage. (FPN): Electronic equipment, electronic/electric-discharge lighting, adjustable speed drive systems, and similar equipment may be nonlinear loads."*

NEC® Section 220-22 (last sentence): *"There shall be no reduction of the neutral capacity for that portion of the load that consists of nonlinear loads supplied from a 4-wire, wye-connected, 3-phase system nor the grounded conductor of a 3-wire circuit consisting of two phase wires and the neutral of a 4-wire, 3-phase, wye-connected system."*

NEC® Section 220-22 FPN No. 2: *"A 3-phase, 4-wire, wye connected power system used to supply power to nonlinear loads may necessitate that the power system design allow for the possibility of high harmonic neutral currents."*

Reprinted with permission from NFPA 70-1999.

Harmonics

In the past, most connected loads were linear, such as resistive heating and lighting, and motors. In a linear circuit, current changes in proportion to a voltage increase or decrease in that circuit. The voltage and current sine waves are "sinusoidal."

The arrival of electronic equipment such as UPS systems, AC-to-DC converters (rectifiers), inverters (AC-to-DC to adjustable frequency AC), computer power supplies, programmable controllers, data processing, electronic ballasts, and similar equipment brought about problems in electrical systems. Unexplained events started to happen such as overheated neutrals, overheating and failure of transformer, overheated motors, hot bus bars in switchboards, unexplained tripping of circuit breakers, incandescent lights blinking, fluorescent lamps flickering, malfunctioning computer, and hot lugs in switches and panelboards, even though the connected loads were found to be well within the conductor and equipment rating. All or the above electronic equipment is considered to be non-linear, when the current in a given circuit does not increase or decrease in proportion to the voltage in that circuit. The resulting distorted voltage and current sine waves are "non-sinusoidal." These problems can feed back into an electrical system and affect circuits elsewhere in the building, not just where the non-linear loads are connected.

The root of the problem can be traced to electronic devices such as thyristors (silicon-controlled rectifiers, SCRs) that can be switched on and off for durations that are extremely small fractions of a cycle. Another major culprit is switching-mode power supplies that "switch" at frequencies of 20,000 to over 100,000 cycles per second. This rapid switching causes distortion of the sine wave. Harmonic frequencies are superimposed on top of the fundamental 60 Hz frequency, creating harmonic distortion.

Harmonics are multiples of a fundamental frequency. In this country, 60 Hz (60 cycles per second) is standard. Other frequencies superimposed on the fundamental frequency can be measured such as:

Odd Harmonies		Even Harmonies	
3rd	180 cycles	2nd	120 cycles
5th	300 cycles	4th	240 cycles
7th	420 cycles	6th	360 cycles
9th	540 cycles	8th	480 cycles
etc.		etc.	

Currents of the 3rd harmonics and odd multiples of the 3rd harmonic (9th, 15th, 21st, etc.) add together in the common neutral conductor of a three-phase system, instead of canceling each other. These odd multiples are referred to as triplens. For example, on an equally balanced three-phase. four-wire wye connected system where each phase carries 100 amperes, the neutral conductor might he called upon to carry 200 amperes or more. If this neutral conductor had been sized to carry 100 amperes, it would become severely overheated. This is illustrated in Figure 8-1.

Unusually high neutral current will cause overheating of the neutral conductor. There might also be excessive voltage drop between the neutral and ground. Taking a voltage reading at a receptacle between the neutral conductor and ground with the loads turned on, and finding more than two volts present probably indicates that the neutral conductor is overloaded because of the connected non-linear load.

Attempting to use a low-cost "average reading" clamp-on meter will not give a true reading of the total current in the neutral conductor, because the neutral current is made up of current values from many frequencies. The readings will be from 30% to 50% lower. To get accurate current readings where harmonics are involved, ammeters referred to as "true rms" must be used.

At least one major manufacturer of Type MC cable manufactures a cable that has an oversized neutral conductor. For cables where the phase conductors are No I2 AWG, the neutral conductor is a No 8 AWG. For cables where the phase conductors are No 10 AWG the neutral conductor is a No 6 AWG. They also provide a Type MC cable that has a separate neutral conductor for each of the phase. conductors. There is still much to be learned about the effects of nonlinear loads. Many experienced electrical consulting engineers are now specifying that three-phase, four-wire branch circuits that supply non-linear loads have a separate neutrals for each phase conductor, instead of using the typical the three-phase four-wire branch circuit using a common neutral. (See Figure 12-15) They also specify that the neutral be sized double that of the phase conductor. The Computer and Business Equipment Manufacturers Association (CBEMA) offers the following recommendation.

Run a separate neutral to 120 volt outlet receptacles on each phase. Avoid using shared neutral conductors for single-phase 120-volt outlets on different phases. Where a shared neutral conductor for a 208Y/120 volt system must be used for multiple phases us a neutral conductor having at least 173% of the phase conductors.

What does this mean to the electrician? It simply means that if over heated neutral conductors, overheated transformers, overheated lugs in switches and panels or other unexplained heating problems are encountered, they might be caused by non-linear loads. Where a commercial or industrial building contains a high number of non-linear loads, it is

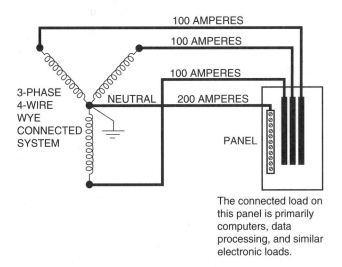

100 AMPERES

100 AMPERES

100 AMPERES

3-PHASE 4-WIRE WYE CONNECTED SYSTEM

NEUTRAL   200 AMPERES

PANEL

The connected load on this panel is primarily computers, data processing, and similar electronic loads.

Figure 8-1 Feeder neutral current resulting from nonlinear loads.

advisable to have the electrical system analyzed and redesigned by an electrical engineer specifically trained and qualified on the subject of non-linear loads

The effects of nonlinear loads such as electric discharge lighting, electronic/data processing equipment and/or variable speed motors is illustrated in following four figures. In each case the power is supplied from a 3-phase 4-wire 208Y/120 system. Figure 8-1 illustrates the possible effect the load could have on a feeder. The neutral could be carrying 200% of the phase conductors. Figure 8-2 illustrates a branch-circuit where the load is fluorescent lighting using core and coil ballast. If the ballast

were electronic the neutral current would be even greater. Figure 8-3 illustrates the recommend arrangement for serving three nonlinear loads from a 4-wire wye system. If the neutral is shared, as is illustrated in figure 8-4, then the neutral should have at least 200% of the carrying capacity as the individual phase conductors. Using separate neutrals is the preferred arrangement.

It is clear from the information given that nonlinear loads can create neutral currents in excess of the phase currents in a 3-phase 4-wire wye system. Considering what is known and unknown it is prudent to take a conservative approach considering the minimal saving involved.

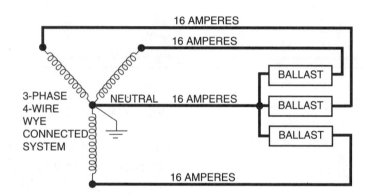

Figure 8-2   Branch-circuit current resulting from nonlinear loads.

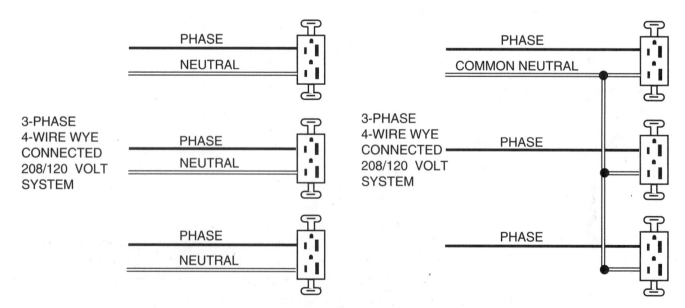

Figure 8-3   Recommended: separate neutrals serving nonlinear loads.

Figure 8-4   Not recommended: shared neutral serving nonlinear loads.

Three stipulations are clear:

- The neutral must be able to carry the maximum unbalanced load.

- On a 3-phase, 4-wire, wye system no reduction is allowed for nonlinear loads.

- In *Section 220-22*, FPN No. 2 a warning is issued concerning the currents created by nonlinear loads.

Three conditions of failure may occur with a 3-phase 4-wire wye system:

- If all three phases are operable. In this, the normal situation, by definition the neutral is part of the feeder and must be sized for the same load and environmental requirements. The minimum size shall comply with *NEC® Section 110-14(c)* and the adjustment and corrections must be applied as with the phase conductors.

- If two phases are operable. In essence this creates a situation similar to a feeder consisting of two phases of a three phase system. A condition where the *NEC®* specifically prohibits a reduction in neutral size.

- Only a single phase is operable, (worse case scenario). It is a tendency for nonlinear loads to be continuous loads and thus likely that in this condition the load in the single phase may be largely a nonlinear load. As previously stated, this is a situation where the neutral load could exceed the phase load.

Giving full consideration to the worse condition scenario, in this commercial building the following strategy will be followed in determining the minimum neutral conductor size:

- the 208 single-phase and three-phase loads will be subtracted from the OCPD selection load (the noncontinuous load plus 125% of the continuous load).

- a load equal to the nonlinear load will be added to the neutral load. This doubles the neutral capacity for the nonlinear load.

Voltage Drop

NEC® Section 215-2(d), FPN No. 2 sets forth a recommendation for limiting the voltage drop on feeders. A similar recommendation is made for branch circuits. The recommendation is for the total voltage drop to be not more than 5%, and not more than 3% for either the feeder or the branch circuit. The author recommends that the feeder be limited to 2% thus leaving 3% for the branch circuit.

As an example the circuit to drugstore window display receptacle outlets is to be loaded to 1500 volt-amperes or 12.5 amperes. The distance from the panelboard to the center outlet is 85 feet and the minimum wire size is No. 12 AWG. Three different methods will be demonstrated to estimate the voltage drop in this circuit.

The first method requires that the resistance per foot of the conductor be known. According to *NEC® Chapter 9, Table 8* the resistance of a No. 12 AWG uncoated copper conductor is 1.93 ohms per 1000 feet. Therefore, the resistance of the circuit is:

$$R = (1.93 \ 3 \ / \ 1000 \ \text{ft}) \times 2 \times 85 \ \text{ft.} = 0.3281 \ \Omega$$

The factor of 2 in the equation is required because both the phase and neutral conductors carry the current. The voltage drop is defined by Ohms Law:

$$VD = I \times R = 12.5 \ A \times 0.3281 \ 3 = 4.1 \ V$$

This is a voltage drop of:

$$4.1 \ V \ / \ 120 \ V = 0.0342 \ \text{or} \ 3.42\%$$

The second method requires that the circular mil area of the conductor be known. From *NEC® Table 310-16* the kcmil area of a No. 12 AWG is 6.53. The equation is:

$$VD = (K \times I \times L \times 2) \ / \ kcmil$$

Where K for copper is 12 and for aluminum 20. Therefore

$$VD = 12 \times 12.5 \times 85 \times 2 \ /6530 = 3.9 \ \text{volts}$$

Another method, is one that considers the power factor, the type of raceway, and if it is a single or three-phase circuit. An abbreviated table is displayed below with values appropriate for use with the circuits in the commercial building. The values given are for installation in steel conduit. A complete table is provided in the Appendix Table 8-1.

To calculate the voltage drop multiply the current, the distance, and the proper factor from the table then move the decimal point 6 places to

TABLE 8-1	Installation in Steel Conduit								
Wire size	12	10	8	6	4	3	2	1	0
90% pf, 1-phase	3659	2214	1460	937	610	501	409	337	263
90% pf, 3-phase	3169	1918	1264	812	528	434	354	292	228
Wire size	00	000	0000	250	300	350	400	500	600
90% pf, 1-phase	227	187	157	142	125	113	105	94	86
90% pf, 3-phase	196	162	136	123	108	98	91	81	75

the left. Again, using the circuit to show window receptacle outlets.

$$VD = 12.5 \text{ A} \times 85 \text{ ft} \times 3659 /10^6 = 3.9 \text{ V}$$

This is a voltage drop of:

$$3.9 \text{ V} / 120 \text{ V} = 0.033 \text{ or } 3.3\%$$

It is recommended that No. 10 AWG conductors be used.

PANELBOARD WORKSHEET SUMMARY

Phase Connection Summary

The volt-ampere loading on each phase is tabulated and compared (Table 8-2). The values should be essentially equal however it is unlikely that they will be exactly equal.

Loads Summary

These values are needed for the design loading calculations. Their use will be discussed in detail in a later unit. The groups are totaled to assure that no loads were omitted.

TABLE 8-2 Drugstore – Load Summary	
Connection Summary:	**VA Total**
Connected load Phase A	4503
Connected load Phase B	5100
Connected load Phase C	4272
Balanced Loads	8604
Connection Total	22479
Loads Summary:	**VA Total**
Continuous	8595
Noncontinuous + Receptacle	5280
Highest Motor	8604
Other Motors	0
Loads Total	22479

FEEDER DETERMINATION, DRUGSTORE

Following is a table, Table 8-3, and an outline illustrating the process of selecting the feeder components for the Drugstore.

Load Summary

- The continuous computed load is taken from the Panelboard Summary. It is multiplied by 1.25.

$$8595 \text{ VA} \times 1.25 = 10744 \text{ VA}$$

- The noncontinuous and receptacle loads are taken from the panelboard schedule. The value remains unchanged.

$$5280 \text{ VA} = 5280 \text{ VA}$$

- The highest motor load is taken from the panelboard summary. It is multiplied by 1.25.

$$8604 \text{ VA} \times 1.25 = 10755 \text{ VA}$$

- There were no other motor loads.

- The four preceding loads are added and the sum multiplied by 0.25 to determine the growth. The growth is increased by 25% on the assumption that it will be all a continuous load.

$$(8595 \text{ VA} + 5280 \text{ VA} + 8604 \text{ VA}) \times 0.25 = 5620 \text{ VA}$$

$$5620 \text{ VA} \times 1.25 = 7025 \text{ VA}$$

- The values in the two columns are summed.

$$8595 \text{ VA} + 5280 \text{ VA} + 8604 \text{ VA} + 5620 \text{ VA} = 28099 \text{ VA}$$

$$10744 \text{ VA} + 5280 \text{ VA} + 10755 \text{ VA} + 7025 \text{ VA} = 33804 \text{ VA}$$

OCPD Selection

- Divide the OCPD selection volt-amperes by 360 to convert to amperes.

$$33804 \text{ VA} \div 360 \text{ V} = 94 \text{ A}$$

TABLE 8-3 Drugstore – Feeder Selection		
Load Summary, 208Y/120, 3 Phase, 4-Wire	Computed	OCPD
Continuous Load	8595	10744
Noncontinuous + Receptacle Load	5280	5280
Highest Motor Load	8604	10755
Other motor Load	0	0
Growth	5620	7025
Computed Load & OCPD Load	28099	33804
OCPD Selection	**Input**	**Output**
OCPD Load Volt-Amperes & Amperes	33804	94
OCPD Rating		100
Minimum Conductor Size		1 AWG
Minimum Derated Ampacity		91
Phase Conductor Selection	**Input**	**Output**
Ambient Temperature & Correction Factor	28°C	1
Current Carrying Conductors & Adjustment Factor	4	0.8
Derating Factor		0.8
Minimum Allowable Ampacity		114
Conductor Size		1 AWG
Conductor Type, Allowable Ampacity	THHN	150
Derated Ampacity		120
Voltage Drop, 0.9 pf, 3 ph, per 100 ft		2.28
Neutral Conductor Selection	**Input**	**Output**
OCPD Load Volt-Amperes	33804	
Balanced Load Volt-Amperes	–8604	
Nonlinear Load Volt-Amperes	4575	
Total Load Volt-Amperes & Amperes	29775	83
Minimum Neutral Size		3 AWG
Minimum Allowable Ampacity		104
Neutral Conductor Type & Size	THHN	3 AWG
Allowable & Derated Ampacity	110	88
Raceway Size Determination	**Input**	**Output**
Feeder Conductors Size & Total Area	1 AWG	0.4686
Neutral Conductor Size & Area	3 AWG	0.0973
Grounding Conductor Size & Area		0
Total Conductor Area		0.5659
Raceway Type & Size		1¼ in

- From *NEC® Section 240-6*, the OCPD ampere rating next higher than 94 is 100-ampere. This becomes the OCPD ampere rating.

- As the OCPD rating is 100-ampere, a No. 1 AWG is the conductor size with an ampacity equal to or next greater than 100 in the 60°C column of *NEC® Table 310-16*.

The minimum derated ampacity must meet two criteria. It must be high enough to carry the computed load. In this case that is: $28099 \div 360 = 78$ A. It also must be high enough to allow the use of the selected OCPD. The next lower rating given in *NEC® Section 240-6* under 100 is 90. An ampacity of 91 would permit the use of an OCPD with a rating of 100-ampere. The higher of these is 91 amperes thus that is the minimum derated ampacity.

Phase Conductor Selection

- The ambient temperature is determined to be 28°C. As this is within the standard range, the correction factor is 1.

- As the harmonic load (fluorescent lighting, etc.) is significant the neutral will be considered as a current carrying conductor. The adjustment factor for 4 conductors is 0.8.

- The correction and adjustment factors are multiplied to yield a reduction factor of 0.8.

- The minimum allowable ampacity is determined by dividing the minimum derated ampacity by the derating factor. In this example $91 \div 0.8 + 114$. This is the lowest ampacity that after derating will be equal to or greater than the minimum derated value.

- The conductor size selection must also meet two criteria. It must be at least the size of the minimum conductor size and it must have an allowable ampacity that is equal or greater than the minimum allowable ampacity. In this example a No. 1 AWG qualifies.

- The conductor type is compliance with the construction specification and the allowable ampacity is taken from *NEC® Table 310-16* for the conductor size.

- The derated ampacity is the allowable ampacity multiplied by the derating factor. This is checked against the minimum allowable ampacity but if the steps have been followed correctly it should never fail.

- The voltage drop is calculated for 90% power factor and a distance of 100 feet by the procedure presented earlier. By using 100 feet, the drop for the actual distance can be easily mentally computed. If after applying the actual length, the drop is in excess of 3% serious consideration should be given to increasing the conductor size.

Neutral Size Determination

As is shown in the feeder selection schedule the neutral size is determined to be:

Total load - Balanced load + linear load
= Neutral load

$$33804 \text{ VA} - 8604 \text{ VA} + 4575 \text{ VA} = 29775 \text{ VA}$$

$$29775 \text{ VA} \div 360 \text{ V} = 83 \text{ A}$$

From the 60°C Column the minimum wire size is a No. 3 AWG.

The minimum allowable ampacity is:

$$83 \text{ A} \div 0.8 = 104 \text{ A}$$

A No. 3 AWG Type THHN conductor has an allowable ampacity of 110 amperes and qualifies for the neutral conductor.

Raceway Size Determination

It has been concluded that the feeder to the drugstore will consist of 3 No. 1 AWG and 1 No. 3 AWG. No equipment grounding conductor is necessary if rigid metal conduit is installed. The specification also permits schedule 80 rigid nonmetallic conduit which would require the installation of an equipment grounding conductor. This grounding conductor would be included in the fill area which may require a larger size raceway.

From *NEC® Chapter 9, Table 5* the area of a No. 1 AWG, Type THHN is 0.1562, the area of a No. 3 AWG, Type THHN, is 0.0973. The total area of the 4 conductors is:

$$(3 \times 0.1562 \text{ sq. in}) + 0.0973 \text{ sq. in.} = 0.565 \text{ sq. in.}$$

Consulting *NEC® Chapter 9, Table 4* for rigid metal conduit the smallest allowable size, for over 2 wires, is 1 1/4-inch.

REVIEW QUESTIONS

Refer to the *National Electrical Code®* or the working drawings when necessary. Where applicable, responses should be written in complete sentences. Write units using unit names, do not use abbreviations or symbols (1 foot, not 1' or 1 ft).

1. Determine the conductor sizes for a feeder to a panelboard. It is a 120/240 Volt single phase system. The OCPD has a rating of 100-ampere. The computed load is 15,600 VA. All the loads are 120 Volt.

_____

_____

_____

For the next three questions, you may use either the graphical or mathematical means to determine the answers.

2. Calculate the neutral current in a 120/240 single-phase system when the current in phase A is 20 amperes and the current in phase B is 40 amperes.

_____

_____

_____

3. Calculate the neutral current in a 208Y/120 three-phase four-wire system when the current in phase A is 0, in phase B is 40, and in phase C is 60 amperes.

_____

_____

_____

4. Calculate the neutral current in a 208Y/120 three-phase four-wire system when the current in phase A is 20, in phase B is 40, and in phase C is 60 amperes.

_____

_____

_____

5. A balanced electronic ballast load connected to a three-phase, four-wire multi-wire branch circuit can result in a neutral current of (zero)(half)(two times) that of the current flowing in the phase conductors. (Circle the correct answer.)

6. Harmonic currents associated with electronic equipment (add together) (cancel out) a shared common neutral on multi-wire branch circuits and feeders. (Circle the correct answer.)

7. Check the current statement:
   For the circuits that supply computer/data processing equipment, it is recommended that
   ☐ a separate neutral be installed for each "hot" phase branch-circuit conductor.
   ☐ a common neutral be installed for each multi-wire branch circuit.

8. Determine the feeder size and other information requested in the following table. The load data is:

Continuous	20,000	VA
Noncontinuous/receptacle	12,000	VA
Highest motor	8,000	VA
Other motor	16,000	VA
Nonlinear	20,000	VA
Balanced	24,000	VA

A 3-phase, 4-wire, 208Y/120 feeder is to be installed. From a distribution panelboard, the feeder is installed underground in rigid PVC, schedule 80 to an adjacent building where intermediate metal conduit is used.

• The IMC is installed in a room where the ambient temperature will be 110°F.

• Type THHN/THWN conductors are used.

• The growth allowance will be 10% of the computed load.

• The power factor will be maintained at 90% or higher.

• The conductor length is 180 feet.

• The voltage drop at maximum load shall not exceed 2%.

Load Summary, 208Y/120, 3 Phase, 4-Wire	Computed	OCPD
Continuous Load		
Noncontinuous + Receptacle Load		
Highest Motor Load		
Other Motor Load		
Growth		
Computed Load & OCPD Load		
OCPD Selection	Input	Output
OCPD Load Volt-Amperes & Amperes		
OCPD Rating		
Minimum Conductor Size		
Minimum Derated Ampacity		
Phase Conductor Selection	Input	Output
Ambient Temperature & Correction Factor		
Current Carrying Conductors & Adjustment Factor		
Derating Factor		
Minimum Allowable Ampacity		
Conductor Size		
Conductor Type, Allowable Ampacity		
Derated Ampacity		
Voltage Drop, 0.9 pf, 3 ph, per 100 ft		
Neutral Conductor Selection	Input	Output
OCPD Load Volt-Amperes		
Balanced Load Volt-Amperes		
Nonlinear Load Volt-Amperes		
Total Load Volt-Amperes & Amperes		
Minimum Neutral Size		
Minimum Allowable Ampacity		
Neutral Conductor Type & Size		
Allowable & Derated Ampacity		
Raceway Size Determination	Input	Output
Feeder Conductors Size & Total Area		
Neutral Conductor Size & Area		
Grounding Conductor Size & Area		
Total Conductor Area		
Raceway Type & Size		

UNIT 9

Special Systems

OBJECTIVES

After studying this unit, the student will be able to

- select and install surface metal raceway.
- select and install multioutlet assemblies.
- compute the loading allowance for multioutlet assemblies.
- select and install a floor outlet system.
- install a branch circuit for a computer.

A number of electrical systems are found in almost every commercial building. Although these systems usually are a minor part of the total electrical work to be done, they are essential systems and it is recommended that the electrician be familiar with the installation requirements of these special systems.

SURFACE METAL RACEWAYS

Surface metal raceways, either metal or nonmetallic, are generally installed as extensions to an existing electrical raceway system, and where it is impossible to conceal conduits, such as in desks, counters, cabinets, and modular partitions. The installation of surface metal raceways is governed by *NEC® Article 352, Part A, NEC® Article 352, Part B* governs the installation of surface nonmetallic raceways. The number and size of the conductors to be installed in surface raceways is limited by the design of the raceway. Catalog data from the raceway manufacturer will specify the permitted number and size of the conductors for specific raceways. Conductors to be installed in raceways may be spliced at junction boxes or within the raceway if the cover of the raceway is removable. See *NEC® Sections 352-7* and *352-29*. It should be noted that the combined size of the conductors, splices, and taps shall not fill more than 75% of the raceway at the point where the splices and taps occur.

Surface raceways are available in various sizes, Figure 9-1, and a wide variety of special fittings, like the supports in Figure 9-2, makes it possible to use surface metal or nonmetallic raceways in almost any dry location. Two examples of the use of surface raceways are shown in Figures 9-3 and 9-4.

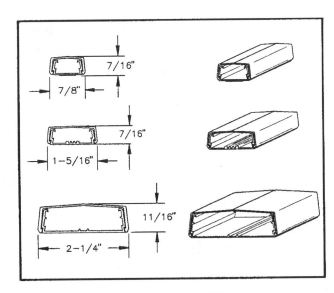

Fig. 9-1 Surface nonmetallic raceways.

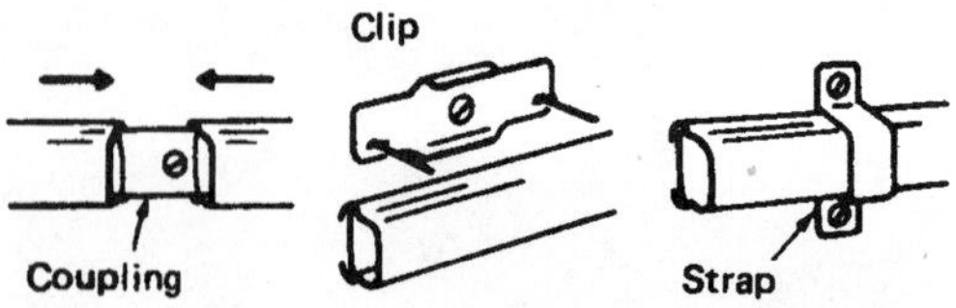

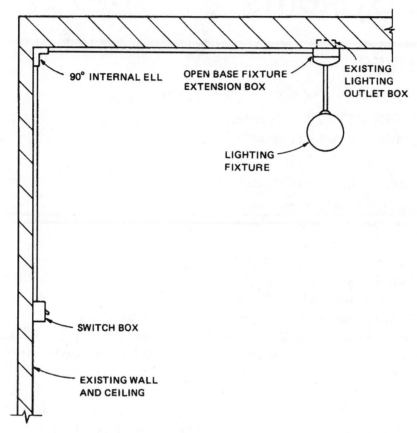

Fig. 9-2   Surface metal raceway supports.

Fig. 9-3   Use of surface raceway to install switch on existing lighting installation.

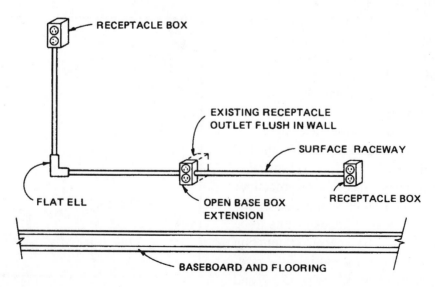

Fig. 9-4   Use of surface metal raceway to install additional receptacle outlets.

MULTIOUTLET ASSEMBLIES

Multioutlet assembly is defined in *NEC® Article 100.* The installation requirements are specified in *NEC® Article 353.* Multioutlet assemblies, Figure 9-5, are similar to surface raceways and are designed to hold both conductors and devices. These assemblies offer a high degree of flexibility to an installation and are particularly suited to heavy use areas where many outlets are required or where there is a likelihood of changes in the installation requirements. The plans for the insurance office specify the use of a multioutlet assembly that will accommodate power, data, telecommunications, security and audio/visual systems, Figure 9-6. This installation will allow the tenant in the insurance office to revise and expand the office facilities as the need arises. The covers are available in steel, aluminum, and PVC with vinyl laminates of different colors and wood veneers.

Loading Allowance

The load allowance for a multioutlet assembly is specified by *NEC® Section 220-3(b)(8)* as:

- 180 volt-amperes for each 5 feet (1.52 m) of assembly when normal loading conditions exist, or

- 180 volt-amperes for each 1 foot (305 mm) of assembly when heavy loading conditions exist.

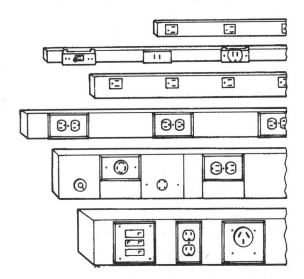

Fig. 9-5   Multioutlet assemblies.

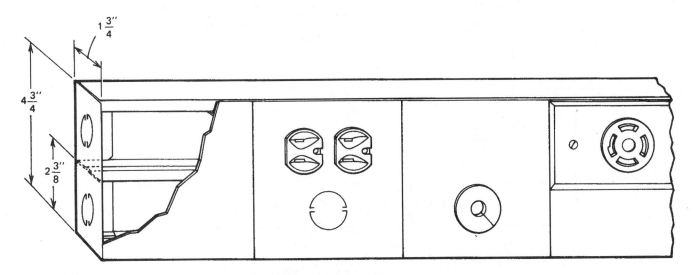

Fig. 9-6   Multioutlet assembly for power and communication systems.

The usage in the insurance office is expected to be intermittent and made up of only small appliances; thus it would qualify for allowable minimum of 180 volt-amperes per five feet. However, the contractor is required to install a duplex receptacle every 18 inches for a total of 56 receptacles. The number of receptacles is multiplied by 180 volt-amperes to determine the connected load, which would be 10,080 volt-amperes. This load is connected to five branch-circuits with a total non-continuous capacity of 12,000 volt-amperes. This allows for some growth, and additional circuits can be easily installed.

When determining the feeder size the requirement for the receptacle load can be reduced by the application of the demand factors given in *NEC® Table 220-13*. The total receptacle loading in the insurance office is 12,780 volt-amperes. The first 10,000 volt-amperes must be included at 100% but the remainder need only be included at 50%. The loading on the feeder due to receptacles would be 11,390 volt-amperes.

Receptacle Wiring

The plans indicate that the receptacles to be mounted in the multioutlet assembly must be spaced 18 inches (457 mm) apart. The receptacles may be connected in either of the arrangements shown in Figure 9-7. That is, all of the receptacles on a phase can be connected in a continuous row, or the receptacles can be connected on alternate phases.

COMMUNICATION SYSTEMS

The installation of the telephone system in the commercial building will consist of two separate installations. The *electrical contractor* will install an empty conduit system according to the speci-

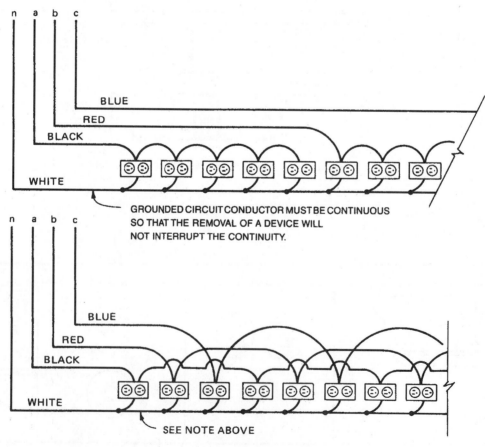

Fig. 9-7   Connection of receptacles.

fications for the commercial building and in the locations indicated on the plans. In addition, the installation will meet the rules, regulations, and requirements of the communications company that will serve the building. Once the conduit system is complete, the electrical contractor, telephone installing company, or the telephone company will install a complete telephone system.

Power Requirements

An allowance of 2500 volt-amperes is made for the installation of the telephone equipment to provide the power required to operate the special switching equipment for a large number of telephones. Because of the importance of the telephone as a means of communication, the receptacle outlets for this equipment are connected to the emergency power system.

The Telephone System

Telephone service requirements vary widely according to the type of business and the extent of the communication convenience desired. In many situations, the telephone lines may be installed in exposed locations. However, improvements in building construction techniques have made it more important to provide facilities for installing concealed telephone wiring. The use of new wall materials and wall insulation, the reduction or the complete omission of trim around windows and doors, the omission of baseboards, and the increased use of metal trim make it more difficult to attach exposed wires. For these reasons (and many others), the wiring is more conspicuous if the installation is exposed. In addition, the unprotected wiring is more subject to subsequent faulty operation.

To solve the problems of exposed wiring, conduits are installed and the proper outlet boxes and junction boxes are placed in position during the construction. The material and construction costs are low when compared to the benefits gained from this type of installation.

Since it is generally more difficult to conceal the telephone wiring to the floors above the first floor after the building construction is completed, it is recommended that telephone conduits be provided to these floors. The materials used in the installation and the method of installing the conduits for the telephone lines are the same as those used for light and power wiring with the following modifications:

- Because of small openings and limited space, junction boxes are used rather than the standard conduit fittings such as ells and tees.

- Since multipair conductors are used extensively in telephone installations, the size of the conduit should be ¾ inch or larger. The current and potential ratings for an installation of this type are not governed by the same conditions as conventional wiring for light and power.

- The number of bends and offsets is kept to a minimum; when possible, bends and offsets are made using greater minimum radii or larger sweeps (the allowable minimum radius is 6 inches, or 152 mm).

- A fishwire is installed in each conduit for use in pulling in the cables.

- When basement telephone wiring is to be exposed, the conduits are dropped into the basement and terminated with a bushing (junction boxes are not required). No more than 2 inches (50.8 mm) of conduit should project beyond the joists or ceiling level in the basement.

- The inside conduit drops need not be grounded unless they can become energized, see *NEC® Section 250-104(c)*.

- The service conduit carrying telephone cables from the exterior of a building to the interior must be permanently and effectively grounded, see *NEC® Section 250-104(a)*.

Installing the Telephone Outlets

The plans for the commercial building indicate that each of the occupancies requires telephone service. As shown previously, the insurance office uses a multioutlet assembly. Telephone outlets are installed in the balance of the occupancies. These outlets are supplied by EMT which runs to the basement, Figure 9-8, where the cable then runs exposed to the main terminal connections. Telephone company personnel will install the cable and connect the equipment.

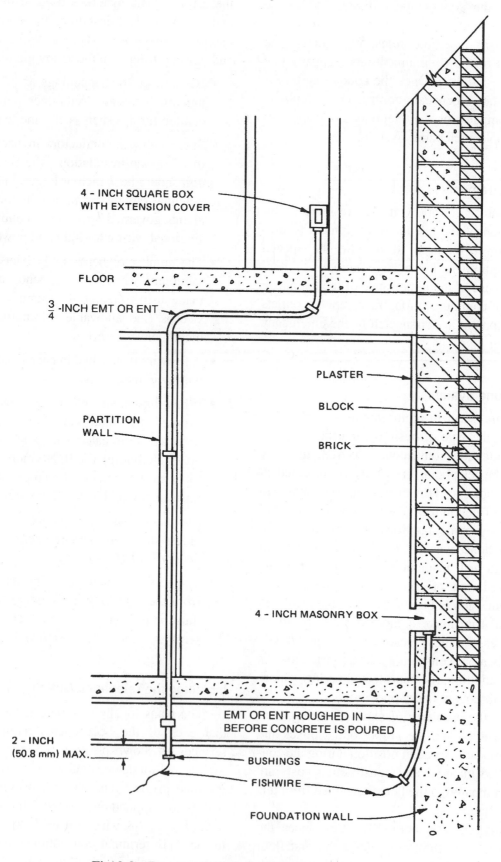

Fig. 9-8   Raceway installation for telephone system.

FLOOR OUTLETS

In an area the size of the insurance office, it may be necessary to place equipment and desks where wall outlets are not available. Floor outlets may be installed to provide the necessary electrical supply to such equipment. Two methods can be used to provide floor outlets: (1) installing underfloor raceway or (2) installing floor boxes.

Underfloor Raceway

The installation requirements for underfloor raceway are set forth in *NEC® Article 354*. It is common for this type of raceway to be installed to provide both power and communication outlets in a dual duct system similar to the one shown in Figure 9-9. The junction box is constructed so that the power and communication systems are always separated from each other. Service fittings are available for the outlets, Figure 9-10.

Floor Boxes

Floor boxes, either metallic or nonmetallic, can be installed using any approved raceway such as rigid conduit, rigid nonmetallic conduit, or electrical nonmetallic tubing. Some boxes must be installed to the correct height by adjusting the leveling screws before the concrete is poured, Figure 9-11. One nonmetallic box can be installed without adjustment and cut to the desired height after the concrete is poured, Figures 9-12 and 9-13.

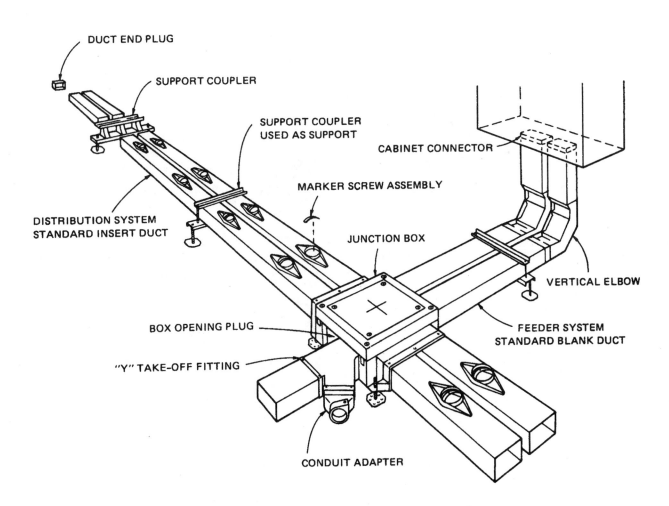

Fig. 9-9 Underfloor raceway.

FOR HIGH-POTENTIAL SERVICE

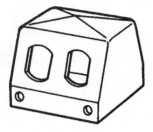
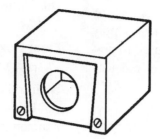
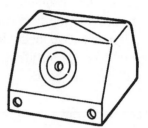

Dimensions: $4\frac{1}{8}''$ (105 mm) long; $4\frac{1}{8}''$ (105 mm) wide; $2\frac{15}{16}''$ (74.6 mm) high.

Fig. 9-10   Service fittings.

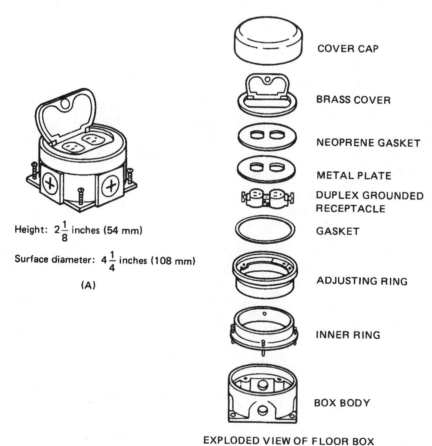

Height: $2\frac{1}{8}$ inches (54 mm)

Surface diameter: $4\frac{1}{4}$ inches (108 mm)

(A)

COVER CAP

BRASS COVER

NEOPRENE GASKET

METAL PLATE

DUPLEX GROUNDED
RECEPTACLE

GASKET

ADJUSTING RING

INNER RING

BOX BODY

EXPLODED VIEW OF FLOOR BOX

Fig. 9-11   A floor box with leveling screws.

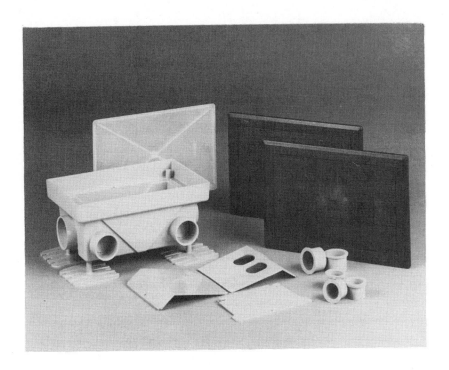

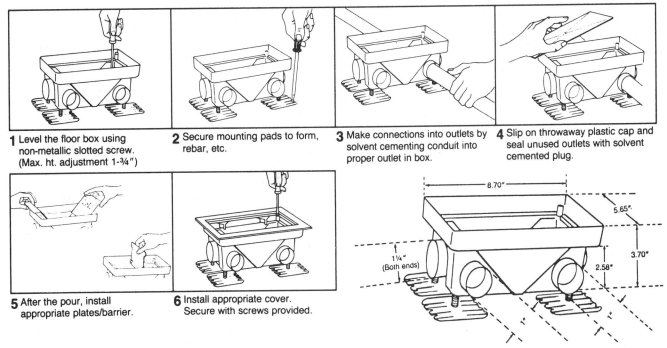

1 Level the floor box using non-metallic slotted screw. (Max. ht. adjustment 1-¾")

2 Secure mounting pads to form, rebar, etc.

3 Make connections into outlets by solvent cementing conduit into proper outlet in box.

4 Slip on throwaway plastic cap and seal unused outlets with solvent cemented plug.

5 After the pour, install appropriate plates/barrier.

6 Install appropriate cover. Secure with screws provided.

Fig. 9-12   Installation of multifunction nonmetallic floor box.

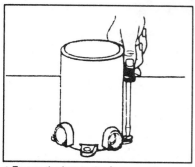

1 Fasten the box to the form or set on level surface.

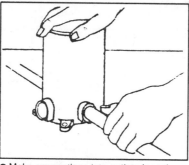

2 Make connections into outlets by solvent cementing conduit into the proper outlet in box.

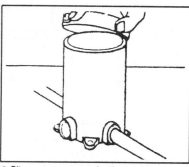

3 Slip on temporary plastic cover and seal unused outlets with solvent cemented plugs.

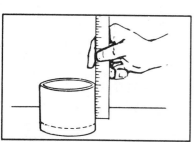

4 Remove temporary plastic cover and determine thickness of flooring to be used. Scribe a line around box at this distance from the floor.

5 Using a handsaw, cut off box at scribed line.

6 Install leveling ring to underside of cover, so that four circular posts extend through slots in outer ring.

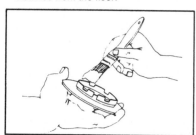

7 Apply PVC cement to leveling ring and to top inside edge of box.

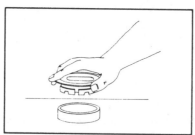

8 Press outer ring cover assembly into floor box for perfect flush fit.

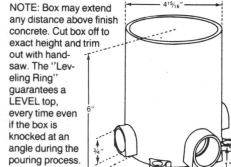

NOTE: Box may extend any distance above finish concrete. Cut box off to exact height and trim out with handsaw. The "Leveling Ring" guarantees a LEVEL top, every time even if the box is knocked at an angle during the pouring process.

4^{15}/$_{16}$"

6"

3/$_4$"

4^1/$_8$"

Fig. 9-13   Installation and trim out of nonmetallic floor box.

REVIEW QUESTIONS

Refer to the *National Electrical Code®* or the working drawings when necessary. Where applicable, responses should be written in complete sentences. Write units using unit names, do not use abbreviations or symbols (1 foot, not 1' or 1 ft).

1. Surface raceways may be extended through dry partitions if _____

    _____ .

2. Where power and communications circuits are to be installed in a combination raceway, the different type circuits must be installed in _____ .

3. If it is necessary to make splices in a multioutlet assembly, the conductors plus the splices shall not fill more than _____ percent of the cross-sectional area at the point where the splices are made.

Complete the following sentences showing required calculations and citing *NEC®* Sections.

4. For three or more conductors the maximum allowable fill area for a metallic surface raceway with a cross-sectional area of 0.25 square inches is _____ .

5. Each compartment of the multioutlet assembly installed in the insurance office has a cross-sectional area of 3.7 square inches. The maximum number of 10 AWG type THWN conductors that may be installed in a compartment of the multioutlet assembly is _____ .

6. Several computer circuits are to be installed in a building where nonmetallic cable and metal boxes are being used. Specify the number of conductors required for a cable serving a computer outlet similar to those in the insurance office. Note any special actions that must be taken to use that cable.

_____

_____

_____

7. A surface metal raceway has a cross-sectional area of 0.2 square inch. Cite Code sections and show calculations. The maximum number of No. 12 AWG, type THHN conductors that may be installed is

_____

UNIT 10

Working Drawings – Upper Level

OBJECTIVES

After studying this unit, the student will be able to

- tabulate materials required to install an electrical rough in.
- select the components to install an electric water heater.
- discuss the advantages/disadvantages between single- and three-phase supply systems.

Before examining the three offices on the upper level of the commercial building, the working drawings, the loading schedules, and the panelboard worksheets should be reviewed. The schedules and worksheets are located in the Appendix.

INSURANCE OFFICE

There are several special wiring systems shown in the insurance office most of which have been discussed in earlier units.

- In-the-floor receptacles are installed in the center of the office area.

- A special electrical raceway is installed in the two exterior walls. This raceway provides for both electrical power and communications.

- Special circuits for personal computers are installed to the computer room and special luminaires are installed in that room.

- A reception area is provided with a combination lighting system to provide a different "look" for the clients.

COMPUTER ROOM CIRCUITS

A room in the insurance office is especially designed to be a computer room, but it is not a computer room by the definitions set forth in *NEC*® Section 645-2. In *NEC*® language it is a room with some computers in it, and the special requirements for a computer room are not applicable. Three actions have been taken to comply with good practice:

- special receptacles are specified,
- a special grounding system is provided,
- and a separate neutral is provided with each of the 120-volt, single-phase circuits.

The special receptacles are discussed in unit 5; provide surge protection. The grounding terminal of each receptacle is connected to an insulated grounding conductor and not directly to the box as is usually the case. These grounding conductors are installed in the conduit with the circuit conductors. The conduit is grounded according to *NEC*® Article 250. The grounding conductors are connected to a special isolated grounding terminal in the panel-board; see *NEC® Section 250-146(d)*. A separate, insulated-grounding conductor is installed with the feeder circuit and is connected to a grounding electrode at the switchgear. This grounding system reduces, if not eliminates, the electromagnetic interference that often is present in the conventional grounding system. The surge protection feature of the receptacle provides protection from lightning and other severe electrical surges that may occur in the electrical system. These surges are often severe enough to cause a loss of computer memory and in some cases damage to the computer. These surge protectors are available in separate units, which can be used to provide protection on any type of sensitive equipment, such as radios and televisions. Three separate neutrals are installed to reduce the possibility of conductor overheating from nonlinear currents. There are size current carrying conductors in the raceway, three neutrals and three phase conductors, thus their ampacity must be adjusted.

A journeyman electrician should be able to look at an electrical drawing and prepare a list of the materials required to install the wiring system. A great deal of time can be lost if the proper materials are not on the job when needed. It is essential that the electrician prepare in advance for the installation so that the correct variety of material is available in sufficient quantities to complete the job.

Experience is by far the best teacher for learning to tabulate materials. Guidance can be given in preparing for the experience.

- Be certain that there is a clear understanding of how the system is to be installed. Decide from where home-runs will be made and establish the sequence of connecting the outlet boxes. The electrician has considerable freedom in making these decisions even when a scheme is shown on the working drawings.

- Take special note of long runs of raceway and determine if voltage-drop is a problem. This can mean an increase in conductor and/or raceway size.

- Be organized and meticulous. Have several sheets of paper (ruled preferred) and a sharp pencil with a good eraser. Start with the home-run and complete that system before starting another.

BEAUTY SALON

Of special interest in the beauty salon are the connection to the water heater, a small laundry, and the lighting for the stations where customers are given special attention.

Water Heater Circuits

A beauty salon (Figure 10-1) uses a large amount of hot water. To accommodate this need the

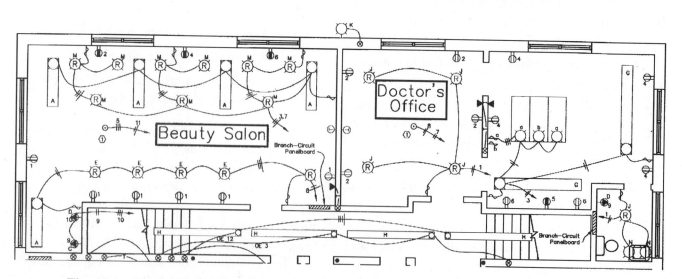

Fig. 10-1   Electrical drawing for a beauty salon and doctor's office. NOTE: For complete blueprint, refer to blueprint E3 in back of text.

specifications indicate that a circuit in the beauty salon is to supply an electric water heater. The water heater is not furnished by the electrical contractor but is to be connected by the electrical contractor, Figure 10-2. The water heater is rated for 3800 watts at 208 volts single phase. In this case it is assumed that the water heater is not marked with the maximum overcurrent protection. The water heater is connected to an individual branch-circuit.

The *branch-circuit rating* must be at least 125% of the appliance's marked rating. The water heater in the Beauty Salon is considered to be "continuously loaded." *NEC® Section 422-10(a)*.

Current rating: $3{,}800/208 = 18.3$ amperes
Minimum BC rating: $18.3 \times 1.25 = 22.9$ amperes
(round up to 23 amperes)

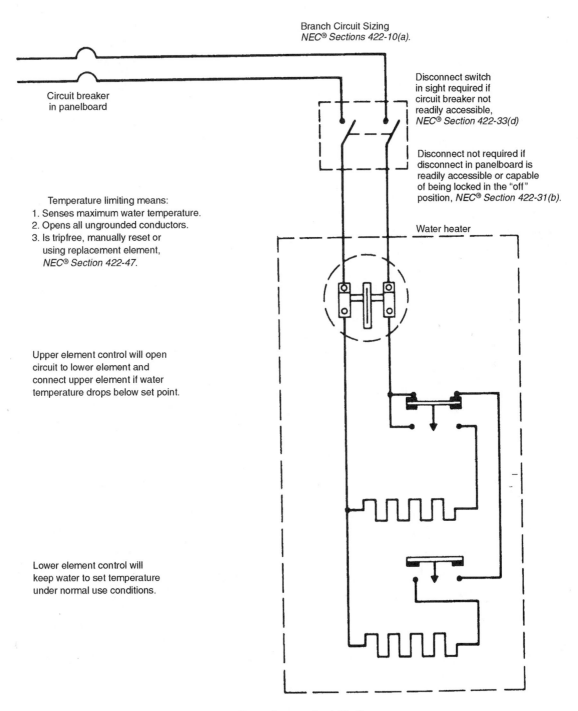

Branch Circuit Sizing
NEC® Sections 422-10(a).

Circuit breaker
in panelboard

Disconnect switch
in sight required if
circuit breaker not
readily accessible,
NEC® Section 422-33(d)

Disconnect not required if
disconnect in panelboard is
readily accessible or capable
of being locked in the "off"
position, *NEC® Section 422-31(b)*.

Temperature limiting means:
1. Senses maximum water temperature.
2. Opens all ungrounded conductors.
3. Is tripfree, manually reset or
   using replacement element,
   NEC® Section 422-47.

Water heater

Upper element control will open
circuit to lower element and
connect upper element if water
temperature drops below set point.

Lower element control will
keep water to set temperature
under normal use conditions.

Fig. 10-2   Water heater installation.

Conductor size is selected from *NEC® Table 310-16*. No. 12 AWG 60°C (Type TW) and 75°C (Type THW) copper conductors have an ampacity of 25 amperes. No. 12 AWG 90°C (Type THHN) conductors have an ampacity of 30 amperes, but must be applied at the 60°C column, see *NEC® Section 110-14(c)(1)*. The exception to these temperature limitations is if the terminals on the equipment are identified for use with the higher-rated conductors.

Note the "dagger" (obelisk) following the ampere ratings for these conductors. This marking references the footnote at the bottom of *NEC® Table 310-16*. The footnote references *Section 240-3*. In *Section 240-3(d)*, it is stated that it is acceptable to provide overcurrent protection for these conductors at greater than the given limits if "otherwise specifically permitted elsewhere in this Code." Several instances of this are given in *NEC® Section 422-11* which addresses overcurrent protection specifically for appliances.

The *branch-circuit overcurrent protection* for single, nonmotor operated appliances might be marked on the nameplate of the appliance. If not marked, then for appliances rated over 13.3 amperes, do not exceed 150% of the appliance's ampere rating . . . or next higher standard ampere rating. If the appliance is rated 13.3 amperes or less, the maximum overcurrent protection is 20 amperes. For the water heater with a current rating of 18.3 amperes the computed maximum is:

$$18.3 \times 1.5 = 27.45 \text{ amperes}$$

NEC® Section 240-6 lists 30-ampere as the next higher standard rating.

Washer-Dryer Combination

One of the uses of the hot water is to supply a washer-dryer combination used to launder the many towels and other items commonly used in a beauty salon. The unit is rated for 4000 VA at 208/120 volts single phase. The branch circuit supplying this unit must have a rating of at least 125 percent of the appliance load, *NEC® Section 422-10(a)*.

- The branch-circuit ampacity must be at least 20 amperes (4000 divided by 208).

- The overcurrent protective device must be rated at least 25 amperes (load in amperes multiplied by 1.25, then raised to the next standard size overcurrent device).

- The conductor ampacity must be at least 21 amperes to allow the use of an overcurrent device with a rating of 25-ampere, *NEC® Section 240-6*.

For appliances that contain one or more motors and other loads, such as heating elements, *NEC® Section 422-2* requires that manufacturers mark appliances with the minimum supply circuit conductor ampacity and the maximum rating of the circuit overcurrent protective device.

Conductor Selection

The two conductors to the water heater receptacle and the three conductors to the washer-dryer combination receptacle are to be installed in a single raceway.

- For 4 or more current-carrying conductors in a raceway the conductor ampacity must be adjusted according to *NEC® Section 310-15(b)(2)*.

- The ampacity of a #10 AWG type THHN conductor will be 32 amperes (the initial ampacity of 40 amperes multiplied by the adjustment factor of 0.8). This will permit the use of a 30-ampere overcurrent protective device. This is the maximum permitted for serving a single 30-ampere receptacle, see *NEC® Section 210-21(b)(2)*.

DOCTOR'S OFFICE

The special feature of the electrical layout for the doctor's office is the single phase three-wire feeder to the panelboard. In comparing the loading schedules of the beauty salon and the doctors office it would be noted that although the beauty salon has a 50% higher load the feeder conductors are smaller.

When a three-phase four-wire system is available in a building all panelboards are usually supplied by a four-wire feeder. In the commercial building this occupancy is supplied by a single phase system to allow for a comparison.

These are the smallest occupancies in the commercial building, and their areas are approximately the same. The total electrical loading in the Doctor's Office is 18,618 volt-ampere and in the Beauty Salon it is 27,809 volt-ampere. The Beauty Salon having a 50% higher load but the feeder for the Beauty Salon (No. 3 AWG) is smaller than the feeder for the Doctor's Office (No. 2 AWG). This reversal is because the Beauty Salon has a three-phase feeder and the Doctor's Office a single-phase feeder. There are several differences in these two systems.

Electrical Power Systems

- Three-phase systems are preferred where there is a large number of motors. Three-phase motors do not require starting windings and thus are less expensive.

- In commercial type buildings the common three phase systems are 208Y/120, 480Y/277 and 240Δ/120. The 480Y provides a higher voltage for motor loads and lighting is available that operates on 277 volts. This system allows the use of smaller (less costly) circuit conductors, raceways, and equipment. It is necessary to install transformers to service 120 volt loads. The 240Δ system provides a slightly higher voltage for motor operation but limits the single phase loading. Only two phases provide 120 volts, the third phase being referred to as the "high-leg" (The conductor color must be, or tagged with, orange.) It is important that all motors have a rating compatible with the system voltage.

- Single phase systems may be taken from a three-phase system or may be from an independent source. If the system has a voltage rating of 208 or 277 it is known that it is a part of a three-phase system. If the higher voltage is 240 then it maybe from a 240Δ system or a 240/120 single phase system.

Number of conductors

- A three-phase circuit requires 3 or 4 conductors depending if a neutral is available. A 208Y/120, and a 480Y/277, requires four conductors; 240Δ and 480Δ circuits require only three.

- Single-phase circuits are either two or three wire. A two wire connection may provide either the higher or lower voltage (120 or 240 from a 120/240 volt source).

Special Problems

- When a single phase circuit is taken from a three-phase wye connected system the neutral is considered a current-carrying-conductor must not be reduced in size.

Special notes

- The Beauty Salon requires an OCPD rated at 90-ampere which places the minimum size conductor selection in the 60°C column. If it were certain that all the terminations were rated for 75°C then the minimum conductor size would be No. 3 AWG, and the raceway size would be 1-inch, assuming that the neutral was reduced.

- The neutral of feeder to the Doctor's Office (see NEC® Section 220-22) shall not be reduced in size.

- In this comparison, the "best choice" is clouded by special rules. Usually, the three-phase system will provide about 50% more power at the same installation cost. In this case the single phase system was a reasonable choice. The message is clear; each circuit should be evaluated on its situation.

TABLE 10-1 Load and Feeder Comparison		
Item	Doctor's Office	Beauty Salon
Total Load	18,618 VA	27,809 VA
Feeder OCPD Rating	110-ampere	90-ampere
Minimum Conductor Size	2 AWG	2 AWG
Neutral Size	2 AWG	6 AWG
Number of Conductors	3	4
Raceway Size	1-inch	1¼-inch

REVIEW QUESTIONS

Refer to the *National Electrical Code*® or the working drawings when necessary. Where applicable, responses should be written in complete sentences. Write units using unit names, use no abbreviations or symbols (1 foot, not 1' or 1 ft).

Refer to the working drawing for the Insurance Office for questions 7 and 8.

1. A receptacle is to be installed for the water heater in the salon. If this is permissible, what section of the Code approves this installation and under what condition?

   _____

   _____

   _____

2. The water heater has two elements. Under what conditions are both elements heating?

   _____

   _____

3. If the water heater is not marked with the maximum overcurrent protection rating, the maximum overcurrent protection is to be sized not to exceed _____ percent of the water heater's ampere rating.

The following questions refer to the branch-circuit panelboard in the doctor's office. The panelboard schedule is shown on Drawing E3 and the panelboard summary on E4.

4. How many single-pole breakers can be added to the panelboard?

   _____

5. How many general-purpose branch circuits are to be installed?

   _____

6. If an appliance that will operate for long periods of time is marked with the rating of the overcurrent device it is permissible to install a device with the next higher rating. (Check the correct answer).

   ☐ Yes

   ☐ No

7. Tabulate the following materials.

   a. Tabulate the luminaires by style and count the number of each.

   b. Tabulate the lamps by types/ratings and count the number of each.

   c. Tabulate the receptacles by type/rating and count the number of each.

   d. Tabulate the switches by type/rating and count the number of each.

8. Determine the conductor count for each ceiling box and the minimum box size. Then tabulate the number of each size required for the installation. Boxes should not be used to support the luminaires and conductors that do not need to be spliced are to be considered as being pulled through.

Suggestions: Examine each box and mark the plans with the required capacities. Check the luminaire styles carefully before counting conductors. The following forms are provided to help and guide the tabulation of information. It is suggested that after the number of conductors in each box has been determined that each unique value be entered in the first column of the first table. Then the required boxes can be identified and a count made.

After all the boxes have been identified and the count established, the second table can be used to summarize by box type and size.

Form for tabulation by number of conductors

Number of conductors	Box type and size	Box count

Form for summary by box type and size

Box type and size	Box count

UNIT 11

Special Circuits (Owner's Circuits)

OBJECTIVES

After studying this unit, the student will be able to

- describe typical connection schemes for photocells and timers.
- list and describe the main parts of an electric boiler control.
- describe the connections necessary to control a sump pump.

Each occupant of the commercial building is responsible for electric power used within that area. It is the owner's responsibility to provide the power to light the public areas and to provide heat for the entire building. The circuits supplying the power to those areas and devices for which the building is responsible will be called the owner's circuits in this unit.

LOADING SCHEDULE

The loading schedule for the owner's circuits, located in the appendix, indicates that these are very special circuits:

- the boiler has by far the greatest electrical demand;
- the circulating pumps have a special power source that will keep them running even if the utility power is interrupted;
- selected lighting can also be supplied by the special power source.

The loading schedule can be divided into three parts:

1. the lighting and receptacles that do not require emergency backup;

2. the lighting, pumps, and receptacles that require emergency backup;

3. The boiler feed, which has a separate 800-ampere main switch in the service equipment.

LIGHTING CIRCUITS

Those lighting circuits which are considered to be the owner's responsibility can be divided into five groups according to the method used to control the circuit:

- continuous operation.
- manual control.
- automatic control.
- timed control.
- photocell control.

Continuous Operation

The luminaires installed at the top of each of the stairways to the second floor of the commercial building are in continuous operation. Since the stairways have no windows, it is necessary to provide artificial light even in the daytime. The power to these luminaires is supplied directly from the panelboard.

Manual Control

A conventional lighting control system supplies the remaining second-floor corridor lights as well as several other miscellaneous lights. All of the lights on manual control can be turned on as needed.

Automatic Control

The sump pump is an example of equipment that is operated by an automatic control system. No human intervention is necessary — the control system starts the pump when the liquid level in the sump rises above the maximum allowable level, then it is pumped down to the minimum level.

Timed Control

The lights at the entrance to the commercial building are controlled by a time clock located near the owner's panel in the boiler room. The circuit is connected as shown in Figure 11-1. When the time clock is first installed, it must be adjusted to the correct time. Thereafter, it automatically controls the lights. The clock motor is connected to the emergency panel so that the correct time is maintained in the event that the utility power fails.

Photocell Control

The three lights located on the exterior of the building are controlled by individual photocells,

Figure 11-2. The photocell control consists of a light-sensitive photocell and an amplifier that increases the photocell signal until it is sufficient to operate a relay which controls the light. The neutral of the circuit must be connected to the photocell to provide power for the amplifier and relay.

SUMP PUMP CONTROL

The sump pump is used to remove water entering the building because of sewage line backups, water main breakage, minor flooding due to natural causes, or plumbing system damage within the building. Since the sump pump is a critical item, it is connected to the emergency panel. The pump motor is protected by a manual motor starter, Figure 11-3,

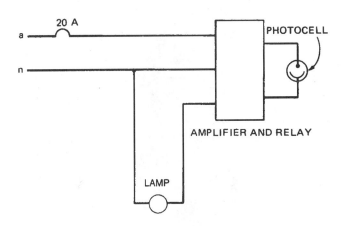

Fig. 11-2   Photocell control of lighting.

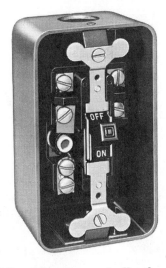

Fig. 11-3   Manual motor controller for sump pump.

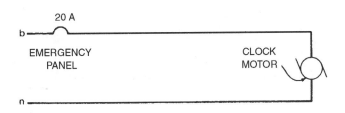

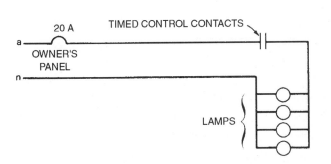

Fig. 11-1   Timed control of lighting.

and is controlled by a float switch, Figure 11-4. When the water in the sump rises, a float is lifted which mechanically completes the circuit to the motor and starts the pump, Figure 11-5. When the water level falls, the pump shuts off.

One type of sump pump that is available commercially has a start-stop operation due to water pressure against a neoprene gasket. This gasket, in turn, pushes against a small integral switch within the submersible pump. This type of sump pump does not use a float.

BOILER CONTROL

An electrically heated boiler supplies heat to all areas of the commercial building. The boiler has a full-load rating of 200 kW. As purchased, the boiler

Fig. 11-4   Float switch.

is completely wired except for the external control wiring, Figure 11-6. A heat sensor plus a remote bulb, Figure 11-7, is mounted so that the bulb is on the outside of the building. The bulb must be mounted where it will not receive direct sunlight and must be spaced at least ½ inch from the brick wall. If these mounting instructions are not followed, the bulb will give inaccurate readings. The heat sensor is adjusted so that it closes when the outside temperature falls below a set point, usually 18°C (65°F). The sensor remains closed until the temperature increases to a value above the differential setting, approximately 21°C (70°F). Once the outside temperature causes the heat sensor to close, the boiler automatically maintains a constant water temperature and is always ready to deliver heat to any zone in the building that requires heat.

A typical heating control circuit is shown in Figure 11-8.

Figure 11-9 is a schematic drawing of a typical connection scheme for the boiler heating elements.

The following points summarize the wiring requirements for a boiler:

- A disconnect means must be installed to disconnect the ungrounded supply conductors.

- The disconnect must be in sight or capable of being locked in the open position.

- The branch-circuit conductor sizes are to be based on 125% of the rated load.

- The overcurrent protective devices are to be sized at 125% of the rated load.

NEC® Article 424 establishes the requirements for the installation of fixed electric space heating equipment.

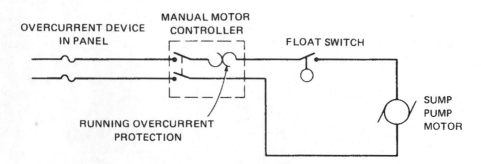

Fig. 11-5   Sump pump control diagram.

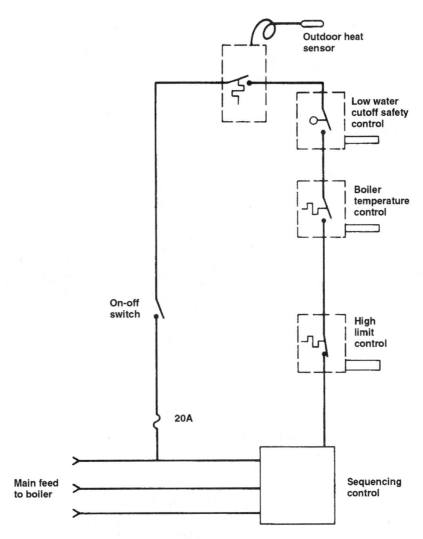

Fig. 11-6   Boiler control diagram.

Fig. 11-7   Heat sensor with remote mounting bulb.

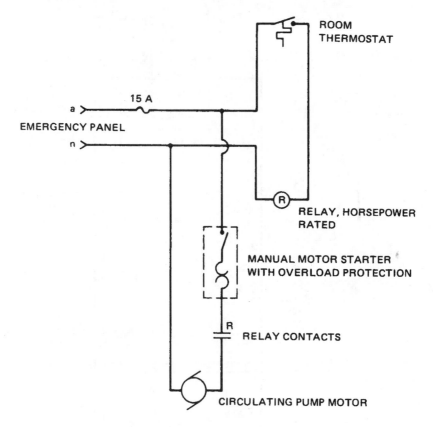

Fig. 11-8   Heating control circuit.

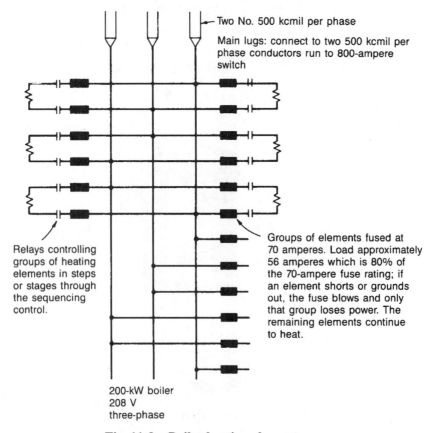

Fig. 11-9   Boiler heating elements.

It should be noted that the boiler for the commercial building is not sized for a particular heat loss; its selection is based on certain electrical requirements. Each of the building occupants has a heating thermostat located within the particular area. The thermostat operates a relay, Figure 11-10, which controls a circulating pump, Figure 11-11, in the hot water piping system serving the area. The circuit to the circulating pump is supplied from the emergency panel so that the water will continue to circulate if the utility power fails, Figure 11-12. In subfreezing weather, the continuous circulation prevents the freezing of the boiler water; such freezing can damage the boiler and the piping system.

Fig. 11-10   Relay in NEMA 1 enclosure.

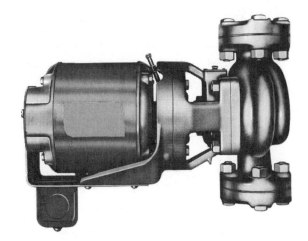

Fig. 11-11   Water circulating pump.

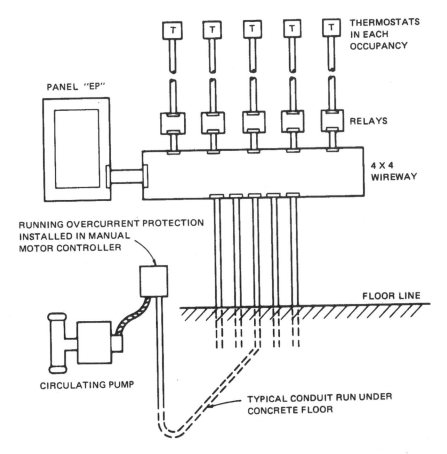

Fig. 11-12   Panel "EP" and installation of devices to connect and control water circulating pumps.

REVIEW QUESTIONS

Refer to the *National Electrical Code®* or the working drawings when necessary. Where applicable, responses should be written in complete sentences. Write units using unit names, do not use abbreviations or symbols (1 foot, not 1' or 1 ft).

The first four questions refer to the installation of the electrical supply to the boiler.

1. What is the proper name of the electrical supply to the boiler?

   _____

2. Describe, in detail, the electrical supply to the boiler, indicating options available and any special precautions.

   _____

   _____

   _____

   _____

   _____

3. Show the calculations to determine the minimum ampacity of the circuit supplying the boiler.

4. Verify, or disprove, the correctness of the sizing of the boiler branch-circuit.

The following questions refer to the installation of the various items of equipment in the equipment room.

5. What is the proper rating of the overload protection to be installed for the sump pump?

_____

_____

_____

6. Detail the installation of the outside heat sensor.

_____

_____

_____

7. What is the difference between a timer and a clock?

_____

_____

_____

UNIT 12

Panelboard Selection and Installation

OBJECTIVES

After studying this unit, the student will be able to

- identify the criteria for selecting a panelboard.
- correctly place and number circuits in a panelboard.
- compute the correct feeder size for a panelboard.
- determine the correct overcurrent protection for a panelboard.

The installation of panelboards is always a part of new building construction such as the commercial building. After a building is occupied it is common for the electrical needs to exceed the capacity of the installed system. In this case the electrician is often expected not only to install a new panelboard, but to select a panelboard that will satisfy the needs of the occupant.

PANELBOARDS

Separate feeders must be run from the main service equipment to each of the areas of the commercial building. Each feeder will terminate in a panelboard which is to be installed in the area to be served, Figure 12-1.

The *National Electrical Code®* defines a panelboard as a single panel or group of panel units designed for assembly in the form of a single panel; such a panel will include buses and automatic overcurrent devices, and may or may not include switches for the control of light, heat, or power circuits. A panelboard is designed to be installed in a

cabinet or cutout box placed in or against a wall or partition. This cabinet (and panelboard) is to be accessible only from the front. Panelboards shall be dead front *(see NEC® Section 384-18)*, which means that no current-carrying parts are exposed to a person operating the panelboard devices.

While a panelboard is accessible only from the front, a switchboard may be accessible from the rear as well. *NEC® Section 384-14* define a lighting and appliance branch-circuit panelboard as a panelboard having more than 10 percent of its overcurrent devices rated at 30 amperes or less with neutral connections provided for these devices. *NEC® Section 384-15* sets the maximum number of overcurrent devices in a lighting and appliance panelboard at 42. *NEC® Section 384-16(c)*, stipulates that if the panelboard contains snap-switches with a rating of 30 amperes or less the panelboard overcurrent protection shall not exceed 200 amperes. Snap-switches are switches that are snapped in place in a panelboard or other similar enclosure. They are additional to the overcurrent devices and must be rated for the load they are to control.

Lighting and appliance branch-circuit panel-boards generally are used on single-phase, three-wire and three-phase, four-wire electrical systems. This type of installation provides individual branch circuits for lighting and receptacle outlets and permits the connection of appliances such as the equipment in the bakery.

Panelboard Construction

In general, panelboards are constructed so that the main feed bus bars run the height of the panel-board. The buses to the branch-circuit protective devices are connected to the alternate main buses as shown in Figures 12-2A and 12-2B. The busing arrangement shown in these figures is typical.

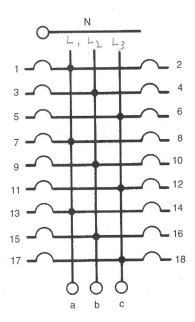

Fig. 12-1   Panelboards.

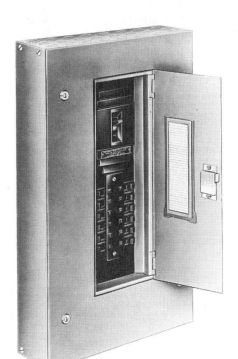

Fig. 12-2A   Lighting and appliance branch-circuit panelboard; single-phase, three-wire.

Fig. 12-2B   Lighting and appliance branch-circuit panelboard; three-phase, four-wire.

Always check the wiring diagram furnished with each panelboard to verify the busing arrangement. In an arrangement of this type, the connections directly across from each other are on the same phase and the adjacent connections on each side are on different phases. As a result, multiple protective devices can be installed to serve the 208-volt equipment. The numbering sequence shown in Figures 12-2 and 12-3 is the common numbering system used for most panels. Figures 12-4 and 12-5 show the phase arrangement requirements of *NEC®* *Section 384-3*.

Numbering of Circuits

The number of the overcurrent devices in a panelboard is determined by the needs of the area being served. Using the bakery as an example, there are 13 single-pole circuits and five three-pole circuits.

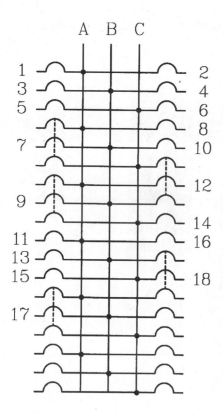

Fig. 12-3 Panelboard circuit numbering scheme.

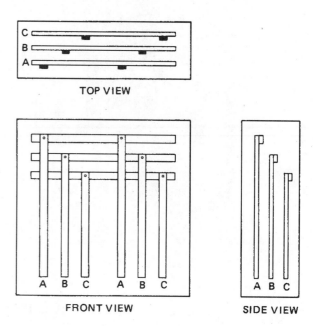

Fig. 12-4 Phase arrangement requirements for switchboards and panelboards.

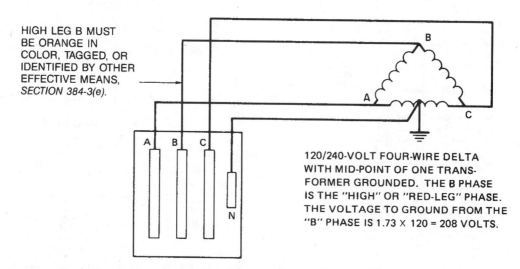

Fig. 12-5 Panelboard or switchboard supplied by four-wire, delta-connected system.

This is a total of 28 poles. When using a three-phase supply, the incremental number is 6 (a pole for each of the three phases on both sides of the panelboard). The minimum number of poles that could be specified for the bakery is 30. This would limit the power available for growth, and it would not permit the addition of a three-pole load. The reasonable choice is to go to 36 poles, which provides consideration flexibility for growth load, Figure 12-3.

Panelboard Sizing

Panelboards are available in various sizes including 100, 225, 400, and 600 amperes. These sizes are determined by the current-carrying capacity of the main bus. After the feeder capacity is determined, the panelboard size is selected to be not smaller than the load computed in accordance with *NEC® Article 220* (refer to *NEC® Section 384-13*).

Panelboard Overcurrent Protection

In many installations, a single feeder may be sized to serve several panelboards, Figure 12-6. For this situation, it is necessary to install an overcurrent device with a trip rating not greater than the rating of the bus bar in each panelboard. In the case of the bakery installation, a main device is not required, as the panelboard is protected by the feeder protective device. Other examples of methods of providing panelboard protection are shown in Figures 12-7, 12-8, and 12-9.

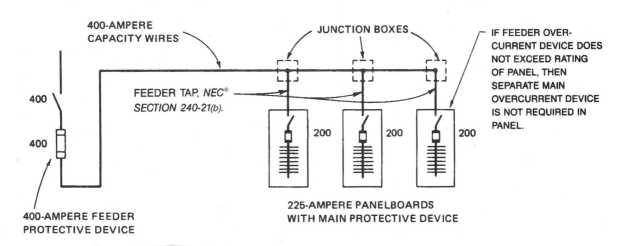

Fig. 12-6   **Panelboards with main,** *NEC® Section 384-16(a), Exception No. 1.*

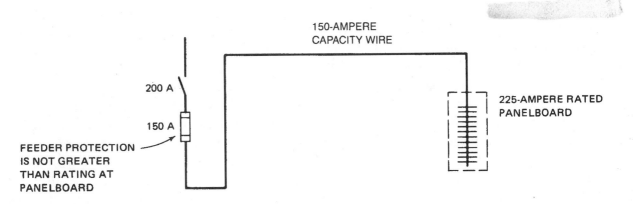

Fig. 12-7   **Panelboard without main.**

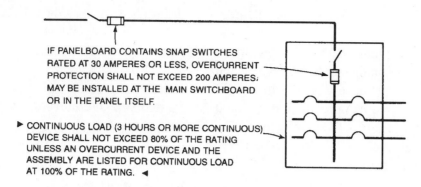

IF PANELBOARD CONTAINS SNAP SWITCHES RATED AT 30 AMPERES OR LESS, OVERCURRENT PROTECTION SHALL NOT EXCEED 200 AMPERES; MAY BE INSTALLED AT THE MAIN SWITCHBOARD OR IN THE PANEL ITSELF.

► CONTINUOUS LOAD (3 HOURS OR MORE CONTINUOUS) DEVICE SHALL NOT EXCEED 80% OF THE RATING UNLESS AN OVERCURRENT DEVICE AND THE ASSEMBLY ARE LISTED FOR CONTINUOUS LOAD AT 100% OF THE RATING. ◄

Fig. 12-8   Panelboard with snap switches.

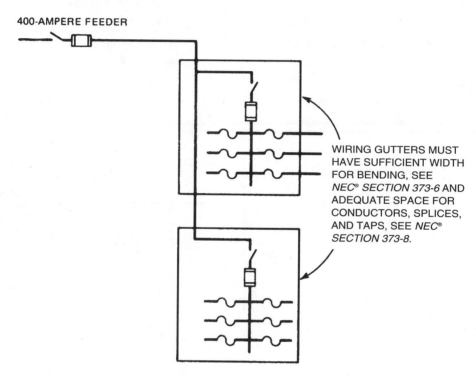

400-AMPERE FEEDER

WIRING GUTTERS MUST HAVE SUFFICIENT WIDTH FOR BENDING, SEE *NEC® SECTION 373-6* AND ADEQUATE SPACE FOR CONDUCTORS, SPLICES, AND TAPS, SEE *NEC® SECTION 373-8.*

Fig. 12-9   Panelboards with adequate space for wiring.

Panelboard Directory

One of the final actions in installing an electrical circuit or system is to update or create the panelboard directory. A holder will be located on the inside of the of the panelboard door. An examination of Figure 12-1 will reveal that the size and shape of these holders will vary. The single column schedule shown on the left has the advantage that the numbers appear numerical order (1, 2, ...x). The advantage of the holder on the right is that the schedule mimics the layout of the overcurrent devices and their numbering (odd numbers on the left, even on the right).

A circuit directory is required by *NEC® Section 384-13*. There is the option of creating a panel directory in your choice of size and shape and secure it to the inside of the panel door. If this is done a sheet of clear plastic covering the directory will lengthen its life. Securing the plastic sheet only at the top will permit lifting to make modifications. The *NEC®* requires that circuits *"be legibly identified as to purpose or use."* Table 12-1 exhibits a directory prepared for the drugstore, all directories are recorded on the working drawings. It is common for them to appear in the specification or on the

	Load/Area Served	OCPD	BC #	Phase	BC #	OCPD	Load/Area Served
TABLE 12-1 Drugstore – Panelboard Directory							
	Sales Area Lighting	20	1	A	2	20	Sales Receptacles, North
	Sales Area Lighting	20	3	B	4	20	Toilet Receptacles
	Toilet Lighting	20	5	C	6	20	Sales Receptacles, South
	Pharmacy Receptacles	20	7	A	8	20	Show Window Receptacles
	Basement Lighting	20	9	B	10	20	Pharmacy Lighting
	Show Window Track	20	11	C	12	20	Basement Receptacles South
	Show Window Track	20	13	A	14	20	Basement Receptacles North
	Exterior Sign	20	15	B			
	Roof Receptacle	20	17	C	16	50	Compressor
				A			
				B			
				C			

working drawings. The directory gives the "load/area" in a wording that would be clear for "non-electrical" people to decipher. In the next column is the size of the overcurrent protective device, then the circuit number, and in the center column the phase connection. The center column would not be considered as "required" but it helps the schedule in mimicking the front of the panelboard. The remaining columns follow the same sequence in reverse.

REVIEW QUESTIONS

Refer to the *National Electrical Code®* or the working drawings when necessary. Where applicable, responses should be written in complete sentences. Write units using unit names, use no abbreviations or symbols (1 foot, not 1' or 1 ft).

The *NEC®* refers to switchboards, distribution boards, and lighting and appliance panelboards. For questions 1, 2, and 3 write the distinguishing characteristics of each. By observing these characteristics a person would know which type of board was being viewed.

1. Switchboard:

   _____

   _____

2. Panelboard:

   _____

   _____

3. Lighting and appliance panelboard.

_____

_____

4. There are several requirements that are the same for distribution boards and panel-boards. List at least two that an electrician should be aware of.

_____

_____

Refer to the Drugstore panelboard directory (Table 12-1) and the Drugstore feeder selection summary in the Appendix for the information necessary to answer questions 5 through 9.

5. The panelboard must be rated for at least _____ ampere.

6. The panelboard has a total of _____ poles.

7. There is space for the addition of _____ three-pole breaker(s).

8. There is space for the addition of _____ two-pole breaker(s).

9. There is space for the addition of _____ single-pole breaker(s).

UNIT 13

The Electric Service

OBJECTIVES

After studying this unit, the student will be able to

- install power transformers to meet the *NEC®* requirements.
- draw the basic transformer connection diagrams.
- recognize different service types.
- connect metering equipment.
- apply ground-fault requirements to an installation.
- install a grounding system.

The installation of the electric service to a building requires the cooperation of the electrician and the local power company and, in some cases, the electrical inspector. The availability of high voltage and the power company requirements determine the type of service to be installed. This unit will investigate several common variations in electrical service installations together with the applicable *NEC®* rules.

TRANSFORMERS

The principle reasons for installing a transformer is to either increase or decrease the voltage. As it is more econmical to transport electricity at a high voltage of transfomers must be installed, by either the utility company or the owner, to step the voltage down to a level that can be utilized by the equipment in the building. Transformers are available in two basic types, liquid-filled or dry.

Liquid Filled Transformers

Many transformers are immersed in a liquid which may be askarel or oil. This liquid performs several important functions: (1) it is part of the required insulation dielectric, and (2) it acts as a coolant by conducting heat from the core and the winding of the transformer to the surface of the enclosing tank, which then is cooled by radiation or fan cooled.

Dry-Type Transformers

Dry-type transformers are widely used because they are lighter in weight than comparable-rated liquid-filled transformers. Installation is simpler because there is no need to take precautions against liquid leaks.

Dry-type transformers are constructed so that the core and coil are open to allow for cooling by the free movement of air. Fans may be installed to increase the cooling effect. In this case, the transformer can be used at a greater load level. A typical dry-type transformer installation is shown in Figure 13-1. An installation of this type is known as a unit substation and consists of three main components: (1) the high-voltage switch, (2) the dry-type transformer, and (3) the secondary distribution section.

High Voltage
Section

Transformer
Section

Secondary Distribution
Section

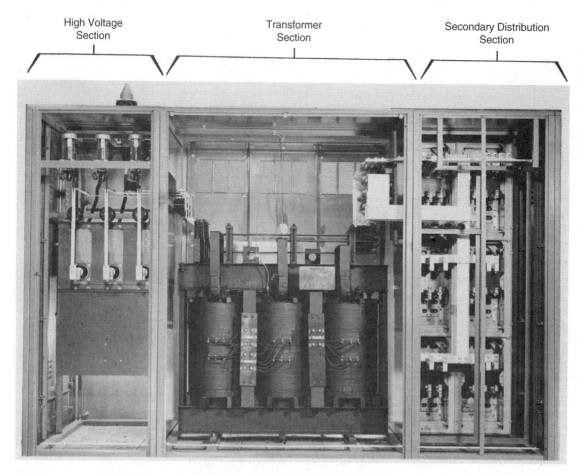

Fig. 13-1   A unit substation.

TRANSFORMER OVERCURRENT PROTECTION

NEC® Article 450 addresses transformer installations and groups transformers into two voltage levels:

1. Over 600 volts   2. 600 volts or less

The Code requirements for the overcurrent protection of transformers rated over 600 volts are separated into two basic types of installations:

NEC® Table 450-3(a) is divided horizontally into two sections. The upper section applies to transformers in any location. The lower section allows some reduction in the device sizes if the location is supervised as defined in Note 3.

Figure 13-2 illustrates five of the more common situations found in commercial building transformer installations. As specified in *NEC® Article 450*, the overcurrent devices protect the transformer only. The conductors supplying or leaving the transformer may require additional overcurrent protection according to *NEC® Articles 240* and *310*. (See units 18 and 19 of this text for information concerning fuses and circuit breakers.)

Figure 13-3 shows one method of installing a dry-type transformer in a commercial or industrial building.

TRANSFORMER CONNECTIONS

A transformer is used in a commercial building primarily to change the transmission line high voltage to the value specified for the building, such as 480Y/277 or 208Y/120 volts. A number of connection methods can be used to accomplish the changing of the voltage. The connection used depends upon the requirements of the building. The following paragraphs describe several of the more commonly used secondary connection methods.

Single-Phase System

Single-phase systems usually provide 120 and/or 240 volts with a two- or three-wire connection, Figure 13-4. The center tap of the transformer secondary shall be grounded in accordance with *NEC® Article 250*, as will be discussed later. Grounding is a safety measure and should be installed with great care.

		I. Primary over 600 volts	II. Secondary over 600 volts	III. Secondary 600 volts or less
Transformers over 600 volts with primary and secondary protection.	For transformers with impedance ▶ not over 6%:	• Maximum fuse-300%* • Maximum breaker-600%*	• Maximum fuse-250% • Maximum breaker-300%*	• Maximum fuse-125%* • Maximum breaker-125%*
	For transformers with impedance ▶ over 6%, but not over 10%:	• Maximum fuse-300%* • Maximum breaker-400%*	• Maximum fuse-225%* • Maximum breaker-250%*	• Maximum fuse-125%* • Maximum breaker-125%*
A	* Where percentages marked with asterisks do not correspond to a standard rating or setting, the next higher standard rating or setting may be used.			

	Primary over 600 volts	Secondary over 600 volts or under 600 volts
Transformers over 600 volts with primary protection only. Supervised installation only.	• Maximum fuse-250%* • Maximum breaker-300%*	• No overcurrent protection required
B	* Where percentages marked with asterisks do not correspond to a standard rating or setting, the next higher standard rating or setting may be used. ** Individual primary overcurrent protection is not required if the primary feeder overcurrent device is sized as shown to the left. This may allow more than one transformer to be connected to one feeder.	

		I. Primary over 600 volts	II. Secondary over 600 volts	III. Secondary 600 volts or less
Transformers over 600 volts with primary and secondary protection. Supervised installations only.	For transformers with impedance ▶ not over 6%:	• Maximum fuse-300%* • Maximum breaker-600%*	• Maximum fuse-250%* • Maximum breaker-300%*	• Maximum fuse-250%* • Maximum breaker-250%*
	For transformers with impedance ▶ over 6%, but not over 10%:	• Maximum fuse-300%* • Maximum breaker-400%*	• Maximum fuse-225%* • Maximum breaker-250%*	• Maximum fuse-250%* • Maximum breaker-250%*
C	* Where percentages marked with asterisks do not correspond to a standard rating or setting, the next higher standard rating or setting may be used. ** Individual primary overcurrent protection is not required if protection on secondary conforms to the values in columns II and III, or if the transformer is equipped by the manufacturer with coordinated thermal overload protection, AND if primary feeder overcurrent protection does not exceed the values in Column I. This may permit more than one transformer to be connected to one feeder.			

	Primary 600 volts or less	Secondary 600 volts or less
Transformers 600 volts and less with primary protection only.	Maximum fuse size is not to exceed 125% of the transformer's rated primary current if secondary protection is not provided. For transformers having rated primary current of less than 9 amperes or more, and if the 125% sizing does not correspond to a standard size, then the next higher standard size of fuse or nonadjustable breaker may be used. For transformers having rated primary current of less than 9 amperes, if the 125% sizing does not correspond to a standard size, then a fuse or nonadjustable trip circuit breaker, not to exceed 167% may be used.	Transformer secondary protection is not required when the primary fuse does not exceed 125% of the rated primary current.
D		

	Primary 600 volts or less	Secondary 600 volts or less
Transformers 600 volts and less with primary and secondary protection.	Maximum fuse or breaker is not to exceed 250% of the transformer's rated primary current when transformer secondary protection does not exceed 125%. If transformer is equipped with coordinated overload protection furnished by the manufacturer, individual primary overcurrent protection is not required if the primary feeder has overcurrent protection: • Set not over 6 times primary current for transformers with not more than 6% impedance. • Set not over 4 times primary current for transformers with more than 6% but not over 10% impedance. This may allow more than one transformer to be connected to one feeder.	Maximum fuse or breaker is not to exceed 125% of the transformer's rated secondary current when the transformer primary protection is not over 250% of the rated primary full-load current. For transformers having rated secondary current of 9 amperes or more, if the 125% sizing does not correspond to a standard size, then the next higher standard size fuse or nonadjustable breaker may be used. For transformers having rated secondary current of less than 9 amperes, if the 125% sizing does not correspond to a standard size, then a fuse or breaker not to exceed 167% may be used.
E		

Fig. 13-2   Typical transformer overcurrent protection requirements. See NEC® Table 450-3(a).

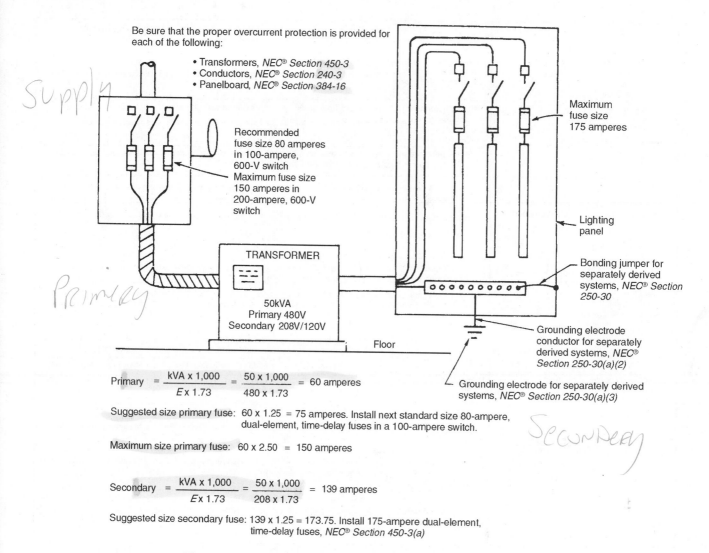

Be sure that the proper overcurrent protection is provided for each of the following:

- Transformers, *NEC® Section 450-3*
- Conductors, *NEC® Section 240-3*
- Panelboard, *NEC® Section 384-16*

Recommended fuse size 80 amperes in 100-ampere, 600-V switch
Maximum fuse size 150 amperes in 200-ampere, 600-V switch

Maximum fuse size 175 amperes

Lighting panel

Bonding jumper for separately derived systems, *NEC® Section 250-30*

Grounding electrode conductor for separately derived systems, *NEC® Section 250-30(a)(2)*

Grounding electrode for separately derived systems, *NEC® Section 250-30(a)(3)*

TRANSFORMER

50kVA
Primary 480V
Secondary 208V/120V

Floor

$$\text{Primary} = \frac{\text{kVA} \times 1{,}000}{E \times 1.73} = \frac{50 \times 1{,}000}{480 \times 1.73} = 60 \text{ amperes}$$

Suggested size primary fuse: 60 × 1.25 = 75 amperes. Install next standard size 80-ampere, dual-element, time-delay fuses in a 100-ampere switch.

Maximum size primary fuse: 60 × 2.50 = 150 amperes

$$\text{Secondary} = \frac{\text{kVA} \times 1{,}000}{E \times 1.73} = \frac{50 \times 1{,}000}{208 \times 1.73} = 139 \text{ amperes}$$

Suggested size secondary fuse: 139 × 1.25 = 173.75. Install 175-ampere dual-element, time-delay fuses, *NEC® Section 450-3(a)*.

Figure 13-3   Diagram of dry-type transformer installation.

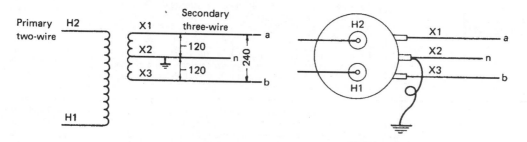

Fig. 13-4   Single-phase transformer connection.

Open Delta System

This connection scheme has the advantage of being able to provide either three-phase or three-phase and single-phase power using only two transformers. It is usually installed where there is a strong probability that the power requirement will increase, at which time a third transformer can be added. The open delta connection is illustrated in Figure 13-5.

In an open delta transformer bank, 86.6% of the capacity of the transformers is available. For example, if each transformer in Figure 13-5 has a

100-kVA rating, then the capacity of the bank is:

$$100 + 100 = 200 \text{ kVA}$$
$$200 \text{ kVA} \times 86.6\% = 173 \text{ kVA}$$

Another way of determining the capacity of an open delta bank is to use 57.7% of the capacity of a full delta bank. Thus, the capacity of three 100-kVA transformers connected in full delta is 300 kVA. Two 100-kVA transformers connected in open delta have a capacity of

$$300 \times 57.7\% = 173 \text{ kVA}$$

When an open delta transformer bank is to serve three-phase power loads only, the center tap is not connected.

Four-Wire Delta System

This connection, illustrated in Figure 13-5, has the advantage of providing both three-phase and single-phase power from either an open (2 transformers) or a closed (3 transformers) delta system. A center tap is brought out of one of the transformers, which is grounded and becomes the neutral conductor to a single-phase three-wire power system. The voltage to ground between phases "A" and "C," which are connected to the transformer that has been tapped, will be equal to ground and additive when measured phase to phase. For example 120 volts to ground and 240 volts between phases. The voltage measured between the grounded tap and the "B" phase will be higher than 120 volts and lower than 240. This phase is called the "high leg" and cannot be used for lighting purposes. This high leg must be identified by using orange colored conductors or by effectively identifying, as the "B" phase conductor, whenever it is present in a box or cabinet with the neutral of the system. See *NEC® Section 384-3(e)*. This "B" phase is to be connected to the center bus bar in panelboards and switchboards.

Three-Wire Delta System

This connection, illustrated in Figure 13-6, provides only three-phase power. One phase of the system may be grounded, in which case it is often referred to a corner grounded delta. The power delivered and the voltages measured between phases remains unchanged. Overcurrent devices are not to be installed in the grounded phase. See unit 17, Figure 17-20.

Three-Phase, Four-Wire Wye System

The most commonly used system for modern commercial buildings is the three-phase, four-wire wye, Figure 13-7. This system has the advantage of being able to provide three-phase power, and also permits lighting to be connected between any of the three phases and the neutral. Typical voltages available with this type of system are 120/208, 265/460, and 277/480. In each case, the transformer connections are the same.

Notes to All Connection Diagrams

All connection diagrams are shown with additive polarity and standard angular displacement. All three-phase connections are shown with the primary connected in delta. For other connections, it is recommended that qualified engineering assistance be obtained.

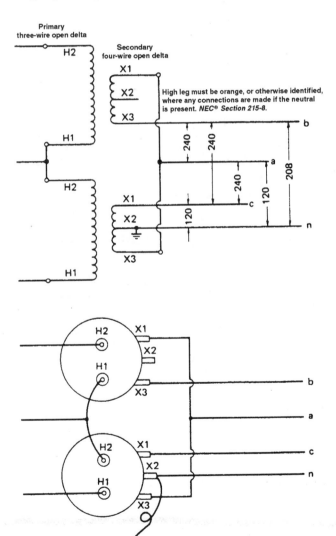

Fig. 13-5 Three-phase open delta connection.

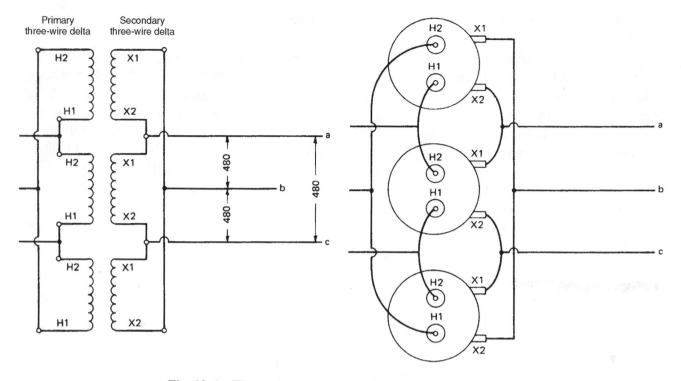

Fig. 13-6   Three-phase delta-delta transformer connection.

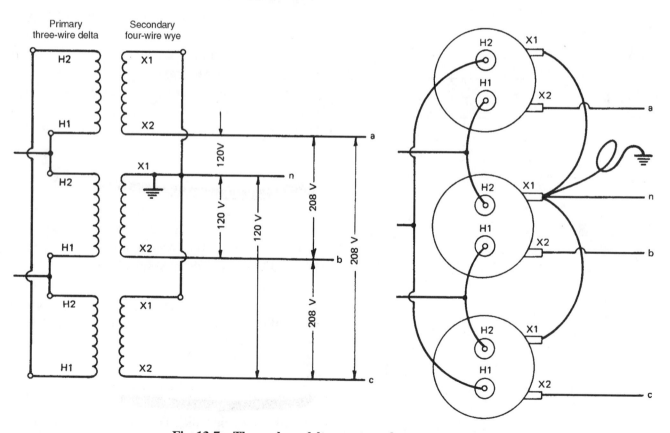

Fig. 13-7   Three-phase delta-wye transformer connection.

THE SERVICE ENTRANCE

The regulations governing the method of bringing the electric power into a building are established by the local utility company. These regula-tions vary considerably between utility companies. Several of the more common methods of installing the service entrance are shown in Figures 13-8 through 13-11.

Except where it has been legally rejected or amended or is specifically not covered, all wiring must conform to the requirements of the *National Electrical Code®* NFPA 70. As stated in *NEC® Section 90-2(b)(5)*, electric utilities are exempt from the *NEC® "Installations, including associated lighting, under the exclusive control of electric utilities for the purpose of communications, metering, generating, control, transformation, transmission, or distribution of electric energy. Such installations shall be located in buildings used exclusively by the utility for such purposes; outdoors on property owned or leased by the utility; on or along public highways, streets, roads, etc., or outdoors on private property by established rights such as easements."* All of the above are governed by the requirements of the *National Electrical Safety Code®* (C2-1992). The *National Electrical Safety Code®* regulations are considerably different than the *National Electrical Code®* and are intended to govern the installation of electrical equipment installed by an electric utility for the purpose of generating, metering, distributing, transforming, etc. as they function as a utility. In the following text and diagrams regarding electrical installations, the requirements of the *National Electrical Code®* apply except as set forth in *NEC® Section 90-2(b)(5)*.

A common example is that of parking lot lighting around shopping centers and industrial and commercial complexes. As the Code is presently written, no matter who makes the installation, an electrical contractor, a building maintenance electrician, or an electrical utility, this type of lighting installation clearly comes under the jurisdiction and requirements of the *National Electrical Code®,* NFPA 70. ◀

Pad-Mounted Transformers

Liquid-insulated transformers as well as dry-type transformers are used for this type of installation. These transformers may be fed by an underground or overhead service, Figure 13-8. The secondary normally enters the building through a bus duct or large cable.

Unit Substation

For this installation, the primary runs directly to the unit substation where all of the necessary

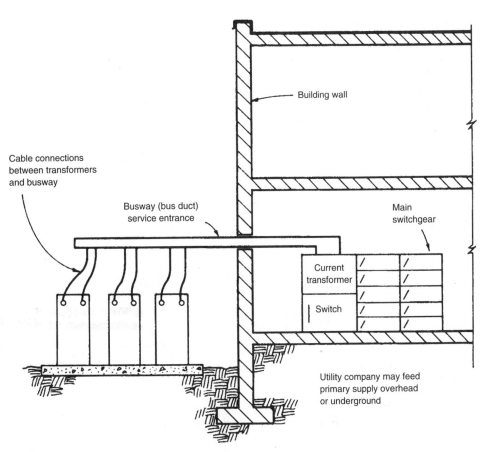

Fig. 13-8   Pad-mounted transformers supplying bus duct service entrance.

equipment is located, Figure 13-9. The utility company requires the building owner to buy the equipment for a unit substation installation.

Pad-Mounted Enclosure

Figure 13-10 illustrates an attractive arrangement in which the transformer and metering equipment are enclosed in a weatherproof cabinet. This type of installation (also known as a *transclosure*) is particularly adapted to smaller service-entrance requirements.

Underground Vault

This type of service is used when available space is an important factor and an attractive site is desired. The metering may be at the utility pole or in the building, Figure 13-11.

METERING

The electrician working on commercial installations seldom makes metering connections. However, the electrician should be familiar with the following two basic methods of metering.

High-Voltage Metering

When a commercial building is occupied by a single tenant, the utility company may elect to meter the high-voltage side of the transformer. To accomplish this, a potential transformer and two current transformers are installed on the high-voltage lines and the leads are brought to the meter as shown in Figure 13-12.

The left-hand meter socket in the illustration is connected to receive a standard socket-type watthour meter; the right-hand meter socket will receive a varhour meter (volt-ampere reactive meter). The two meters are provided with 15-minute demand attachments which register kilowatt (kW) and kilovolt-ampere reactive (kVAR) values respectively. These demand attachments will indicate the maximum usage of electrical energy for a 15-minute period during the interval between the readings made by the utility company. The rates charged by the utility company for electrical energy are based on the maximum demand and the power factor as determined from the two meters. A high demand or a low power factor will result in higher rates.

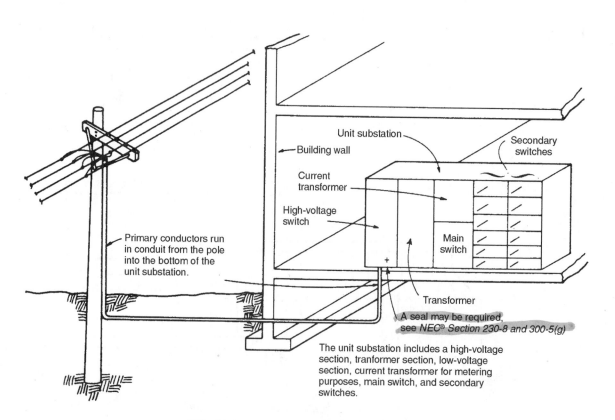

Unit substation

Secondary switches

Building wall

Current transformer

High-voltage switch

Main switch

Primary conductors run in conduit from the pole into the bottom of the unit substation.

Transformer

A seal may be required, see NEC® Section 230-8 and 300-5(g)

The unit substation includes a high-voltage section, tranformer section, low-voltage section, current transformer for metering purposes, main switch, and secondary switches.

Fig. 13-9   High-voltage service entrance.

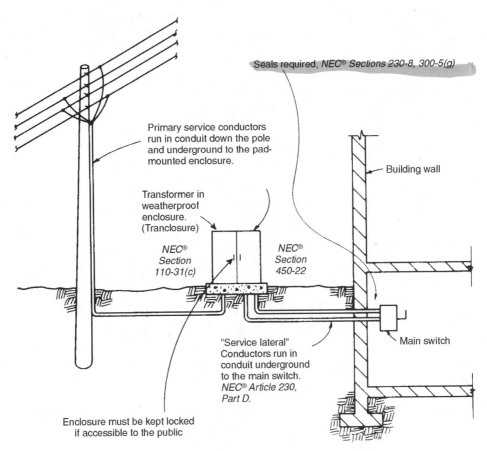

Fig. 13-10   Pad-mounted enclosed transformer supplying underground service entrance.

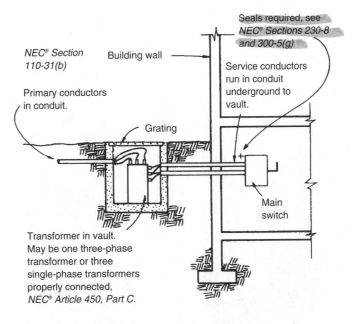

Fig. 13-11   Transformer in underground vault supplying underground service entrance.

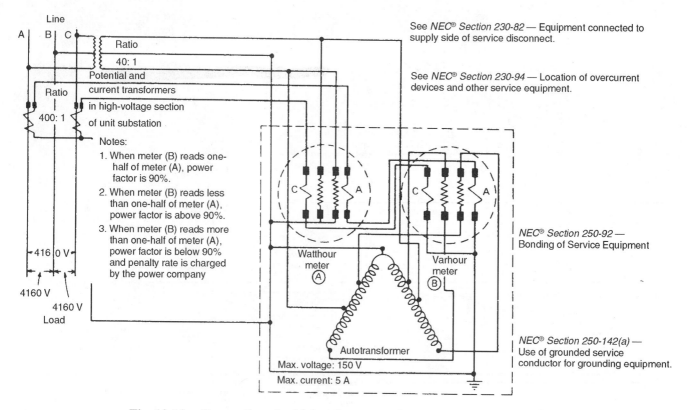

Fig. 13-12   Connections for high-voltage watt-hour and var-hour meters.

Low-Voltage Metering

Low-voltage metering of loads greater than 200 amperes is accomplished in the same manner as high-voltage metering. In other words, potential and current transformers are used.

Fig. 13-13   Meter socket.

For loads of 200 amperes or less, the feed wires from the primary supply are run directly to the meter socket, figure 13-13. For multiple occupancy buildings, such as the commercial building discussed in this text, the meters are usually installed as a part of the service-entrance equipment, called a *switchboard*.

SERVICE-ENTRANCE EQUIPMENT

When the transformer is installed at a location far from the building, the service-entrance equipment consists of the service-entrance conductors, the main switch or switches, the metering equipment, and the secondary distribution switches, Figure 13-14. The commercial building shown in the plans is equipped in this manner and will be used as an example for the following paragraphs.

The Service

The service for the commercial building is similar to the service shown in Figure 13-10. A pad-mounted, three-phase transformer is located outside the building and rigid conduit running underground serves as the service raceway.

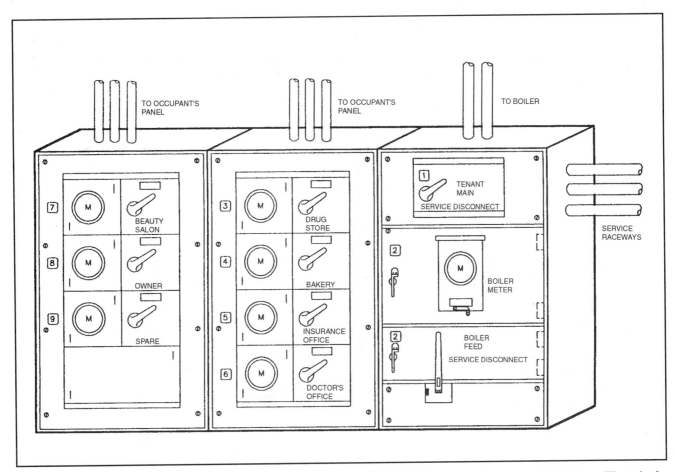

Fig. 13-14 Commercial building service-entrance equipment, pictorial view. (*Courtesy* of Erickson Electrical Equipment Co.)

The service-entrance equipment is located in the basement of the commercial building. This equipment and the service and feeder raceways are shown in Figure 13-14. It will be noticed there is a switch and meter for each of the panelboards. In addition, the building main disconnect and, meter and a boiler disconnect and meter is shown. The exact configuration of this equipment would be developed by the manufacturer.

Figure 13-15 provides a diagram of the connections made within the equipment. For each occupant there is an overcurrent protection device (a set of fuses), a feeder disconnect switch, a meter, and terminals for connection of the feeder. In addition there are two main switches, which together would disconnect the utility electrical power from the building. As many as six disconnects may be installed and be in compliance with *NEC® Section 230-71(a)*. It is also permissible, see *NEC® Section 230-90(a)*

Exception No. 3, for the total of the ratings of the feeder overcurrent protection devices to be greater than the rating of the main. Adding the feeder fuse sizes in the commercial building will yield a sum of 695-ampere. The main disconnect is fused at 600-ampere. This is because the requirements for the main are somewhat different from the feeders as is illustrated in the following calculations. It is the responsibility of the building owner, or a representative, to provide the manufacturer with the required sizes or the information necessary for determining the sizes.

In Unit 8 the process for sizing the feeder to the drugstore panelboard was explained in detail. Similar information for the other feeders is recorded in the appendix. A tabulation of these values is recorded in Tables 13-1 and 13-2. In Table 13-1, the sums of the various load types are shown in the main totals column. It will be noticed there is no total in

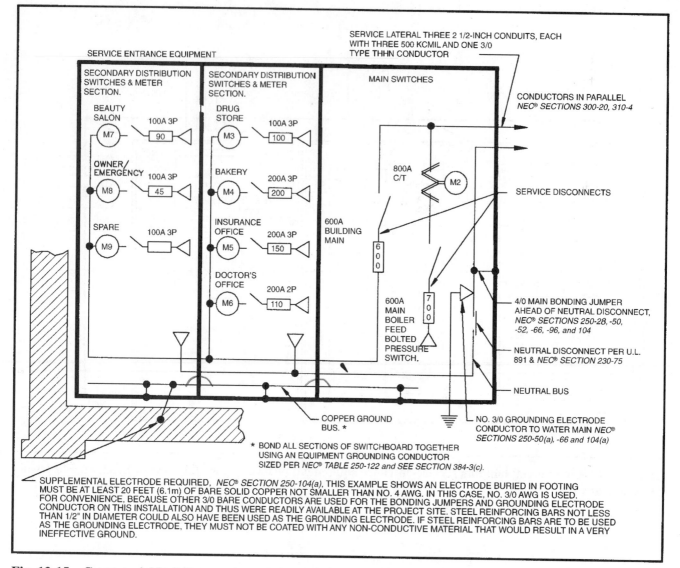

SERVICE LATERAL THREE 2 1/2-INCH CONDUITS, EACH
WITH THREE 500 KCMIL AND ONE 3/0
TYPE THHN CONDUCTOR

SERVICE ENTRANCE EQUIPMENT

SECONDARY DISTRIBUTION
SWITCHES & METER
SECTION.

BEAUTY
SALON   100A 3P
M7   90

OWNER/
EMERGENCY   100A 3P
M8   45

SPARE   100A 3P
M9

SECONDARY DISTRIBUTION
SWITCHES & METER
SECTION.

DRUG
STORE   100A 3P
M3   100

BAKERY   200A 3P
M4   200

INSURANCE
OFFICE   200A 3P
M5   150

DOCTOR'S
OFFICE   200A 2P
M6   110

MAIN SWITCHES

CONDUCTORS IN PARALLEL
NEC® SECTIONS 300-20, 310-4

800A
C/T   M2

SERVICE DISCONNECTS

600A
BUILDING
MAIN

600

600A
MAIN
BOILER
FEED
BOLTED
PRESSURE
SWITCH.

700

4/0 MAIN BONDING JUMPER
AHEAD OF NEUTRAL DISCONNECT,
NEC® SECTIONS 250-28, -50,
-52, -66, -96, and 104

NEUTRAL DISCONNECT PER U.L.
891 & NEC® SECTION 230-75

NEUTRAL BUS

COPPER GROUND
BUS. *

NO. 3/0 GROUNDING ELECTRODE
CONDUCTOR TO WATER MAIN NEC®
SECTIONS 250-50(a), -66 and 104(a)

* BOND ALL SECTIONS OF SWITCHBOARD TOGETHER
USING AN EQUIPMENT GROUNDING CONDUCTOR
SIZED PER NEC® TABLE 250-122 and SEE SECTION 384-3(c).

SUPPLEMENTAL ELECTRODE REQUIRED, NEC® SECTION 250-104(a), THIS EXAMPLE SHOWS AN ELECTRODE BURIED IN FOOTING
MUST BE AT LEAST 20 FEET (6.1m) OF BARE SOLID COPPER NOT SMALLER THAN NO. 4 AWG. IN THIS CASE, NO. 3/0 AWG IS USED.
FOR CONVENIENCE, BECAUSE OTHER 3/0 BARE CONDUCTORS ARE USED FOR THE BONDING JUMPERS AND GROUNDING ELECTRODE
CONDUCTOR ON THIS INSTALLATION AND THUS WERE READILY AVAILABLE AT THE PROJECT SITE. STEEL REINFORCING BARS NOT LESS
THAN 1/2" IN DIAMETER COULD ALSO HAVE BEEN USED AS THE GROUNDING ELECTRODE. IF STEEL REINFORCING BARS ARE TO BE USED
AS THE GROUNDING ELECTRODE, THEY MUST NOT BE COATED WITH ANY NON-CONDUCTIVE MATERIAL THAT WOULD RESULT IN A VERY
INEFFECTIVE GROUND.

Fig. 13-15   Commercial building service-entrance equipment, oneline diagram. (*Courtesy* of Erickson Electrical Equipment Co.)

TABLE 13-1 Commercial Building — Load Summary by Type						
Occupancies Load Types	Drugstore	Bakery	Insurance Office	Doctor's Office	Beauty Salon	Owner
Continuous	8595	22566	8974	5330	5605	4096
Noncontinuous + Receptacle	5280	6380	15780	5300	7080	2556
Highest Motor	8604	8604	8604	4264	5762	1128
Total Motor	8604	17333	8604	4264	9562	3768
Noncoincident	8604	0	8604	4264	5762	0
Balanced	8604	34985	8604	8064	13562	0
Nonlinear	4575	4446	21864	660	1155	2022

Occupancies Load Connections	Drugstore	Bakery	Insurance Office	Doctor's Office	Beauty Salon	Owner	Totals
TABLE 13-2 Commercial Building — Load Summary by Phase							
Phase A	4503	4213	7886	0	3387	3420	23409
Phase B	5100	2893	8998	3950	1950	3560	26451
Phase C	4272	4188	7870	2880	3348	3440	25998

the noncoincident row. This value only applies after the boiler is included. The boiler branch circuit load is added and the noncoincident load subtracted, to obtain the service requirement.

In Table 13-2 the loads are listed by the phase connection. It is required that these be reasonably equal as is the case in the commercial building. It is unnecessary to arrange the loads to be exactly equal because there will always be a variation in actual loads either over or under the computed load.

Building Main

The internal connections between the feeder switches and the building main will be determined by the manufacturer. These connections are usually made with copper or aluminum bars sized for the load. The calculations for the overcurrent protection are shown in Tables 13-3 and 13-5. The computed load values are taken from Table 13-1. They are adjusted to determine the OCPD selection load just as was done for the feeders. There is one important difference, the OCPD is a bolted pressure switch and

is listed for operation at 100% of rating, thus the conductors must be adjusted for the continuous load but the OCPD may be sized for the computed load. In Table 13-3 the internal connections of the building main are shown as needing an ampacity of 621 amperes. The minimum OCPD rating is 600-ampere.

Boiler Branch Circuit

The operation of the boiler is discussed in Unit 11. The supply is 3-phase and no neutral is required. There is a requirement for a grounding conductor to be installed in the feeder raceway because of the necessity of installing a short section of flexible metal conduit to reduce vibration transmissions. The sizing of this conductor is discussed later in this unit. See Figure 13-24 (page 177). The branch circuit to the boiler must have an ampacity and a rating of 125% of the load to be in compliance with *NEC® Section 424-82*. As shown in Table 13-4 the minimum ampacity is 694 amperes, a value higher than those listed in *NEC® Table 310-16* for 75°C (167°F) conductors. By using two sets of parallel conductors the ampacity for each set is reduced to 347 amperes which requires a 500 kcmil. Two paralleled sets of 500 kcmil Type THHN conductors will have an allowable ampacity of 860 amperes. When paralleling conductors, all of the phase conductors must be identical in length, type, size, material and terminations. It is customary for electricians to lay all the conductors, in this case six, on a floor where there exact lengths can be assured. See *NEC® Section 310-4*.

The main overcurrent protective device is a bolted pressure switch and is listed for operation at 100% of rating. A switch rated at 600-ampere, may be used with the load of 600 amperes. (If the 25% increase was required a 700-ampere OCPD and an 800-ampere switch would be needed).

TABLE 13-3 Commercial Building Load Summary for Building Main		
Load Summary 208Y/120 3-Phase 4-Wire	**Computed**	**Conductor**
Continuous load	55166	68958
Noncontinuous + Receptacle Load	42376	42376
Highest Motor Load Allowance	8604	10755
Total Motor Load	52135	52135
Growth	39570	49463
Total Loads (Computed) (OCPD Selection)	197851 ~~19781~~	223687
Component Selection	**Input**	**Output**
Selection Amperes	550	621
Ampere Rating	600	

TABLE 13-4 Commercial Building Boiler Branch-Circuit Calculation		
Load Summary, 208Y 3 Phase, 4-Wire	Computed	Selection
Continuous Load	200000	250000
OCPD Selection	Input	Output
OCPD Selection Amperes	556	694
OCPD Ampere Rating	600	
Minimum Conductor Size		2-500 kcmil
Minimum Derated Ampacity		347
Phase Conductor Selection	Input	Output
Ambient Temperature & Correction	38	0.91
Current Carrying Conductors & Adjustment	3	1
Reduction Factor		0.91
Number of Parallel Sets	2	
Load per set Amperes		278
Minimum Allowable Ampacity		381
Conductor Size		500 kcmil
Conductor Type & Allowable Ampacity	THHN	430
Derated Ampacity		391
Raceway Size Determination	Input	Output
Circuit Conductors Size & Area	500 kcmil	2.1219
Grounding Conductor Size & Area	1/0 AWG	0.1855
Total Conductor Area		2.3074
Minimum Raceway Type & Size	RMC	3 in

TABLE 13-5 Commercial Building Service Calculations		
Load Summary — Service	Computed	
Building Main	197851	
Boiler	200000	
Total Load	397851	
Service Conductor Selection	Input	Output
Computed Load Total	397851	
Noncoincident Load	−27234	
Adjusted Load Volt-Amperes & Amperes	370617	1029
Number of Parallel Sets & Amperes per Set	3	343
Minimum Conductor Size		500 kcmil
Minimum Derated Ampacity		343
Phase Conductor Sizing	Input	Output
Ambient Temperature, Correction	28°C	1
Current Carrying Conductors (#) (Adjustment)	3	1
Reduction Factor		1
Minimum Allowable Ampacity		343
Conductor Size		500 kcmil
Conductor Type & Ampacity	THHN	430
Total Derated Ampacity (3 Sets)		1290
Neutral Sizing	Input	Output
Computed Load Total	397851	
Balanced Load	−31092	
Boiler	−200000	
Nonlinear Load	34722	
Neutral Load	201481	560
Neutral Sets & Load per Set	3	187
Minimum Neutral Size		3/0 AWG
Minimum Allowable Ampacity		187
Conductor Type & Size	THHN	3/0 AWG
Allowable & Derated Ampacity	225	225
Raceway Size Determination	Input	Output
Feeder Conductor Size & Area	500 kcmil	2.1219
Neutral Conductor Size & Area	3/0 AWG	0.2679
Grounding Conductor Size & Area		0
Total Conductor Area		2.3898
Raceway Type & Size	RMC	3 in

Service Entrance Conductor Size

As previously stated the two main overcurrent protective devices are bolted pressure switches listed for operation at 100% of rating. This allows the service entrance conductors to be sized to the computed load. This sum is reduced by the noncoincident load as the boiler load greatly exceeds that of the air conditioning, resulting in a total load of 1000 amperes. Three parallel sets of 500 kcmil, Type THHN, conductors will be installed for the service. The total derated ampacity will be 1290 amperes. The same procedure is followed to size the neutral. As with a feeder neutral, the balanced load is subtracted and the nonlinear load is added. Two additional *NEC®* requirements must be applied when sizing neutrals.

- *NEC® Section 250-24(b)(2)* requires that the grounded conductor, i.e. the neutral, in a service must not be smaller than the required grounding electrode conductor. Grounding electrode conductor sizes are given in *NEC® Table 250-66.*

- In *NEC® Section 250-24(b)(2)* the grounded conductor shall not be smaller than 1/0 AWG, when part of a parallel set.

When applied to the commercial building service, both of these require a minimum conductor size of 1/0 AWG.

WORKING SPACE AROUND THE MAIN SWITCHBOARD

▶ Electricians or other personnel working on or needing access to electrical equipment must have enough room and adequate lighting to safely move around and work on the equipment to perform the function they are doing, such as examining, adjusting, servicing, or maintaining the equipment. Many serious injuries and deaths have occurred as a result

of "arc blasts" and electrical shock. Should an electrical arcing fault occur, personnel must have enough space to get away safely and quickly from the faulted electrical equipment.

NEC® Section 110-26 of the *National Electrical Code®* covers the **minimum** working space, access, headroom, and lighting requirements for electrical equipment such as switchboards, panels, boards and motor control centers operating at 600 volts or less. Here is a brief recap of *Section 110-26.*

General Considerations

When discussing working space, the major concern is ACCESS TO and ESCAPE FROM the required working space (Figure 13-16A and Figure 13-16B).

NEC® Section 110-26 addresses "working space," not the room itself. Don't confuse the term "working space" with "room." The concern for

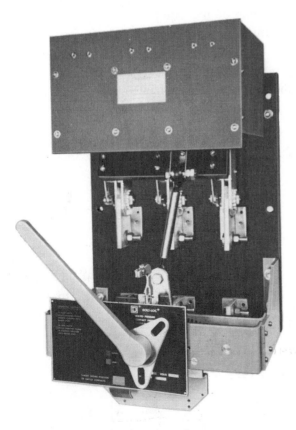

Fig. 13-16A Bolted pressure contact switch for use with high-capacity fuses.

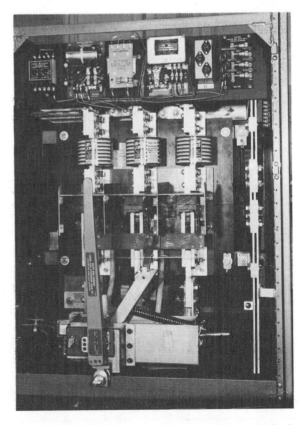

Fig. 13-16B Bolted pressure contact switch has ground-fault protection phase failure relay, shunt tripping and antisingle phasing indicator in addition to high-capacity fuses.

safety requires adequate working space around the equipment, but not necessarily the entire room in which the equipment is located. In some cases, the required working space might be the entire room.

For simplicity, the diagrams on the following pages show the required working space in front of electrical equipment. If access is needed on the sides and/or rear of the electrical equipment, adequate working space must be provided on the sides and/or rear of the equipment.

In addition to metal water or steam pipes, electrical conduits, electrical equipment, metal ducts, and any other metal grounded surfaces, consider concrete, brick, tile, or similar conductive building materials as "grounded."

Working space requirements are classified by *Conditions.*

Condition 1 Exposed live parts on one side of the working space . . . and no live or grounded parts on the other side, or exposed live parts on both sides if effectively guarded by suitable insulating material.

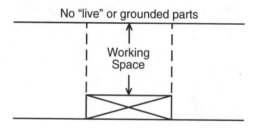

No "live" or grounded parts

Working Space

Condition 2 Exposed live parts on one side of the working space . . . and grounded parts on the other side.

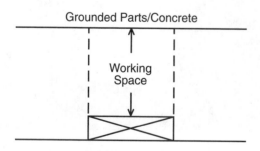

Grounded Parts/Concrete

Working Space

Condition 3 Exposed live parts on both sides of the working space having no guarding of the live parts through the use of suitable insulating material (i.e., wood).

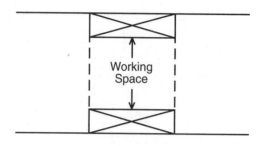

Working Space

Working Space "Depth"

Working space is the space in the direction where access to live parts is necessary. Generally, this is in front of the electrical equipment, but it could also be on the sides or back of the equipment, depending upon the design of the equipment. To keep the diagrams simple, work space requirements are shown from the front of the equipment only. If there is a need to access live parts from the sides and/or rear of the equipment, minimum working space must also be provided from the sides and/or rear. If rear access to de-energized parts is necessary, a minimum working space of 30 inches is required in the rear of the equipment.

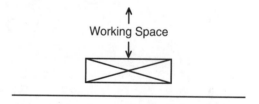

Working Space

All working spaces must allow for the opening of equipment doors and hinged panels to at least a 90° position.

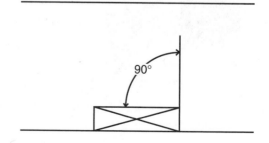

90°

Working spaces must be kept clear, and must not be used for storage.

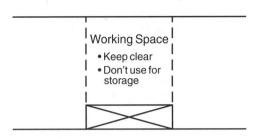

Working space distances are measured from exposed live parts, or from the enclosed front or opening for enclosed "dead front" switchboards. When enclosed equipment is "opened" to work on, there will be exposed live parts.

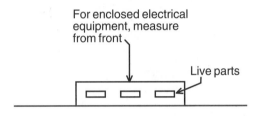

Different Voltages to Ground . . . Different Working Space Requirements

For system voltages of 0 to 150 volts to ground, the minimum working space requirement is 3 feet. An example of this is the 208/120-volt wye-connected system in the commercial building discussed in this text, where the voltage to ground is 120 volts.

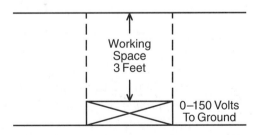

For system voltages of 151 to 600 volts to ground, the minimum working space requirement varies from 3 feet to 4 feet, depending upon the *Condition*. An example of this would be a 480/277-volt wye-connected system where the voltage to ground is 277 volts.

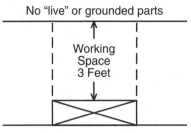

Condition 1:

No "live" or grounded parts

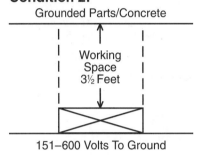

Condition 2:

Grounded Parts/Concrete

151–600 Volts To Ground

Condition 3:

Live Parts Both Sides

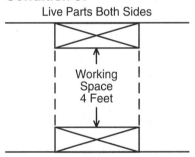

Access and Entrance to Working Space

Access and *Entrance* refer to the actual working space requirements, not to the room itself. However, in instances where the electrical equipment is located in tight quarters of a small room, the working space could in fact be the entire room.

For electrical equipment, at least one entrance of sufficient area to the working space is required.

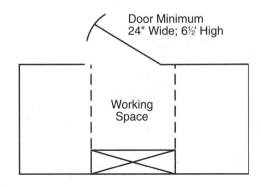

For electrical equipment rated 1200 or more *and* over 6 feet wide (1.83 m), two entrances to the working space are required. The entrances to the working space shall be 24 inches (610 mm) wide minimum and 6½ feet (1.98 m) high minimum, one on each end of the equipment.

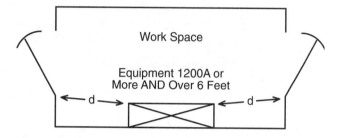

Exception No. 1: If there is an unobstructed, continuous way of exit travel from the working space, only one means of egress is permitted.

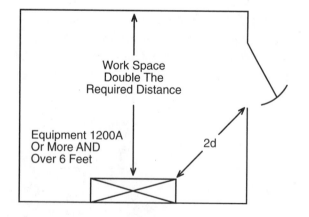

Exception: If the minimum working space requirements listed in *NEC® Section 110-26(a)* are doubled, only one entrance to the working space is required.

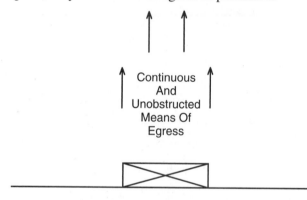

Working Space "Width"

For equipment less than 30 inches wide, provide working space *width* of not less than 30 inches.

This will enable a person to stand and work on the equipment.

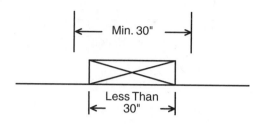

For equipment 30 inches wide or greater, provide working space *width* of not less than the width of the equipment. This will enable a person to stand anywhere in front of the equipment.

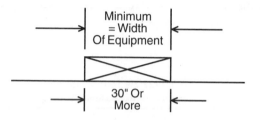

Working Space "Height"

Working space height must not be less than 6½ feet (1.98 m). For equipment more than 6½ feet high, working space height must be at least as high as the equipment.

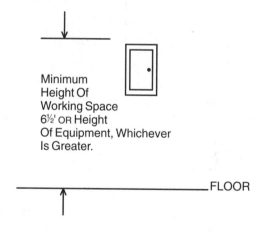

Keep the "zone" for the space *"equal to the width and depth of the equipment and extending from the floor to a height of 6 feet (1.83 m) above the equipment or to the structural ceiling, whichever is lower, shall be dedicated to the electrical installation."* This space must be kept clear of any and all piping, ducts, or other equipment that is *not* related to the electrical installation. This "zone" is the width and depth of the electrical equipment, and is dedicated for electrical use only. See *NEC® Section 110-26(f)(1)(a)*.

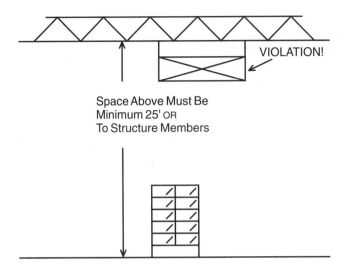

Space Above Must Be Minimum 25' OR To Structure Members

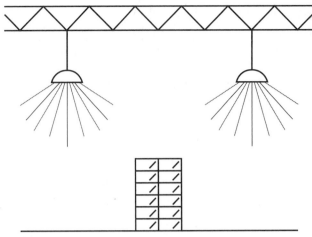

Headroom

The minimum headroom for the working space around electrical equipment is 6½ feet (1.98 m), or the height of the electrical equipment, whichever height is greater.

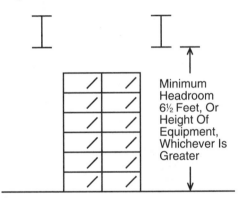

Minimum Headroom 6½ Feet, Or Height Of Equipment, Whichever Is Greater

Illumination

Provide illumination for all working spaces around service equipment, switchboards, panelboards, or motor control centers. This required illumination shall not be controlled by automatic means only (motion sensors), as this would introduce a hazard. For example, if the person working on the equipment did not physically move around enough to keep the motion detector's sensor in the "on" position, the lights would shut off. The present Code does not specify how much illumination is required. It's a judgment call. In some cases, adjacent lighting might be adequate to provide the required illumination. Good examples of this are the fluorescent fixtures located on the ceiling above and to the front of the main electrical equipment in the commercial building discussed in this text. ◀

GROUNDING

The principal reason that electrical systems, circuits, equipment, and conductive materials enclosing the above are grounded is to facilitate, i.e., enable, the immediate response of overcurrent protective devices to ground faults. This action removes the hazard, to people and animals, that would otherwise exist. See *NEC® Section 250-2.*

Electrical systems, such as a three-phase four-wire wye, and circuits, such as those supplied by a transformer connected to a system that exceeds 150 volts to ground, are grounded to stabilize and limit the voltage to ground. This is the grounding of current-carrying conductors. *NEC® Article 250, Part B* identifies the systems and circuits to be grounded.

Electrical equipment and conductive materials enclosing this equipment are grounded to limit the voltage to ground on these items. See *NEC® Article 250, Parts D and G,* which detail the items to be bonded to ground. Ideally, except in special cases where isolation is necessary, comprehensive bonding of all conductive materials should be the goal. This includes all metallic piping, duct work, framing, partitions, siding, and all other items that could come in contact with electrical system wiring or circuits and with a person or animal.

The system/circuit grounding and the equipment/materials grounding are to be joined at only one location on a premise. Failure to maintain the separation will result in currents extraneous from the electrical conductors, i.e., through the pipes, ducts, etc. These currents would increase the level of electro-magnetic radiation, which, many have alleged, has a harmful effect on the human body.

It is very important that the ground connections and the grounding electrode system be properly installed. To achieve the best possible ground system, the electrician must use the recommended procedures and equipment when making the installation. Figure 13-17 illustrates some of the terminology used in *NEC® Article 250, Grounding*.

If the grounding is installed in accordance with *NEC® Article 250*, a good path for ground currents will exist if a ground fault occurs. The reason for this is that the lower the impedance of the grounding path (ac resistance to the current), the greater is the ground-fault current. This increased ground-fault current causes the overcurrent device protecting the circuit to respond faster than it otherwise would for low-value currents. For example,

$$I = \frac{E}{Z} = \frac{277}{0.1} = 2770 \text{ amperes}$$

but, with a lower impedance,

$$I = \frac{E}{Z} = \frac{277}{0.01} = 27{,}700 \text{ amperes}$$

The amount of ground-fault damage to electrical equipment is related to (1) the response time of the overcurrent device and (2) the amount of current. One common term used to relate the time and current to the ground-fault damage is *ampere squared seconds (I^2t)*:

$$\text{Amperes} \times \text{Amperes} \times \text{Time in seconds} = I^2t$$

It can be seen in the expression for I^2t that when the current *(I)* and time *(t)* in seconds are kept to a minimum, a low value of I^2t results. Lower values of I^2t mean that less ground-fault damage will occur. Units 17, 18, and 19 provide detailed coverage of overcurrent protective devices, fuses, and circuit breakers.

Comprehensive Grounding

The concept of comprehensive grounding dictates that all conductive material such as water pipes, gas pipes, duct work, siding, and framing be bonded to all electrical equipment and connected by a main bonding jumper to the system ground. The interconnection of the equipment/material makes it highly unlikely that any of the items could become isolated, for even if one connection failed, the integrity of the grounding would be maintained through other connections. The value of this concept is illustrated in Figure 13-18 and described in the following example.

1. A live wire contacts the gas pipe. The bonding jumper Ⓐ is not installed originally.

2. The gas pipe now has 120 volts through it.

3. The insulating joint in the gas pipe results in a poor path to ground; assume the resistance is 8 ohms.

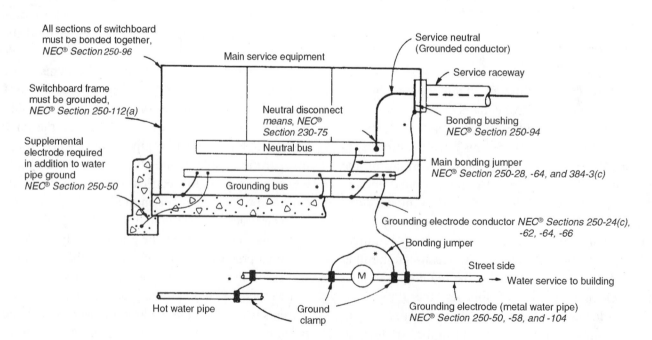

Fig. 13-17   Terminology of service grounding and bonding.

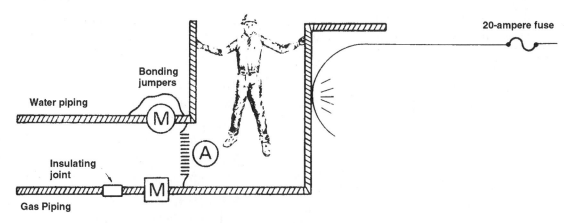

Fig. 13-18 System grounding.

4. The 20-ampere fuse will not open.

$$I = \frac{E}{R} = \frac{120}{8} = 15 \text{ amperes}$$

5. If a person touches the hot gas pipe and the water pipe at the same time, a current passes through the person's body. If the body resistance is 12,000 ohms, then the current is:

$$I = \frac{E}{R} = \frac{120}{12,000} = 0.01 \text{ ampere}$$

This value of current passing through a human body can cause death.

6. The fuse is now "seeing" 15 + 0.01 = 15.01 amperes; however, it still does not blow.

7. If comprehensive grounding had been achieved, bonding jumper Ⓐ would have kept the voltage difference between the water pipe and the gas pipe at zero, and the fuse would have opened. If 10 feet (3.05 m) of No. 4 AWG copper wire were used as the jumper, then the resistance of the jumper would be 0.00308 ohm per *NEC® Chapter 9, Table 8*. The current is:

$$I = \frac{E}{R} = \frac{120}{0.00308} = 38,961 \text{ amperes}$$

(In an actual system, the impedance of all of the parts of the circuit would be much higher. Thus, a much lower current would result. The value of current, however, would be enough to cause the fuse to open.)

The advantages of comprehensive system grounding are:

- the potential voltage differentials between the different parts of the system are kept at a minimum, thereby reducing shock hazard.

- impedance of a ground path is kept at a minimum, which results in higher current flow in the event of a ground fault. The lower the impedance, the greater is the current flow, and the faster the overcurrent device opens.

GROUNDING ELECTRODE SYSTEMS

In Figures 13-14 and 13-17, the main service equipment, the service raceways, the neutral bus, the grounding bus, and the hot and cold water pipes have been bonded together to form a *Grounding Electrode System*, as required by *NEC® Section 250-50*.

For discussion purposes of the Commercial Building, at least 20 feet (6.1 m) of No. 3/0 bare copper conductor is installed in the footing to serve as the additional supplemental electrode that is required by *NEC® Section 250-50*. The minimum size permitted is No. 4 copper. This conductor, buried in the concrete footing, is supplemental to the water pipe ground. *NEC® Section 250-50(a)* permits the supplemental electrode to be bonded to the grounding electrode conductor, the grounded service-entrance conductor, the grounded service raceway, or the grounded service enclosure.

For commercial or industrial installations the supplemental electrode may be bonded to the interior metal water piping only where the entire length of water piping that will serve as the conductor is exposed and will be under qualified maintenance conditions. See *NEC® Section 250-50(a)*.

Had No. 3/0 bare copper conductor not been selected for installation in the concrete footing, the supplemental grounding electrode could have been:

- the metal frame of the building where effectively grounded.

- at least 20 feet (6.1 m) of steel reinforcing bars (re-bars) ½-inch (12.7 mm) minimum diameter, encased in concrete at least 2 inches (50.8 mm) thick, in direct contact with the earth, near the bottom of the foundation or footing.

- at least 20 feet (6.1 m) of bare copper wire, encircling the building, minimum size No. 2 AWG, buried directly in the earth at least 2½ feet (762 mm) deep.

- ground rods.

- ground plates.

All of these factors are discussed in *NEC® Article 250, Parts C, E,* and *F*.

Many Code interpretations are probable as a result of the grounding electrode systems concept. Therefore, the local code authority should be consulted. For instance, some electrical inspectors may not require the bonding jumper between the hot and cold water pipes, as shown in Figure 13-17. They may determine that an adequate bond is provided through the water heater itself of either type, electric or gas. Other electrical inspectors may require that the hot and cold water pipes be bonded together because some water heaters contain insulating fittings that are intended to reduce corrosion inside the tank caused by electrolysis. According to their reasoning, even though the water heater originally installed contains no fittings of this type, such fittings may be included in a future replacement heater, thereby necessitating a bond between the cold and hot water pipes. The local electrical inspector should be consulted for the proper requirements. However, bonding the pipes together, as shown in figure 13-17, is the recommended procedure.

Grounding Requirements

When grounding service-entrance equipment, the following Code rules must be observed:

- The electrical system must be grounded when maximum voltage to ground does not exceed 150 volts, see *Section 250-20(b)(1)*.

- The electrical system must be grounded when the neutral is used as a circuit conductor (example: a 480/277-volt, wye-connected, three-phase, four-wire system), *Section 250-20(b)(2)*. This is similar to the connection of the system shown in Figure 13-6 (page 156).

- The electrical system must be grounded where the midpoint of one transformer is used to establish the grounded neutral circuit conductor, as on a 240/120-volt, three-phase, four-wire delta system, see *Section 250-20(b)(3)* and refer to Figure 13-5 (page 155).

- The earth must not be used as the sole means of accomplishing an effective grounding means, see *NEC® Section 250-2(d)*.

- *NEC® Section 250-2(d)* requires that an effective grounding path be established. The path to ground from all circuits, equipment, and metal enclosures (1) must be permanent and continuous, (2) must have adequate capacity to conduct safely any fault current that it might be called upon to carry, and (3) must have impedance low enough to limit the voltage to ground and to facilitate operation of the overcurrent device. Details of this subject are discussed in Units 17, 18, and 19.

- All grounding schemes shall be installed so that no objectionable currents will exist in the grounding conductors and other grounding paths, see *NEC® Section 250-6(a)*.

- The grounding electrode conductor must be connected to the supply side of the service disconnecting means, see *NEC® Section 250-24(a)*.

- The identified neutral conductor is the conductor that must be grounded, see *NEC® Section 250-26*.

- Tie (bond) everything together, see *NEC® Sections 250-50, -104(a), -104(b)*.

- Do not use interior metal water piping to serve as a means of interconnection steel framing members, concrete-encased electrodes, and ground ring for the purposes of establishing the grounding electrode system as required by *NEC® Section 250-104*. Should any PVC nonmetallic water piping be interposed in the metal piping runs, the continuity of the bonding-grounding system is broken.

The first five feet of metal water piping entering the building is NOT considered to be "interior." The proper location to connect the grounding electrode conductor, the bonding conductors associated with the metal framing members, concrete-encased electrodes, and ground ring is anywhere on the first five feet of metal water piping after it enters the building. The first five feet may include a water meter.

The only exception to this is for industrial and commercial buildings, where the entire length of the water piping being used as the conductor to interconnect the various electrodes is exposed, and where only qualified people will be doing maintenance on the installation; see *NEC® Section 250-50*.

- The grounding electrode conductor is to be sized as given in *NEC® Table 250-66*.

- The metal frame of the building shall be bonded to the grounding electrode system where it is effectively grounded, *NEC® Section 250-50(b)* and *(FPN)*.

- The hot and cold water metal piping system shall be bonded to the service equipment enclosure, to the grounded conductor at the service, and to the grounding electrode conductor, *NEC® Section 250-104*.

- *NEC® Section 250-64(b)* states that grounding electrode conductors shall not be spliced. There are exceptions for commercial and industrial installations where the grounding electrode conductor may be spliced by either exothermic welding or irreversible compression type connectors that are listed for that purpose.

- The grounding electrode conductor shall be connected to the metal underground water pipe when 10 feet (3.05 m) long or longer, including the metal well casing, *NEC® Section 250-50*.

- In addition to grounding the service equipment to the underground water pipe, one or more additional electrodes are required, such

as a bare conductor in the footing, a grounding ring, rod or pipe electrodes, or plate electrodes. All of these items must be bonded if they are available on the premises. See *NEC® Section 250-50* and -*104*.

- *NEC® Section 250-52(a)* prohibits using a metal underground gas piping system as the grounding electrode. However, where metal gas piping comes into a building, it must be bonded to all of the other metal piping that is required to be bonded, see *NEC® Section 250-104(c)*.

- The grounding electrode conductor shall be copper, aluminum, or copper-clad aluminum, *NEC® Section 250-62*.

- The grounding electrode conductor may be solid or stranded, uninsulated, covered or bare, and must not be spliced, see *NEC® Section 250-62 and 250-64(c)*.

- Bonding is required around all insulating joints or sections of the metal piping system that might be disconnected, *NEC® Section 250-68(b)*.

- The connection to the grounding electrode must be accessible, *NEC® Section 250-68(a)*. A connection to a concrete-encased, drive, or buried grounding electrode does not have to be accessible.

- Make sure that the length of bonding conductor is long enough so that if the equipment is removed, the bonding will still be acceptable, *NEC® Section 250-68(b)*. The bonding jumper around the water meter is a good example where this requirement applies.

- The grounding electrode conductor must be tightly connected by using proper lugs, connectors, clamps, or other approved means, *NEC® Section 250-70*. One type of grounding clamp is shown in Figure 13-19.

Fig. 13-19   Grounding clamp.

Sizing the Grounding Electrode Conductors

The grounding electrode conductor connects the grounding electrode to the equipment grounding conductor and to the systems grounding conductor. In the commercial building, this means that the grounding electrode conductor connects the main water pipe *(grounding electrode)* to the grounding bus *(equipment grounding conductor)* and to the neutral bus *(systems grounding conductor)*. See Figure 13-17.

NEC® Table 250-66 is referred to when selecting grounding electrode conductors for services where there is no overcurrent protection ahead of the service-entrance conductors other than the utility companies' primary or secondary overcurrent protection.

In *NEC® Table 250-66*, the service conductor sizes are given in the wire size and not the ampacity values. To size a grounding electrode conductor for a service with three parallel No. 500 kcmil service conductors, it is necessary to total the wire size and select a grounding electrode conductor for an equivalent No. 1500 kcmil service conductor. In this case, the grounding electrode conductor is No. 3/0 copper (see Figure 13-14). Figure 13-20 shows another example.

The main bonding jumper, see Figures 13-21 and 13-22, connects the neutral bus bar to the equipment grounding bus bar. The jumper is sized according to *NEC® Section 250-28(d)*. The service to the commercial building consists of three 500 kcmil conductors in parallel. This adds up to a total conductor area of 1500 kcmil. Multiplying by 0.125 yields a minimum size for the main bonding jumper of 187.5 kcmil. Referring to *NEC® Chapter 9, Table 8*, this requires a No. 4/0 AWG conductor.

The computed neutral load is 560 amperes . . . 187 amperes per conductor. The 90°C column of *NEC® Table 310-16* indicates that No. 3/0 AWG conductors have an ampacity of 200 amperes. This would seem to be adequate. However, there are other Code sections that must be checked.

Referring to *NEC® Section 250-24(b)*, the requirement is set forth that the minimum size for the neutral is determined by the requirement for a grounding electrode conductor. When the service conductors are paralleled, the neutral shall be routed with the phase conductors and shall be sized no smaller than the required grounding electrode conductor, *NEC® Table 250-66*. The three 500 kcmil service-entrance conductors in each service raceway will require that the neutral conductor be no smaller than 1/0, see *NEC® Section 310-4*. So, although the calculated neutral load may allow a smaller conductor, 1/0 is the minimum size because this conductor would carry high-level fault currents, if a ground fault occurs.

Table 250-66. Grounding Electrode Conductor for Alternating-Current Systems

Size of Largest Service-Entrance Conductor or Equivalent Area for Parallel Conductors[1]		Size of Grounding Electrode Conductor	
Copper	Aluminum or Copper-Clad Aluminum	Copper	Aluminum or Copper-Clad Aluminum[2]
2 or smaller	½ or smaller	8	6
1 or 1/0	2/0 or 3/0	6	4
2/0 or 3/0	4/0 or 250 kcmil	4	2
Over 3/0 through 350 kcmil	Over 250 kcmil through 500 kcmil	2	1/0
Over 350 kcmil through 600 kcmil	Over 500 kcmil through 900 kcmil	1/0	3/0
Over 600 kcmil through 1100 kcmil	Over 900 kcmil through 1750 kcmil	2/0	4/0
Over 1100 kcmil	Over 1750 kcmil	3/0	250 kcmil

Notes:

1. Where multiple sets of service-entrance conductors are used as permitted in Section 230-40, Exception No. 2, the equivalent size of the largest service-entrance conductor shall be determined by the largest sum of the areas of the corresponding conductors of each set.

2. Where there are no service-entrance conductors, the grounding electrode conductor size shall be determined by the equivalent size of the largest service-entrance conductor required for the load to be served.

[1]This table also applies to the derived conductors of separately derived ac systems.

[2]See installation restrictions in Section 250-64(a).

FPN: See Section 250-24(b) for size of ac system conductor brought to service equipment.

Reprinted with permission from NFPA 70-1999.

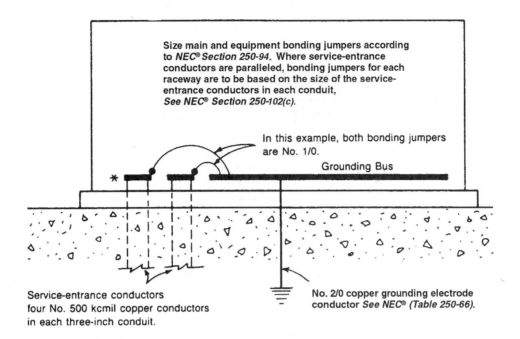

Size main and equipment bonding jumpers according to *NEC® Section 250-94*. Where service-entrance conductors are paralleled, bonding jumpers for each raceway are to be based on the size of the service-entrance conductors in each conduit, *See NEC® Section 250-102(c).*

In this example, both bonding jumpers are No. 1/0.

Grounding Bus

*

Service-entrance conductors four No. 500 kcmil copper conductors in each three-inch conduit.

No. 2/0 copper grounding electrode conductor *See NEC® (Table 250-66).*

*Conduits shall rise not more than three inches (76 mm) above bottom of enclosure, *See NEC® Section 384-10.*

Fig. 13-20   Typical service entrance (grounding and bonding).

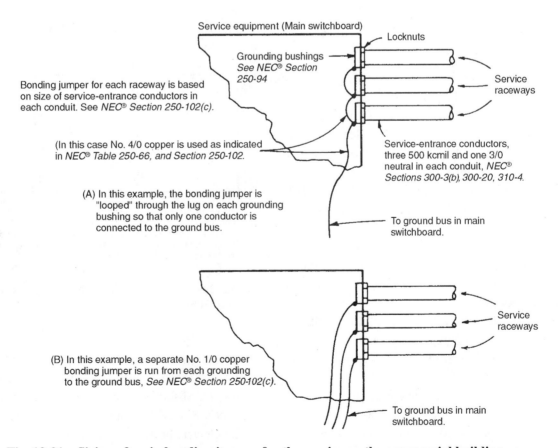

Service equipment (Main switchboard)

Locknuts

Grounding bushings *See NEC® Section 250-94*

Service raceways

Bonding jumper for each raceway is based on size of service-entrance conductors in each conduit. See *NEC® Section 250-102(c).*

(In this case No. 4/0 copper is used as indicated in *NEC® Table 250-66, and Section 250-102.*

Service-entrance conductors, three 500 kcmil and one 3/0 neutral in each conduit, *NEC® Sections 300-3(b), 300-20, 310-4.*

(A) In this example, the bonding jumper is "looped" through the lug on each grounding bushing so that only one conductor is connected to the ground bus.

To ground bus in main switchboard.

Service raceways

(B) In this example, a separate No. 1/0 copper bonding jumper is run from each grounding to the ground bus, *See NEC® Section 250-102(c).*

To ground bus in main switchboard.

Fig. 13-21   Sizing of main bonding jumper for the service on the commercial building.

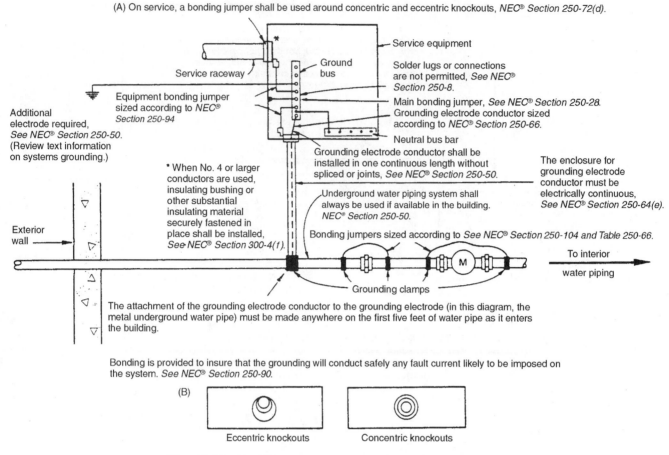

(A) On service, a bonding jumper shall be used around concentric and eccentric knockouts, *NEC® Section 250-72(d).*

Service equipment

Service raceway

Ground bus

Solder lugs or connections are not permitted, See *NEC® Section 250-8.*

Equipment bonding jumper sized according to *NEC® Section 250-94*

Main bonding jumper, *See NEC® Section 250-28.*
Grounding electrode conductor sized according to *NEC® Section 250-66.*

Additional electrode required, *See NEC® Section 250-50.* (Review text information on systems grounding.)

Neutral bus bar

Grounding electrode conductor shall be installed in one continuous length without spliced or joints, *See NEC® Section 250-50.*

The enclosure for grounding electrode conductor must be electrically continuous, *See NEC® Section 250-64(e).*

* When No. 4 or larger conductors are used, insulating bushing or other substantial insulating material securely fastened in place shall be installed, *See NEC® Section 300-4(f).*

Underground water piping system shall always be used if available in the building. *NEC® Section 250-50.*

Exterior wall

Bonding jumpers sized according to *See NEC® Section 250-104 and Table 250-66.*

To interior water piping

Grounding clamps

The attachment of the grounding electrode conductor to the grounding electrode (in this diagram, the metal underground water pipe) must be made anywhere on the first five feet of water pipe as it enters the building.

Bonding is provided to insure that the grounding will conduct safely any fault current likely to be imposed on the system. *See NEC® Section 250-90.*

(B)

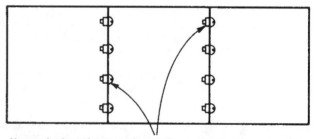

Eccentric knockouts   Concentric knockouts

Fig. 13-22   Bonding and grounding of service equipment.

Remember that *NEC® Sections 250-2(d), -90,* and *-96* require that the grounding and bonding conductors be capable of carrying any fault current that they might be called upon to carry under fault conditions. Also refer to *NEC® Section 250-2.* This subject is covered in Units 17, 18, and 19.

Refer to Figures 13-21 through 13-23 for the proper procedures to be used in bonding electrical equipment.

NEC® Table 250-122 is referred to when selecting equipment grounding conductors when there is overcurrent protection ahead of the conductor supplying the equipment.

NEC® Table 250-122 is based on the current setting or rating, in amperes, of the overcurrent device installed ahead of the equipment (other than service-entrance equipment) being supplied.

The electric boiler branch circuit in the commercial building consists of two No. 500 kcmil conductors per phase protected by 700-ampere fuses in the main switchboard. The use of flexible metal conduit

Nonconductive paint, enamel, or similar coating must be removed at contact points when sections of electrical equipment are bolted together to insure that the sections are effectively bonded. Tightly driven metal bushings and locknuts generally will bite through paints and enamels, thus making the removal of the paint unnecessary. *NEC® Section 250-96.*

Fig. 13-23   Insuring bonding when electrical equipment is bolted together.

at the boiler means that a bonding wire must be run through each conduit as shown in the figure. A detailed discussion on flexible connections is given in unit 8. An illustration of a typical application for the electric boiler feeder in the commercial building is shown in Figure 13-24.

Table 250-122. Minimum Size Equipment Grounding Conductors for Grounding Raceway and Equipment

Rating or Setting of Automatic Overcurrent Device in Circuit Ahead of Equipment, Conduit, etc., Not Exceeding (Amperes)	Size (AWG or kcmil)	
	Copper	Aluminum or Copper-Clad Aluminum*
15	14	12
20	12	10
30	10	8
40	10	8
60	10	8
100	8	6
200	6	4
300	4	2
400	3	1
500	2	1/0
600	1	2/0
800	1/0	3/0
1000	2/0	4/0
1200	3/0	250
1600	4/0	350
2000	250	400
2500	350	600
3000	400	600
4000	500	800
5000	700	1200
6000	800	1200

Note: Where necessary to comply with Section 250-2(d), the equipment grounding conductor shall be sized larger than this table.

*See installation restrictions in Section 250-120.

Reprinted with permission from NFPA 70-1999.

GROUND-FAULT PROTECTION

The *NEC®* requires the use of ground-fault protection (GFP) devices on services which meet the conditions outlined in *NEC® Section 230-95*. Thus, ground-fault protection devices are installed:

- on solidly grounded wye services above 150 volts to ground, but not over 600 volts between phases on service disconnects rated at 1000 amperes or more (for example, on 480Y/277-volt systems).

- to operate at 1200 amperes or less.

- so that the maximum time of opening the service switch or circuit breaker does not exceed one (1) second for ground-fault currents of 3000 amperes or more.

- to limit damage to equipment and conductors on the *load side* of the service disconnecting means. GFP will *not* protect against damage caused by faults occurring on the *line side* of the service disconnect.

These ground-fault protection requirements do not apply on services for a continuous industrial process where a nonorderly shutdown will introduce additional or increased hazards, *NEC® Sections 230-95, Exception No. 1*, and *240-12*.

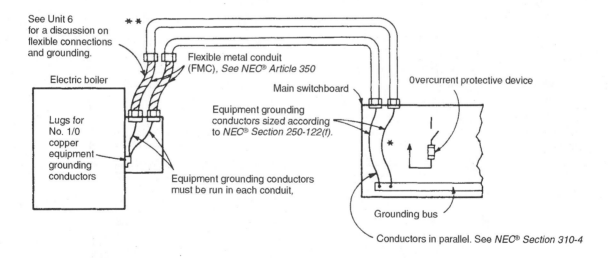

** Each conduit contains three No. 500 kcmil "hot" conductors and one No. 1/0 equipment grounding conductor solidly connected to the grounding bus in the switchboard and to the lugs in the terminal box on the boiler. The No. 1/0 equipment grounding conductor may be insulated or bare, *NEC® Section 250-118*

Fig. 13-24   Sizing equipment grounding conductors.

When a fuse-switch combination serves as the service disconnect, the fuses must have adequate interrupting capacity to interrupt the available fault current *(NEC® Section 110-9)*, and must be capable of opening any fault currents that exceed the interrupting rating of the switch during any time when the ground-fault protective system will not cause the switch to open, see *NEC® Section 230-95(b)*.

Ground-fault protection is not required on:

- delta-connected three-phase systems.

- ungrounded wye-connected three-phase systems.

- single-phase systems.

- 120/240-volt single-phase systems.

- 208Y/120-volt three-phase, four-wire systems.

- systems over 600 volts; for example, 2400/4160 volts.

- service disconnecting means rated at less than 1000 amperes.

- systems where the service is subdivided; for example, a 1600-ampere service may be divided between two 800-ampere switches.

The time of operation of the device as well as the ampere setting of the GFP device must be considered carefully to insure that the continuity of the electrical service is maintained. The time of operation of the device includes: (1) the sensing of a ground fault by the GFP monitor, (2) the monitor signaling the disconnect switch to open, and (3) the actual opening of the contacts of the disconnect

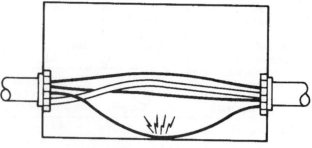

Arcing ground fault can occur when a phase wire and the conduit grounding system contact each other.

Fig. 13-25   Ground fault.

device (either a switch or a circuit breaker). The total time of operation may result in a time lapse of several cycles or more (units 17 and 18).

GFP circuit devices were developed to overcome a major problem in circuit protection: the low-value phase-to-ground arcing fault, Figure 13-25. The amount of current that flows in an arcing phase-to-ground fault can be low when compared to the rating or setting of the overcurrent device. For example, an arcing fault can generate a current of 600 amperes. Yet, a main breaker rated at 1,600 amperes will allow this current without tripping, since the 600-ampere current appears to be just another *load* current. The operation of the GFP device assumes that under normal conditions the total instantaneous current in all of the conductors of a circuit will exactly balance, Figure 13-26. Thus, if a current coil is installed so that all of the circuit conductors run through it, the normal current

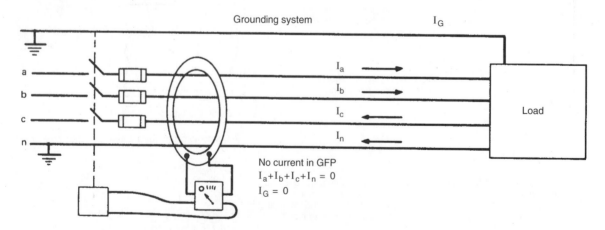

Fig. 13-26   Normal condition.

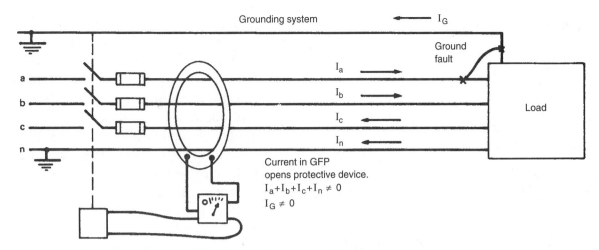

Fig. 13-27   Abnormal condition.

measured by the coil will be zero. If a ground fault occurs, some current will return through the grounding system and an unbalance will result in the conductors. This unbalance is then detected by the GFP device, Figure 13-27.

The purpose of ground-fault protection devices is to sense and protect equipment against *low-level ground faults*. GFP monitors do not sense phase-to-phase faults, three-phase faults, or phase-to-neutral faults. These monitors are designed to sense phase-to-ground faults only.

Large (high magnitude) ground-fault currents can cause destructive damage even though a GFP is installed. The amount of arcing damage depends upon (1) how much current flows and (2) the length of time that the current exists. For example, if a GFP device is set for a ground fault of 500 amperes and the time setting is six cycles, then the device will need six cycles to signal the switch or circuit breaker to open the circuit whether the ground fault is slightly more than 500 amperes or as large as 20,000 amperes. The six cycles needed to signal the circuit breaker plus the operation time of the switch or breaker may be long enough to result in damage to the switchgear.

The damaging effects of high-magnitude ground faults, phase-to-phase faults, three-phase faults, and phase-to-neutral faults can be reduced substantially by the use of current-limiting overcurrent devices *NEC® Section 240-11*. These devices reduce both the peak let-through current and the time of opening once the current is sensed. For example, a ground fault of 20,000 amperes will open a current-limiting fuse in less than one-half cycle. In addition, the peak let-through current is reduced to a value much less than 20,000 amperes (see units 17 and 18).

Ground-fault protection for equipment (GFP) is not to be confused with personal ground-fault protection (GFCI). In commercial buildings GFCI devices are only required in bathrooms and on rooftops, see *NEC® Section 210-8(b)*, and are discussed in Unit 5.

The GFP is connected to the normal fused switch or circuit breaker which serves as the circuit protective device. The GFP is adjusted so that it will signal the protective device to open under abnormal ground-fault conditions. The maximum setting of the GFP is 1200 amperes. In the commercial building in the plans, the service voltage is 208Y/120 volts. The voltage to ground on this system is 120 volts. This value is not large enough to sustain an electrical arc. Therefore, it is not required (according to the *NEC®*) that ground-fault protection for the service disconnecting means be installed for the commercial building. The electrician can follow a number of procedures to minimize the possibility of an arcing fault. Examples of these procedures follow.

- Insure that conductor insulation is not damaged when the conductors are pulled into raceway.

- Insure that the electrical installation is properly grounded and bonded.

- Locknuts and bushings must be tight.

- All electrical connections must be tight.

- Tightly connect bonding jumpers around concentric and/or eccentric knockouts.

- Insure that conduit couplings and other fittings are installed properly.

- Check insulators for minute cracks.

- Install insulating bushings on all raceways.

- Insulate all bare bus bars in switchboards when possible.

- Conductors must not rest on bolts, or other sharp metal edges.

- Electrical equipment must not be allowed to become damp or wet either during or after construction.

- All overcurrent devices must have an adequate interrupting capacity.

- Do not work on *hot* panels.

- Be careful when using *fish tape* since the loose end can become tangled with the electrical panelboard.

- Be careful when working with *live* parts; do not drop tools or other metal objects on top of such parts.

- Avoid large services; for example, it is usually preferable to install two 800-ampere service disconnecting means rather than to install one 1600-ampere service disconnecting means.

SAFETY IN THE WORKPLACE

Many injuries have occurred by' individuals working electrical equipment "hot." A fault, whether line-to-line, line-to-line-to-line, or line-to-ground can develop tremendous energy. Splattering of melted copper and steel, the heat of the arc blast, the blinding light of the arc, and electric shock are hazards that are present. The temperatures of an electrical arc, such as might occur on a 480/277-volt solidly grounded wye-connected system, are hotter than the surface of the sun. The enormous pressures of an "arc blast" can blow a person clear across the room. These awesome pressures will vent (dis-

charge) through openings, such as the open cover of a panel, just where the person is standing.

Federal laws are in place to protect workers from injury while on the job. When working on or near electrical equipment, certain safety practices must be followed.

In the Occupational Safety and Health Act (OSHA), Sections 1910.331 through 1910.360 are devoted entirely to Safety Related Work Practices. Proper training in work practices, safety procedures, and other personnel safety requirements are described. Such things as turning off the power, locking the switch off, or properly tagging the switch are discussed in the OSHA Standards.

The fundamental rule is **NEVER WORK ON EQUIPMENT WITH THE POWER TURNED ON!**

In the few cases where the equipment absolutely must be left "on," proper training regarding electrical installations, proper training in established safety practices, proper training in first aid, proper insulated tools, safety glasses, hard hats, protective insulating shields (rubber blankets to cover "live" parts), rubber gloves, nonconductive and nonflammable clothing, and working with more than one person are required.

Proper training is required to enable a person to become qualified. The *National Electrical Code®* describes a Qualified Person as "One familiar with the construction and operation of the equipment and the hazards involved."

The National Fire Protection Association Standard NFPA 70B is entitled *Electrical Equipment Maintenance*. This Code also discusses safety issues and safety procedures, and mirrors the OSHA regulations in that *No Electrical Equipment Should Be Worked On While It Is Energized*.

The National Electrical Manufacturers Association (NEMA) in Publication PB 2.1-1991 entitled *General Instructions for Proper Handling, Installation, Operation, and Maintenance of Deadfront Distribution Switchboards Rated 600 Volts or Less* repeat the safety rules discussed above. This standard states that "The installation, operation, and maintenance of switchboards should be conducted only by qualified personnel." This standard discusses the *"Lock-Out, Tag-Out"* procedures. It clearly states that:

WARNING: HAZARDOUS VOLTAGES IN ELECTRICAL EQUIPMENT CAN CAUSE SEVERE PERSONNEL INJURY OR DEATH. UNLESS OTHERWISE SPECIFIED, INSPECTION AND PREVENTATIVE MAINTENANCE SHOULD ONLY BE PERFORMED ON SWITCHBOARDS, AND EQUIPMENT TO WHICH POWER HAS BEEN TURNED OFF, DISCONNECTED AND ELECTRICALLY ISOLATED SO THAT NO ACCIDENTAL CONTACT CAN BE MADE WITH ENERGIZED PARTS.

There have been lawsuits where serious injuries have occurred on equipment that had GFP protection. The claims were that the GFP protection for equipment also provides protection for personnel. **This is not true!** Throughout the *National Electrical Code®* references are made to **Ground Fault Protection For Personnel (GFCI)** and **Ground Fault Protection For Equipment (GFP)**.

Many texts have been written about the hazards of electricity. All of these texts say **TURN THE POWER OFF!**

IT'S THE LAW!

REVIEW QUESTIONS

Refer to the *National Electrical Code®* or the working drawings when necessary. Where applicable, responses should be written in complete sentences. Write units using unit names, use no abbreviations or symbols (1 foot, not 1' or 1 ft).

A 300 kVA transformer bank has a three-phase, 480-volt delta primary and a three-phase, 208Y/120 secondary.

1. What is the full load current in the primary? _____ *361* _____

2. What is the full load current in the secondary? _____ *833* _____

3. What is the proper rating of fuse to install in the secondary? _____ *1200* _____

4. What is the kVA of the bank if one of the transformers is removed? _____ *174 KVA* _____

5. Sketch the secondary connections and indicate where a 120-volt load could be connected.

300 × .58 = 174 KVA

The next question refers to the accompanying drawing.

6. A panelboard is added to an existing panelboard installation. Two knockouts, one near the top of the panelboard and the other near the bottom, are cut in the adjoining sides of the boxes. Indicate on the drawing the proper way of extending the phase and neutral conductors to the new panelboard. Line side lugs are suitable for two conductors.

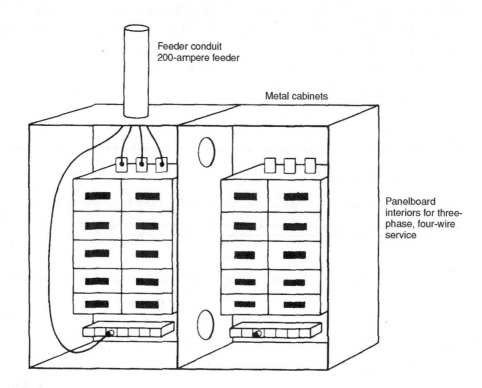

Feeder conduit
200-ampere feeder

Metal cabinets

Panelboard interiors for three-phase, four-wire service

The following questions refer to grounding requirements.

7. What is the proper size copper grounding electrode conductor for a 100-ampere service that consists of 3/0 AWG phase conductors? _____

8. What is the minimum size copper grounding electrode conductor for a service with three sets of parallel 350 kcmil conductors (three conductors per phase)? _____

9. If they are available, what four items must be bonded together to form a grounding electrode system? _____

_____

_____

_____

_____

10. If none of the four items listed in the previous question are available, what three items can be used? _____

_____

_____

11. The engineering calculations for an 800-ampere service entrance call for two 500-kcmil copper conductors per phase, connected in parallel. The neutral calculations show that the neutral conductors need only be No. 3 copper conductors. Yet, the specifications and riser diagram show that the neutral conductor is a 1/0 AWG copper conductor, run in parallel. The riser diagram and specifications call for two conduits, each containing three 500-kcmil phase conductors, plus one 1/0 AWG neutral conductor. Why is the neutral conductor sized as a 1/0 AWG?

_____

_____

_____

_____

_____

_____

_____

12. Electric utility installations and equipment under the exclusive control of the utility do not have to conform to the *National Electrical Code.®* However, they do have to conform to the _____.

13. How much working space must be provided in front of a 480/277-volt dead front switchboard according to *Table 110-26(a)*? The switchboard faces a concrete block wall. (3 feet) (3½ feet) (4 feet). Circle the correct answer.

14. A wall-mounted panel has a number of electrical conduits running out of the top and bottom of the panel. A sheet metal worker started to run a cold air return duct through the wall approximately 8 feet high directly above the panel. The electrician stated that this is not permitted, and cited *Section* _____ of the *National Electrical Code®* to support his argument.

15. You are an electrical apprentice. The electrical foreman on the job tells you to connect some equipment grounding conductors to the ground bus that runs along the front bottom of a dead front switchboard. The system is 480/277-volt solidly grounded wye. The building is still under construction, and a number of other trades are working in the building. The switchboard has already been energized, and it is totally enclosed except for two access covers near the front bottom where the ground bus runs. The ground bus already has the proper number of and size of lugs for the equipment grounding conductors. You ask permission to turn "off" the main power so you can work on connecting the equipment grounding conductors to the ground bus without having to worry about accidentally touching the phase buses with your hands or with a screwdriver or wrench. The foreman says, "No." What would you do?

_____

_____

_____

_____

_____

16. For the main switchboard in the commercial building, what "condition" describes the installation? Refer to *NEC® Section 110-16*.

_____

_____

_____

_____

_____

UNIT 14

Lamps for Lighting

OBJECTIVES

After studying this unit, the student will be able to

- define the technical terms associated with lamp selection.
- list the lamps scheduled to be used in the commercial building.
- explain the application of lamps used in the commercial building.
- identify the parts of the three most popular types of lamps.
- order the lamp types according to certain characteristics.
- define lamps by their designations.

For most construction projects, the electrical contractor is required to purchase and install lamps in the luminaires (lighting fixtures). Thomas Edison provided the talent and perseverance that led to the development of the incandescent lamp in 1879 and the fluorescent lamp in 1896. Peter Cooper Hewitt produced the mercury lamp in 1901. All three of these lamp types have been refined and greatly improved since they were first developed.

NEC® Article 410 contains the provisions for the wiring and installation of lighting fixtures, lampholders, lamps, receptacles, and rosettes.

Several types of lamps are used in the commercial building. In the *NEC®* these lamps are referred to as either incandescent or electric discharge lamps. See *NEC® Article 410, Part K*. In the industry, the incandescent lamp is also referred to as a filament lamp because the light is produced by a heated wire filament. The electric discharge lamps include a variety of types but all require a ballast. The most common types of electric discharge lamps are fluorescent, mercury, metal halide, high-pressure sodium, and low-pressure sodium. Mercury, metal halide, and high-pressure sodium lamps are also classed as high-intensity discharge (HID) lamps.

LIGHTING TERMINOLOGY

Candela (cd): The luminous intensity of a source, when expressed in candelas, is the candlepower (cp) rating of the source.

Lumen (lm): The amount of light received in a unit of time on a unit area at a unit distance from a point source of one candela, Figure 14-1. The surface area of a sphere with unit radius is 12.57 times the radius; therefore, a one-candela source produces 12.57 lumens. When the measurement is in English units, the unit area is 1 square foot and the unit distance is 1 foot. If the units are in SI, then the unit area is 1 square meter and the unit distance is 1 meter.

Illuminance: The measure of illuminance on a surface is the lumen per unit area expressed in footcandles (fc) or lux (lx). The recommended illuminance levels vary greatly, depending on the task to be performed and the ambient lighting conditions. For example, while 5 footcandles (fc), or 54 lux (lx), is often accepted as adequate illumination for a dance hall, 200 fc (2,152 lx) may be necessary on a drafting table for detailed work.

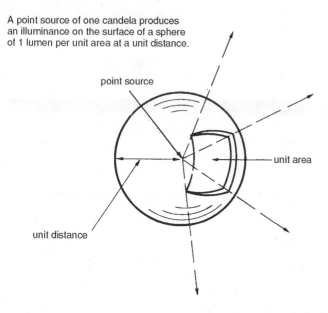

A point source of one candela produces an illuminance on the surface of a sphere of 1 lumen per unit area at a unit distance.

point source

unit area

unit distance

Fig. 14-1   Pictorial presentation of unit sphere.

Lumen per watt (lm/W):   The measure of the effectiveness (efficacy) of a light source in producing light from electrical energy. A 100-watt incandescent lamp producing 1670 lumens has an effective value of 16.7 lumens per watt.

Kelvin (K):   The kelvin (sometimes incorrectly called degree Kelvin) is measured from absolute zero; it is equivalent to a degree value in the Celsius scale plus 273.16. The color temperature of lamps is given in Kelvin. The lower the number, the warmer is the light (more red content). The higher the number, the cooler is the light (more blue content).

Color Rendering Index (CRI):   This value is often given for lamps so the user can have an idea of the color rendering probability. The CRI uses filament light as a base for 100 and the warm white fluorescent for 50. The CRI can be used only to compare lamps that have the same color temperature. The only sure way to determine if a lamp will provide good color rendition is to see the material in the lamplight.

Refer to Table 14-1 for a comparison of the various types of electric lamps.

INCANDESCENT LAMPS

The incandescent lamp has the lowest efficacy of the types listed in Table 14-1. However, incandescent lamps are very popular and account for more than 50 percent of the lamps sold in the United States. This popularity is due largely to the low cost of incandescent lamps and luminaires.

Federal energy legislation that became effective in April 1994 removed several of the more commonly used lamps from the marketplace. No longer can the standard reflector lamps, like the R40, or standard PAR lamps, like the PAR38, be manufactured. The use of halogen and krypton filled, ER, and low voltage lamps is encouraged. The ER lamp (elliptical reflector) is designed primarily for use as a replacement and is not recommended for new installations. Typical of the recommended substitutions are: a 50-watt PAR30 halogen lamp with a 65° spread to replace the 75-watt R30 reflector lamp, and a 60- or 90-watt PAR halogen flood with a 30° spread to replace the 150-watt PAR38. A full list of substitutions is available at major hardware stores and electrical distributors.

	Filament	Fluorescent	Mercury	Metal halide	HPS	LPS
Lumen per watt	6 to 23	25 to 100	30 to 65	65 to 120	75 to 140	130 to 180
Wattage range	40 to 1500	4 to 215	40 to 1000	175 to 1500	35 to 1000	35 to 180
Life	750 to 8k	9k to 20k	16k to 24k	5k to 15k	20k to 24k	18k
Color temperature	2400 to 3100	2700 to 7500	3000 to 6000	3000 to 5000	2000	1700
Color rendition Index	90 to 100	50 to 110	25 to 55	60 to 70	20 to 25	0
Potential for good color rendition	high	highest	fair	good	color discrimination	no color discrimination
Lamp cost	low	moderate	moderate	high	high	moderate
Operational cost	high	good	moderate	moderate	low	low

TABLE 14-1   Characteristics of Electric Lamps

The values given above are generic for general service lamps. A survey of lamp manufacturers catalogs should be made before specifying or purchasing any lamp.

Construction

The light-producing element in the incandescent lamp is a tungsten wire called the *filament*, Figure 14-2. This filament is supported in a glass envelope or bulb. The air is evacuated from the bulb and is replaced with an inert gas such as argon. The filament is connected to the base by the lead-in wires. The base of the incandescent lamp supports the lamp and provides the connection means to the power source. The lamp base may be any one of the base styles shown in Figure 14-3.

Characteristics

Incandescent lamps are classified according to the following characteristics.

Voltage Rating. Incandescent lamps are available with many different voltage ratings. When installing lamps, the electrician should be sure that a lamp

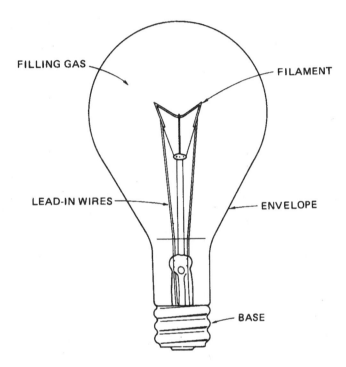

Fig. 14-2 Incandescent lamp.

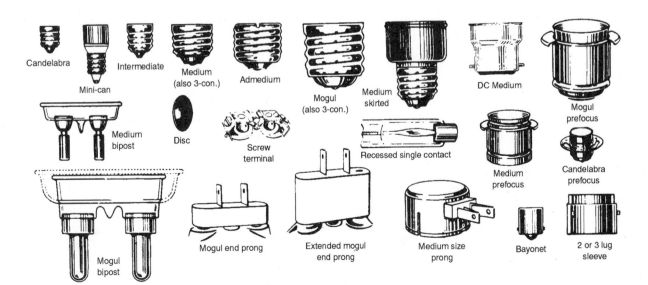

Fig. 14-3 Incandescent lamp bases.

with the correct rating is selected since a small difference between the rating and the actual voltage has a great effect on the lamp life and lumen output, Figure 14-4.

Wattage.   Lamps are usually selected according to their wattage rating. This rating is an indication of the consumption of electrical energy but is not a true measure of light output. For example, at the rated voltage, a 60-watt lamp produces 840 lumens and a 300-watt lamp produces 6000 lumens; therefore, one 300-watt lamp produces more light than seven 60-watt lamps.

Shapes.   Figure 14-5 illustrates the common lamp configurations and their letter designations.

Size.   Lamp size is usually indicated in eighths of an inch and is the diameter of the lamp at the widest place. Thus, the lamp designation A19 means that the lamp has an arbitrary shape and is $^{19}/_8$ inches or $2^3/_8$ inches (60.3 mm) in diameter, Figure 14-6.

Operation

The light-producing filament in an incandescent lamp is a resistance load that is heated to a high temperature by the electric current through the

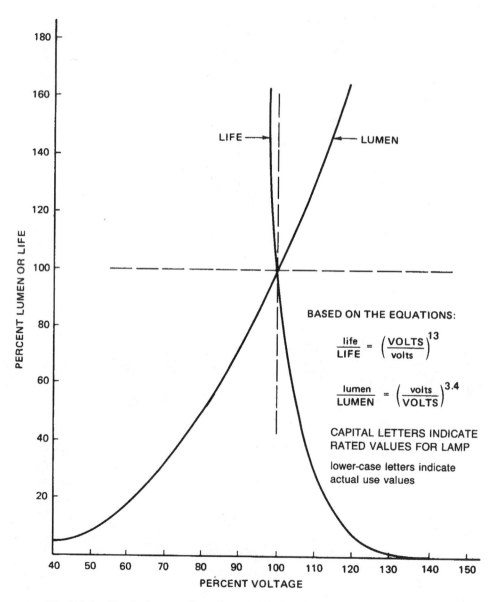

BASED ON THE EQUATIONS:

$$\frac{life}{LIFE} = \left(\frac{VOLTS}{volts}\right)^{13}$$

$$\frac{lumen}{LUMEN} = \left(\frac{volts}{VOLTS}\right)^{3.4}$$

CAPITAL LETTERS INDICATE RATED VALUES FOR LAMP

lower-case letters indicate actual use values

Fig. 14-4   Typical operating characteristics of an incandescent lamp.

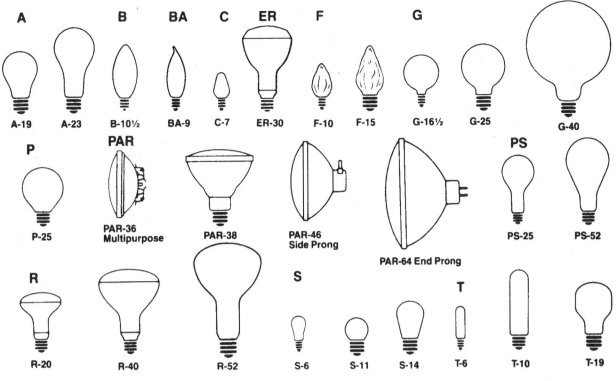

Fig. 14-5   Incandescent lamps.

lamp. This filament is usually made of tungsten which has a melting point of 3655 K. At this temperature, the tungsten filament produces light with an efficacy of 53 lumens per watt. However, to increase the life of the lamp, the operating temperature is lowered, which also means a lower efficacy. For example, if a 500-watt lamp filament is heated to a temperature of 3000 K, the resulting efficacy is 21 lumens per watt.

Catalog Designations

Catalog designations for incandescent lamps usually consist of the lamp wattage followed by the shape and ending with the diameter and other special designations as appropriate. Common examples are:

60A19   60 watt, arbitrary shape, 19/8 inches diameter

60PAR/HIR/WFL55°   60 watt, parabolic reflector, halogen, wide flood with 55° spread

LOW-VOLTAGE INCANDESCENT LAMPS

In recent years, low-voltage (usually 12-volt) incandescent lamps have become very popular for accent lighting. Many of these lamps are tungsten

←———— 19/8″ (60.3 mm) ————→

Fig. 14-6   An A19 lamp.

halogen lamps. They have a very small source size, as shown in Figure 14-7. This feature allows very precise control of the light beam. A popular size is the MR16, which has a reflector diameter of just two inches. These lamps provide a whiter light than do regular incandescent lamps. When dimming tungsten halogen lamps, a special dimmer is required because of the transformer that is installed to reduce the voltage. The dimmer is installed in the line voltage circuit supplying the transformer. Dimmed lamps will darken if they are not occasionally operated at full voltage.

Catalog Designations

Catalog designations for low-voltage incandescent lamps are similar to other incandescent lamps except for some special cases such as:

MR16   mirrored reflector, $^{16}/_8$ inches
              diameter

FLUORESCENT LAMPS

Luminaires using fluorescent lamps are considered to be electric-discharge lighting. Fluorescent lighting has the advantages of a high efficacy and long life.

Construction

A fluorescent lamp consists of a glass bulb with

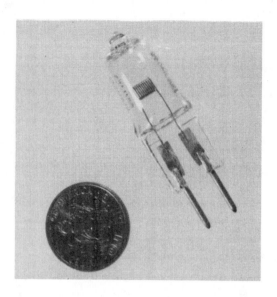

Fig. 14-7   A 100-watt, 12-volt tungsten halogen lamp.

an electrode and a base at each end, Figure 14-8. The inside of the bulb is coated with a phosphor (a fluorescing material), the air is evacuated, and an inert gas plus a small quantity of mercury is released into the bulb. The base styles for fluorescent lamps are shown in Figure 14-9.

Characteristics

Fluorescent lamps are classified according to type, length or wattage, shape, and color. See Figure 14-10.

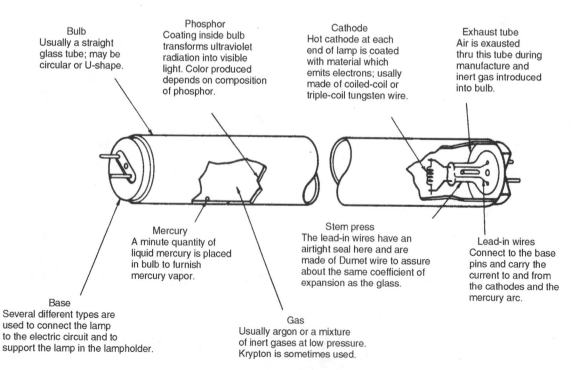

Bulb
Usually a straight glass tube; may be circular or U-shape.

Phosphor
Coating inside bulb transforms ultraviolet radiation into visible light. Color produced depends on composition of phosphor.

Cathode
Hot cathode at each end of lamp is coated with material which emits electrons; usally made of coiled-coil or triple-coil tungsten wire.

Exhaust tube
Air is exausted thru this tube during manufacture and inert gas introduced into bulb.

Mercury
A minute quantity of liquid mercury is placed in bulb to furnish mercury vapor.

Stem press
The lead-in wires have an airtight seal here and are made of Dumet wire to assure about the same coefficient of expansion as the glass.

Lead-in wires
Connect to the base pins and carry the current to and from the cathodes and the mercury arc.

Base
Several different types are used to connect the lamp to the electric circuit and to support the lamp in the lampholder.

Gas
Usually argon or a mixture of inert gases at low pressure. Krypton is sometimes used.

Fig. 14-8   Basic parts of a typical hot cathode fluorescent lamp.

Type. The lamps may be preheat, rapid start, or instant start depending upon the ballast circuit.

Length or Wattage. Depending on the lamp type, either the length or the wattage is designated. For example, both the F40 preheat and the F48 instant start are 40-watt lamps, 48 inches (1.22 m) long. The bases of these two lamps are different, however.

Shapes. The fluorescent lamp usually has a straight tubular shape. Exceptions are the circline lamp, which forms a complete circle, and the U-shaped lamp, which is an F40T12 lamp having a 180° bend in the center to fit a 2-foot (610 mm) long luminaire, and the PL lamp, which has two parallel tubes with a short connecting bridge at the ends opposite the base.

Catalog Designation and Color

An earlier reference was made to federal energy legislation; this legislation also affects the manufacturing of fluorescent lamps. The popular cool white and warm white lamps will no longer be available. Instead, a series of lamps with superior color and efficiency will be available as a substitution. The color of the light to be produced is usually indicated by the Kelvin color temperature. A lamp with a 30 in its nomenclature indicates it produces light with a color temperature of 3000 K, which would provide good rendition to warm colors. Lamps with a 50 or 65 would provide good rendition to cool colors.

Common examples of recommended lamps are:

F40/41U/RS   A substitute for the F40CW, it is a T12, 34-watt lamp producing 2900 initial lumen with a CRI of 85 at a color temperature of 4100 K.

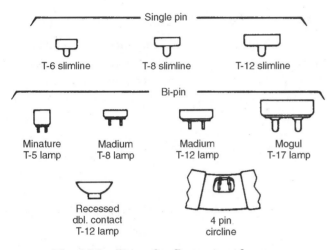

Fig. 14-9 Bases for fluorescent lamps.

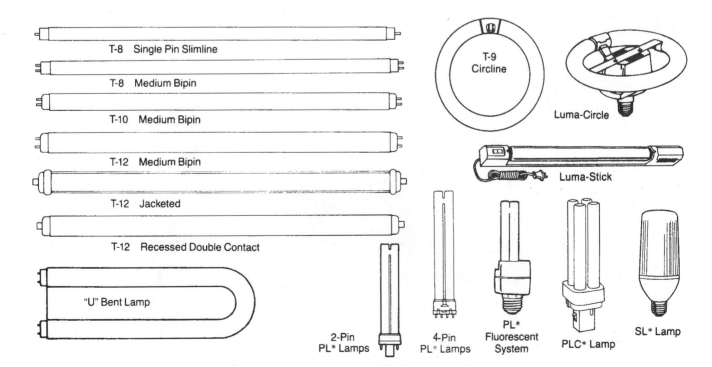

Fig. 14-10 Fluorescent lamps.

F96T12/SPEC30/HO   A substitute for the F96T12/HO/CW, it is a 95-watt lamp, producing 8350 initial lumen with a CRI of 70 at a color temperature of 4100 K.

PL*18/27   Compact fluorescent, 18 watts, 2700 K color temperature

Operation

If a substance is exposed to such rays as ultraviolet and X rays and emits light as a result, then the substance is said to be fluorescing. The inside of the fluorescent lamp is coated with a phosphor material which serves as the light-emitting substance. When sufficient voltage is applied to the lamp electrodes, electrons are released. Some of these electrons travel between the electrodes to establish an electric discharge or arc through the mercury vapor in the lamp. As the electrons strike the mercury atoms, radiation is emitted by the atoms. This radiation is converted into visible light by the phosphor coating on the tube, Figure 14-11.

As the mercury atoms are ionized, the resistance of the gas is lowered. The resulting increase in current ionizes more atoms. If allowed to continue, this process will cause the lamp to destroy itself. As a result, the arc current must be limited. The standard method of limiting the arc current is to connect a reactance (ballast) in series with the lamp.

Ballasts and Ballast Circuits

Although inductive, capacitive, or resistive means can be used to ballast fluorescent lamps, the most popular ballast is an assembly of a core and coil, a capacitor, and a thermal protector installed in a metal case, Figure 14-12. Once the assembled parts are placed in the case, it is filled with a potting compound to improve the heat dissipation and reduce ballast noise. Electronic ballasts are also available for many applications. Electronic ballasts are quieter, from 25-40% more efficient, and weigh less but are more expensive and increase the nonlinear load. Nonlinear loads may require larger neutral conductors. Ballast are required for high-intensity discharge lamps. They serve two basic functions:

- to provide the proper voltage for starting and
- to control the current during operation.

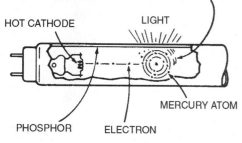

ULTRAVIOLET RADIATION STRIKES PHOSPHOR COATING CAUSING IT TO FLUORESCE

HOT CATHODE   LIGHT

MERCURY ATOM

PHOSPHOR   ELECTRON

Fig. 14-11   How light is produced in a typical hot cathode fluorescent lamp.

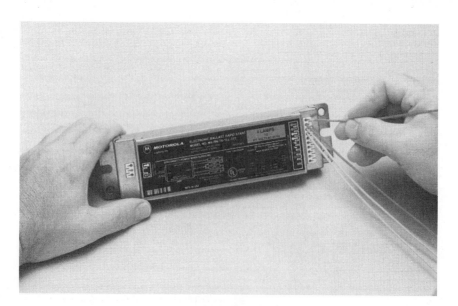

Fig. 14-12   An electronic Class P ballast. Ⓜ and Motorola are registered trademarks of Motorola Inc. All rights reserved.

The installation requirements for ballast are enumerated in *NEC® Article 410, Part P*.

Preheat Circuit. The first fluorescent lamps developed were of the preheat type and required a starter in the circuit. This type of lamp is now obsolete and is seldom found except in smaller sizes which may be used for items such as desk lamps. The starter serves as a switch and closes the circuit until the cathodes are hot enough. The starter then opens the circuit and the lamp lights. The cathode temperature is maintained by the heat of the arc after the starter opens. Note in Figure 14-13 that the ballast is in series with the lamp and acts as a choke to limit the current through the lamp.

Rapid-Start Circuit. In the rapid-start circuit, the cathodes are heated continuously by a separate winding in the ballast, Figure 14-14, with the result that almost instantaneous starting is possible. This type of fluorescent lamp requires the installation of a continuous grounded metal strip within an inch of the lamp. The metal wiring channel or the reflector of the luminaire can serve as this grounded strip. The standard rapid-start circuit operates with a lamp

current of 430 mA. Two variations of the basic circuit are available; the high-output (HO) circuit operates with a lamp current of 800 mA and the very high-output circuit has 1500 mA of current. Although high-current lamps are not as efficacious as the standard lamp, they do provide a greater concentration of light, thus reducing the required number of luminaires.

Instant-Start Circuit. The lamp cathodes in the instant-start circuit are not preheated. Sufficient voltage is applied across the cathodes to create an instantaneous arc, Figure 14-15. As in the preheat circuit, the cathodes are heated during lamp operation by the arc. The instant-start lamps require single-pin bases, Figure 14-16, and are generally called *slimline lamps*. Bi-pin base fluorescent lamps

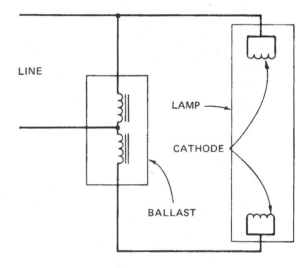

Fig. 14-15   Instant-start circuit.

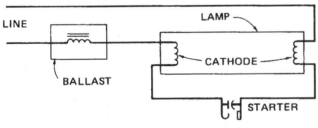

Fig. 14-13   Preheat circuit.

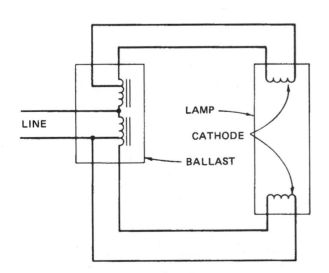

Fig. 14-14   Rapid-start circuit.

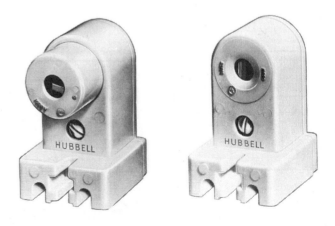

Fig. 14-16   Single-pin base for instant-start fluorescent lamp.

are available, such as the 40-watt F40T12/CW/IS lamp. For this style of lamp, the pins are shorted together so that the lamp will not operate if it is mistakenly installed in a rapid-start circuit.

Special Circuits

Most fluorescent lamps are operated by one of the circuits just covered: the preheat, rapid-start, or instant-start circuits. Variations of these circuits, however, are available for special applications.

Dimming Circuit.   The light output of a fluorescent lamp can be adjusted by maintaining a constant voltage on the cathodes and controlling the current passing through the lamp. Devices such as thyratrons, silicon-controlled rectifiers, and autotransformers can provide this type of control. The manufacturer of the ballast should be consulted about the installation instructions for dimming circuits.

Flashing Circuit.   The burning life of a fluorescent lamp is greatly reduced if the lamp is turned on and off frequently. Special ballasts are available that maintain a constant voltage on the cathodes and interrupt the arc current to provide flashing.

High-Frequency Circuit.   Fluorescent lamps operate more efficiently at frequencies above 60 hertz. The gain in efficacy varies according to the lamp size and type. However, the gain in efficacy and the lower ballast cost generally are offset by the initial cost and maintenance of the equipment necessary to generate the higher frequency.

Direct-Current Circuit.   Fluorescent lamps can be operated on a dc power system if the proper ballasts are used. A ballast for this type of system contains a current-limiting resistor that provides an inductive kick to start the lamp.

Special Ballast Designation

Special ballasts are required for installations in cold areas such as out-of-doors. Generally, these ballasts are necessary for installations in temperatures lower than 10°C (50°F). These ballasts have a higher open-circuit voltage and are marked with the minimum temperature at which they will operate properly.

Class P. Ballast   The National Fire Protection Association reports that the second most frequent cause of electrical fires in the United States is the overheating of fluorescent ballasts. To lessen this hazard, Underwriters Laboratories, Inc. has established a standard for a thermally protected ballast which is designated as a Class P ballast. This type of ballast has an internal protective device which is sensitive to the ballast temperature. This device opens the circuit to the ballast if the average ballast case temperature exceeds 90°C when operated in a 25°C ambient temperature. The requirements for the thermal protection of ballasts are established in *NEC® Section 410-73(e)*. After the ballast cools, the protective device is automatically reset. As a result, a fluorescent lamp with a Class P ballast is subject to intermittent off-on cycling when the ballast overheats.

It is possible for the internal thermal protector of a Class P ballast to fail. The failure can be in the "welded shut" mode or it can be in the "open" mode. Since the "welded shut" mode can result in overheating and possible fire, it is recommended that the ballast be protected with in-line fuses as indicated in Figure 14-17.

External fuses can be added for each ballast so that a faulty ballast can be isolated to prevent the shutdown of the entire circuit because of a

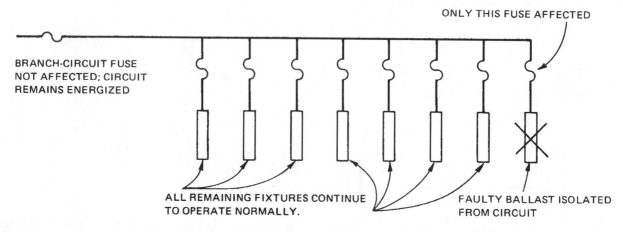

ONLY THIS FUSE AFFECTED

BRANCH-CIRCUIT FUSE
NOT AFFECTED; CIRCUIT
REMAINS ENERGIZED

ALL REMAINING FIXTURES CONTINUE
TO OPERATE NORMALLY.

FAULTY BALLAST ISOLATED
FROM CIRCUIT

Fig. 14-17   Ballast protected by fuse.

single failure, Figure 14-17. The ballast manufacturer normally provides information on the fuse type and its ampere rating. The specifications for the commercial building require that all ballasts shall be individually fused; the fuse size and type are selected according to the ballast manufacturer's recommendations, Figure 14-18.

Sound Rating. All ballasts emit a hum that is caused by magnetic vibrations in the ballast core. Ballasts are given a sound rating (from A to F) to indicate the severity of the hum. The quietest ballast has a rating of A. The need for a quiet ballast is determined by the ambient noise level of the location where the ballast is to be installed. For example, the additional cost of the A ballast is justified when

it is to be installed in a doctor's waiting room. In the bakery work area, however, a ballast with a C rating is acceptable; in a factory, the noise of an F ballast probably will not be noticed.

Power Factor. The ballast limits the current through the lamp by providing a coil with a high reactance in series with the lamp. An inductive circuit of this type has an uncorrected power factor of from 40% to 60%; however, the power factor can be corrected to within 5% of unity by the addition of a capacitor. In any installation where there are to be a large number of ballasts, it is advisable to install ballasts with a high power factor.

Compact Fluorescent Lamps

One of the newest items is the compact fluorescent lamp. These lamps usually consist of a twin tube arrangement that has a connecting bridge at the end of the tubes, Figure 14-19. Lamps are available that have two sets of tubes; these are called double twin tube or quad tube lamps, Figure 14-20.

This type of lamp has a rated life ten times that of an incandescent lamp and provides about three

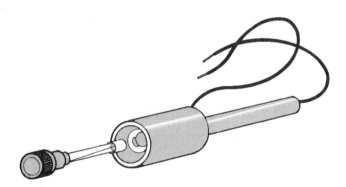

Fig. 14-18   In-line fuseholder for ballast protection.

Fig. 14-19   A 5-watt, twin tube, compact fluorescent lamp.

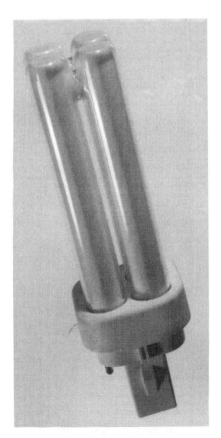

Fig. 14-20   A 10-watt, quad tube compact fluorescent lamp.

times the light per watt of power. A typical socket, along with a low-power factor ballast, is shown in Figure 14-21. Some lamps have a medium screw base, with the ballast in the base, Figure 14-22. This type of lamp can directly replace an incandescent lamp. The bases for these lamps often have a

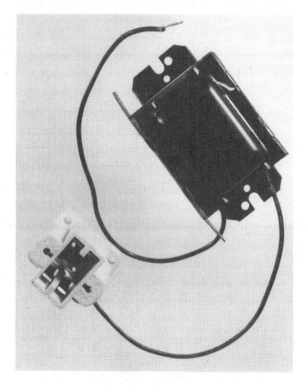

Fig. 14-21   Socket and ballast for compact fluorescent lamp.

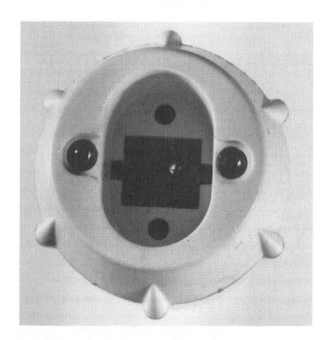

Fig. 14-22   Medium base socket with ballast for compact fluorescent lamp.

retractable pin for the base connection, Figure 14-23. This allows the positioning of the lamp.

Class E Ballast

Manufacturers of certain fluorescent ballasts are required to meet a Ballast Efficiency Factor (BEF). Initially, this applied only to F40T12 one- and two-lamp ballasts and to two-lamp ballasts for F96T12 and F96T12HO lamps. The BEF for the two-lamp F40T12, as used in the commercial building, requires that the lamp produce at a minimum 84.8 lumens per watt efficacy.

HIGH-INTENSITY DISCHARGE LAMPS (HID)

Two of the lamps in this category, mercury and metal halide, are similar in that they use mercury as an element in the light-producing process. The other HID lamp, high-pressure sodium, uses sodium in the light-producing process. In all three lamps, the light is produced in an arc tube that is enclosed in an outer glass bulb. This bulb serves to protect the arc tube from the elements and to protect the elements from the arc tube. An HID lamp will continue to give light after the bulb is broken, but it should be promptly removed from service. When the outer bulb is broken, people can be exposed to harmful ultraviolet radiation. Mercury and metal halide lamps produce a light with a strong blue

Fig. 14-23   Medium base screw-in socket with retractable pin for positioning twin tube compact fluorescent lamp.

content. The light from sodium lamps is orange in color.

Mercury Lamps

Many people consider the mercury lamps to be obsolete. They have the lowest efficacy of the HID family, which ranges from 30 lumens per watt for the smaller-wattage lamps to 65 lumens per watt for the larger-wattage lamps. Some of the positive features of mercury lamps are that they have a long life, with many lamps still functioning at 24,000 hours. With a clear bulb, they give a greenish light that makes them popular for landscape lighting.

Catalog Designations

Catalog designations vary considerably for HID lamps depending on the manufacturer. In general the designation for a mercury lamp will begin with an "H," for metal halide lamps it will begin with an "M," and either an "L" or "C" for high-pressure sodium lamps. See Figure 14-24.

MH250/C/U	metal halide, 250 watt, phosphor coated, base up (Philips)
MVR250/C/U	same as above (General Electric)
M250/C/U	same as above (Sylvania)
H38MP-100/DX	mercury, type 38 ballast, 100 watt, deluxe white (Philips)
HR100DX38/A23	same as above (General Electric)
H38AV-100/DX	same as above (Sylvania)

Metal Halide Lamps

The metal halide lamp has the disadvantages of a relatively short life and a rapid drop-off in light output as the lamp ages. Rated lamp life varies from 5000 hours to 15,000 hours. During this period the light output can be expected to drop by 30% or more. These lamps are considered to have good color-rendering characteristics and are often used in retail clothing and furniture stores. The lamp has a high efficacy rating, which ranges from 65 to 120 lumens per watt. Only a few styles are available with ratings below 175 watts. Operating position (horizontal or vertical) is critical with many of these

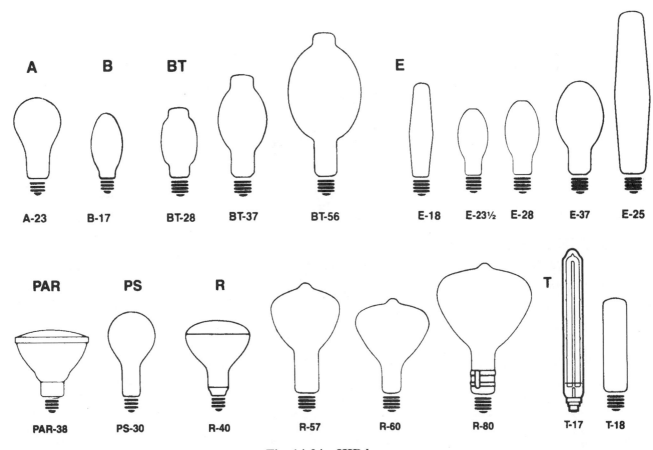

Fig. 14-24   HID lamps.

lamps and should be checked before a lamp is installed.

High-Pressure Sodium Lamps (HPS)

This type of lamp is ideal for applications in warehouses, parking lots, and similar places where color recognition is necessary but high-quality color rendition is not required. The light output is rather orange in color. The lamp has a life rating equal to or better than that of any other HID lamp and has very stable light output over the life of the lamp. The efficacy is very good, ranging as high as 140 lumens per watt.

Low-Pressure Sodium Lamps

This lamp has the highest efficacy of any of the lamps, ranging from 130 to above 180 lumens per watt. The light is monochromatic, containing energy in only a very narrow band of yellow. This lamp is usually used only in parking and storage areas where no color recognition is required. The lamp has a good life rating of 18,000 hours. It maintains a very constant light output throughout its life. The lamp is physically longer than HID lamps but generally is shorter than fluorescent lamps.

REVIEW QUESTIONS

Refer to the *National Electrical Code*® or the working drawings when necessary. Where applicable, responses should be written in complete sentences. Write units using unit names, do not use abbreviations or symbols (1 foot, not 1' or 1 ft).

Describe the lamps specified by the following lamp designations:

1. 100G40 _____
2. 150A23 _____
3. 120PAR/FL _____
4. F40SPEC35/RS _____
5. MH70 _____

Give a brief definition of each of the following:

6. Color rendering index _____
7. Efficacy _____
8. Footcandle _____
9. Illuminance _____
10. Lumen _____
11. Luminous intensity _____

Rank the following lamp types according to their efficacy. Assume that the best efficacy is used in each case. Give a 1 to the lamp with the highest efficacy and a 6 to the one with the lowest efficacy.

| 12. | _____ Filament | 14. | _____ HPS | 16. | _____ Mercury |
| 13. | _____ Fluorescent | 15. | _____ LPS | 17. | _____ Metal halide |

Define the light-producing element in each of the following lamp types.

18. Fluorescent _____
19. Incandescent _____
20. Metal halide _____

UNIT 15

Luminaires

OBJECTIVES

After studying this unit, the student will be able to

- locate luminaires in a space.
- properly select and install luminaires.
- discuss the attributes of several types of luminaires.
- select and locate a luminaire in a clothes closet.
- compute the lighting watts per square foot for a space.

DEFINITIONS

The terms "luminaire" and "lighting fixture" are used interchangeably. The Illuminating Engineering Society recommends the use of "luminaire" but the *National Electrical Code®* uses either "lighting fixture" or just "fixture." There is a danger in using the term "fixture" alone, for in different contexts it may have different meanings to the reader. In a toilet facility it is common to hear of the fixtures when this means the water closet or the lavatory. For the sake of clarity, and to follow good practice, in this text the word "luminaire" will be used except when addressing *NEC®* requirements, in which case "lighting fixture" will be used.

The following definition from the IES *Handbook* applies to either of the terms. A luminaire (lighting fixture) is a complete lighting unit consisting of a lamp or lamps together with the parts designed to distribute the light, to position and protect the lamps, and connect the lamps to the power supply.

The *NEC®* defines a lighting outlet as an outlet intended for the direct connection of a lampholder, a lighting fixture, or a pendant cord terminating in a lampholder.

INSTALLATION

The installation of luminaires is a frequent part of the work required for new building construction and for remodeling projects where customers are upgrading the illumination of their facilities. To execute work of this sort, the electrician must know how to install luminaires and, in some cases, select the luminaires.

The luminaires required for the commercial building are described in the specifications and indicated on the plans. The installation of luminaires, lighting outlets, and supports is covered in *NEC® Sections 370-23, 410-4*, and *Article 410, Parts C* and *D*. These *NEC®* Parts and Sections set forth the basic requirements for the installation of the outlets and supports in what is commonly referred to as the rough-in.

The rough-in must be completed before the ceiling material can be installed. The exact location of the luminaires is rarely dimensioned on the electrical plans and, in some remodeling situations, there are no plans. In either case the electrician should be able to rough in the outlet boxes and supports so that the luminaires will be correctly spaced in the area. If a single luminaire is to be installed in an area, the center of the area can be found by drawing diagonals from each corner, Figure 15-1. When more than one luminaire is required in an area, the procedure shown in Figure 15-2 should be followed. Uniform light distribution is achieved by spacing the luminaires so that the distances between the luminaires and between the luminaires and the walls follow these recommended spacing guides. The spacing ratios for specific luminaires are given in the data sheets published by each manufacturer. This number, usually between 0.5 and 1.5, when multiplied by the mounting height, gives the maximum distance that the luminaires may be separated and provide uniform illuminance on the work surface.

While doing luminaire layouts it maybe necessary to apply good judgement in making the final decision. Consider the following situation where uniform illuminance is desired.

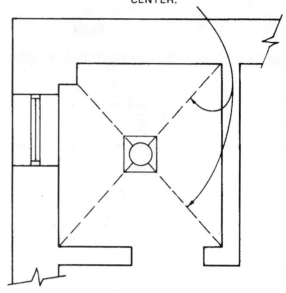

STRINGS STRETCHED FROM CORNERS TO LOCATE EXACT CENTER.

Fig. 15-1   Fixture location.

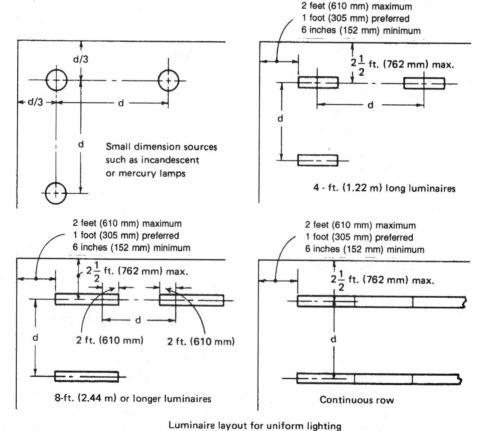

Luminaire layout for uniform lighting
Note: d should not exceed spacing ratio times mounting height. Mounting height is from luminaire to work plane for direct, semidirect, and general diffuse luminaires and from ceiling to work plane for semi-indirect and indirect luminaires.

Fig. 15-2   Fixture spacing.

- The space is 20-foot square.
- One-foot square ceiling tile will be installed.
- The floor to ceiling height is 9 feet.
- The work plane height is 2.5 feet.
- The luminaire is one-foot square
- The luminaire spacing ratio is 1.1.

The work plane to luminaire distance is:

$$9 \text{ ft} - 2.5 \text{ ft} = 6.5 \text{ ft}$$

The maximum center to center spacing distance is:

$$6.5 \text{ ft} \times 1.1 = 7.15 \text{ feet}$$

Since the ceiling is a one foot square grid the maximum spacing becomes 7 feet.

The maximum distance from the center of the luminaire to the wall should not be greater than:

$$7 \text{ ft} / 3 = 2.3 \text{ feet}$$

Working with the installer of the ceiling system it is a given that the ceiling will be installed in a uniform layout. There will be either a full or a half grid at the wall. This limits the luminaire placement at 1.5, 2 or 2.5 feet from center to wall.

Usually it will be decided, when there is an odd number of rows, that the center row of luminaires will be installed in the center of the space. This would place a 6-inch grid along each wall. For a uniform luminaire layout, the luminaires would be centered at 2.5, 10, and 17.5 feet from the wall.

This layout exceeds both the recommended spacing distance and the distance from the wall. The alternatives are to exchange the luminaires for a style with a higher spacing ratio or to install additional luminaires. This type of compromise is a common occurrence in luminaire layout where there is a grid ceiling system.

Supports

Both the lighting outlet and the luminaire must be supported from a structural member of the building. To provide this support, a large variety of clamps and clips are available, Figures 15-3 and 15-4. The selection of the type of support depends upon the way in which the building is constructed.

Surface-Mounted Luminaires

For surface-mounted and pendant-hung luminaires, the lighting outlets and supports must be roughed in so that the luminaire can be installed after the ceiling is finished. Support rods should be placed so that they extend about one inch (25.4 mm) below the finished ceiling. The support rod may be either a threaded rod or an unthreaded rod, Figure 15-5. If the luminaires are not available when the rough-in is necessary, luminaire construction information should be requested from the manufacturer. The manufacturer can provide drawings that will indicate the exact dimensions of the mounting holes in the back of the luminaire, Figure 15-6.

Recessed Luminaires

For recessed luminaires, the lighting outlet box will be located above the ceiling. This box must be accessible to the luminaire opening by means of a metal raceway that is at least four feet (1.22 mm) long, but not more than six feet (1.83 m) long.

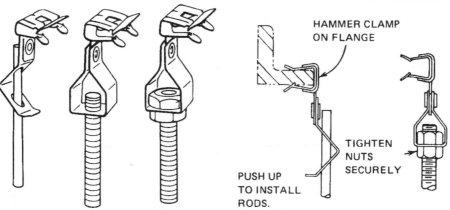

INSTALLATION INSTRUCTIONS

HAMMER CLAMP
ON FLANGE

TIGHTEN
NUTS
SECURELY

PUSH UP
TO INSTALL
RODS.

Fig. 15-3 Rod hangers for connection to flange.

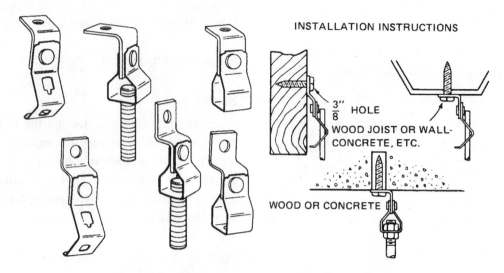

Fig. 15-4   Rod hanger supports for flat surfaces.

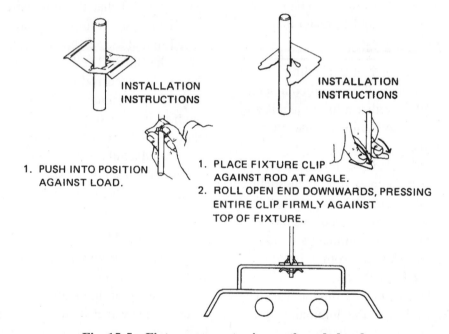

Fig. 15-5   Fixture support using unthreaded rod.

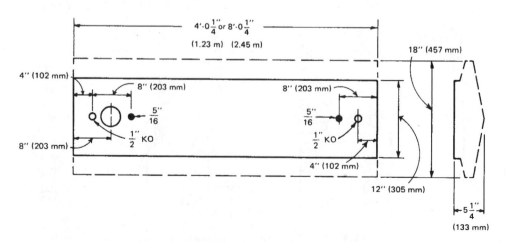

Fig. 15-6   Shop drawing indicating mounting holes for fixture.

Conductors suitable for the temperatures encountered are necessary. This information is marked on the luminaire. Branch-circuit conductors may be run directly into the junction box on listed prewired luminaires (*NEC® Section 410-67*), Figure 15-7.

Recessed luminaires are usually supported by rails installed on two sides of the rough-in opening. The rails can be heavy lather's channel or another substantial type of material.

LABELING

Always carefully read the label(s) on the luminaires. The labels will provide, as appropriate, information similar to the following:

- wall mount only
- ceiling mount only
- maximum lamp wattage
- type of lamp
- access above ceiling required
- suitable for air handling use
- for chain or hook suspension only
- suitable for operation in ambient not exceeding _____ degrees F (degrees C)
- suitable for installation in poured concrete
- for installation in poured concrete only
- for line volt-amperes, multiply lamp wattage by _____
- suitable for use in suspended ceilings
- suitable for use in uninsulated ceilings
- suitable for use in insulated ceilings
- suitable for damp locations (such as bathrooms and under eaves)
- suitable for wet locations
- suitable for use as a raceway
- suitable for mounting on low-density cellulose fiberboard

The information on these labels, together with conformance with *NEC® Article 410* should result in a safe installation.

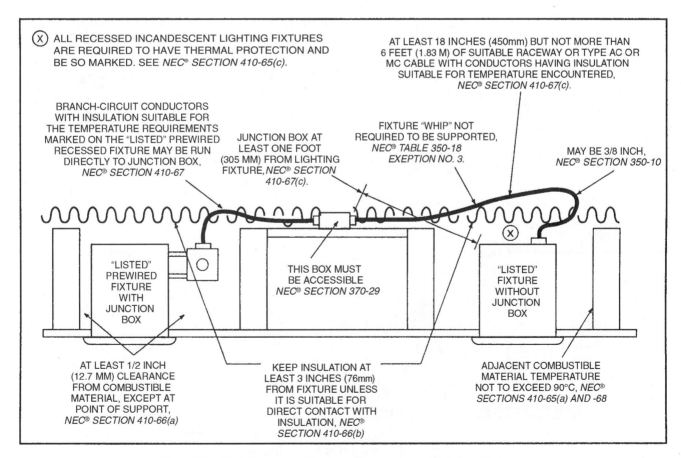

Fig. 15-7   Requirements for installing recessed lighting fixtures.

The Underwriters Laboratories *Electrical Construction Materials Directory* (known as the Green Book), the *General Information Directory* (known as the White Book), and the luminaire manufacturers catalogs and literature are also excellent sources of information on how to install luminaires properly. The Underwriters Laboratories lists, tests, and labels luminaires and publishes annual revisions to these directories.

Code Requirements for Installing Recessed Luminaires

The electrician must follow very carefully the Code requirements given in *NEC® Article 410 Parts N* and *M* for the installation and construction of recessed luminaires. Of particular importance are the Code restrictions on conductor temperature ratings, luminaire clearances from combustible materials, and maximum lamp wattage.

Recessed luminaires generate a considerable amount of heat within the enclosure and are a definite fire hazard if not wired and installed properly, Figures 15-7, 15-8, and 15-9. In addition, the excess heat will have an adverse effect on lamp and ballast life and performance.

If other than a type IC luminaire is being installed the electrician should work closely with the installer of the insulation to be sure that the clearances, as required by the *NEC*,® are followed.

To be absolutely sure that clearances are maintained, a boxlike device, Figure 15-10, is available that snaps together around a recessed luminaire, thus preventing the insulation from coming into contact

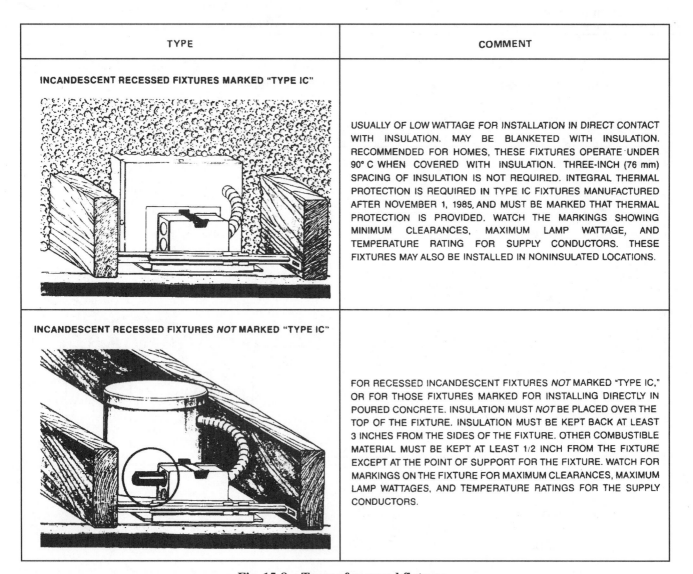

TYPE	COMMENT
INCANDESCENT RECESSED FIXTURES MARKED "TYPE IC"	USUALLY OF LOW WATTAGE FOR INSTALLATION IN DIRECT CONTACT WITH INSULATION. MAY BE BLANKETED WITH INSULATION. RECOMMENDED FOR HOMES, THESE FIXTURES OPERATE UNDER 90° C WHEN COVERED WITH INSULATION. THREE-INCH (76 mm) SPACING OF INSULATION IS NOT REQUIRED. INTEGRAL THERMAL PROTECTION IS REQUIRED IN TYPE IC FIXTURES MANUFACTURED AFTER NOVEMBER 1, 1985, AND MUST BE MARKED THAT THERMAL PROTECTION IS PROVIDED. WATCH THE MARKINGS SHOWING MINIMUM CLEARANCES, MAXIMUM LAMP WATTAGE, AND TEMPERATURE RATING FOR SUPPLY CONDUCTORS. THESE FIXTURES MAY ALSO BE INSTALLED IN NONINSULATED LOCATIONS.
INCANDESCENT RECESSED FIXTURES *NOT* MARKED "TYPE IC"	FOR RECESSED INCANDESCENT FIXTURES *NOT* MARKED "TYPE IC," OR FOR THOSE FIXTURES MARKED FOR INSTALLING DIRECTLY IN POURED CONCRETE. INSULATION MUST *NOT* BE PLACED OVER THE TOP OF THE FIXTURE. INSULATION MUST BE KEPT BACK AT LEAST 3 INCHES FROM THE SIDES OF THE FIXTURE. OTHER COMBUSTIBLE MATERIAL MUST BE KEPT AT LEAST 1/2 INCH FROM THE FIXTURE EXCEPT AT THE POINT OF SUPPORT FOR THE FIXTURE. WATCH FOR MARKINGS ON THE FIXTURE FOR MAXIMUM CLEARANCES, MAXIMUM LAMP WATTAGES, AND TEMPERATURE RATINGS FOR THE SUPPLY CONDUCTORS.

Fig. 15-8   Types of recessed fixtures.

with the luminaire, as required by the Code. The material used in these boxes is fireproof. Although this box assures *NEC®* compliance, it does violate the integrity of the ceiling insulation and the heat loss around the luminaire will be high.

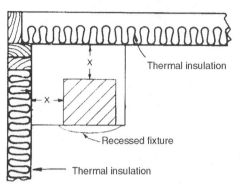

X distance at least 3 inches. Insulation above fixture must not trap heat; fixture and insulation must be installed to permit free air circulation unless the fixture is identified for installation within thermal insulation.

Fig. 15-9 Clearances for recessed lighting fixture.

Fig. 15-10 Box-like device prevents insulation from coming into contact with lighting fixture.

According to *NEC® Section 410-67(c)*, a tap conductor must be run from the fixture terminal connection to an outlet box. For this installation, the following conditions must be met:

- The conductor insulation must be suitable for the temperatures encountered.

- The outlet box must be at least one foot (305 mm) from the fixture.

- The tap conductor must be installed in a suitable raceway or cable such as Type AC or MC.

- The raceway shall be at least 18 inches (450 mm) but not more than six feet (1.83 m) long.

The branch-circuit conductors are run to the junction box. Here they are connected to conductors from the luminaire. These fixture wires have an insulation suitable for the temperature encountered at the lampholder. Locating the junction box at least one foot (305 mm) from the luminaire insures that the heat radiated from the luminaire cannot overheat the wires in the junction box. The conductors must run through at least 18 inches (450 mm) of the metal raceway (but not to exceed six feet, or 1.83 m) between the luminaire and the junction box. Thus, any heat that is being conducted in the metal raceway will be dissipated considerably before reaching the junction box. Many recessed luminaires are factory equipped with a flexible metal raceway containing high-temperature wires that meet the requirements of *NEC® Section 410-67*.

Some recessed luminaires have a box mounted on the side of the luminaire so that the branch-circuit conductors can be run directly into the box and then connected to the conductors entering the luminaire.

Additional wiring is unnecessary with these pre-wired luminaires, Figure 15-11. It is important to note that *NEC® Section 410-11* states that branch-circuit wiring shall not be passed through an outlet box that is an integral part of an incandescent luminaire unless the luminaire is identified for through wiring.

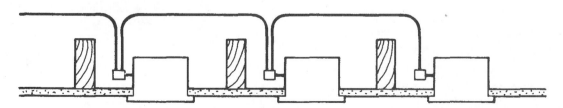

Fig. 15-11 Installation permissible with prewired lighting fixtures.

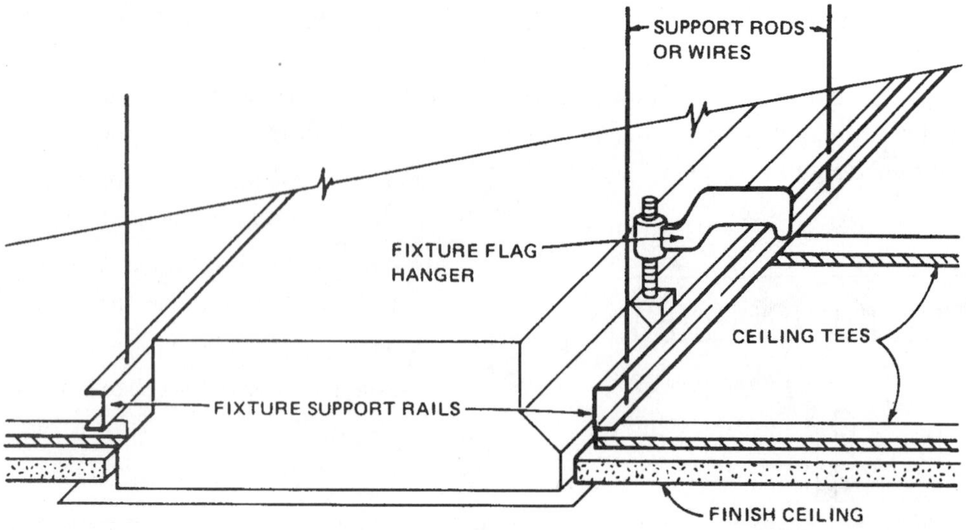

SUPPORT RODS OR WIRES

FIXTURE FLAG HANGER

CEILING TEES

FIXTURE SUPPORT RAILS

FINISH CEILING

Fig. 15-12   Recessed lighting fixture supported with flag hanger and support rails.

A lighting fixture may not be used as a raceway unless it is identified for that use, see *NEC® Section 410-31.*

If a recessed luminaire is not prewired, the electrician must check the luminaire for a label indicating what conductor insulation temperature rating is required.

Recessed luminaires are inserted into the rough-in opening and fastened in place by various devices. One type of support and fastening method for recessed luminaires is shown in Figure 15-12. The flag hanger remains against the luminaire until the screw is turned. The flag then swings into position and hooks over the support rail as the screw is tightened.

Thermal Protection

All recessed incandescent luminaires must be equipped with factory-installed thermal protection, Figure 15-7. Marking on these luminaires must indicate this thermal protection. The ONLY exceptions to the Code rule are: (1) if the luminaire is identified for use and installed in poured concrete, or (2) if the construction of the luminaire is such that temperatures would be no greater than if the lighting fixtures had thermal protection. See *NEC® Section 410-65(c)* for the exact phrasing of the exceptions.

Thermal protection, Figure 15-13, will cycle *on-off-on-off* repeatedly until the heat problem is removed. These devices are factory-installed by the manufacturer of the luminaire.

Both incandescent and fluorescent recessed luminaires are marked with the temperature ratings required for the supply conductors if over 60 degrees C. In the case of fluorescent luminaires, branch-circuit conductors within three inches of a ballast must have a temperature rating of at least 90 degrees C. All fluorescent ballasts, including replacement ballasts, installed indoors must have integral thermal protection to be in compliance with *NEC® Section 410-73(e)(1).* These thermally protected ballasts are called "Class P ballasts." This device will provide protection during normal operation, but it should not be expected to provide protection from the excessive heat that will be created by covering the luminaire with insulation. For this reason, fluorescent luminaires, just as incandescent luminaires, must have one-half inch clearance from combustible materials and three inches clearance from thermal insulation.

Because of the inherent risk of fires due to the heat problems associated with recessed luminaires, always read the markings on the luminaire and any instructions furnished with the luminaire and consult *NEC® Article 410.*

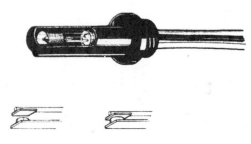

Fig. 15-13   A thermal protector.

Wiring

It was stated previously that it is very important to provide an exact rough-in for surface-mounted luminaires. *NEC® Section 410-14(b)* emphasizes this fact by requiring that the lighting outlet be accessible for inspection (without removing the surface-mounted luminaire). The installation of the outlet meets the requirements of *NEC® Section 410-14(b)* if the lighting outlet is located so that the large opening in the back of the luminaire can be placed over it, Figure 15-14.

To meet the requirements of *NEC® Section 410-31*, branch-circuit conductors with a rating of 90 degrees Celsius may be used to connect luminaires. However, these conductors must be of the single branch circuit supplying the luminaires. Therefore, all of the conductors of multiwire branch circuits can be installed as long as these conductors are the grounded and ungrounded conductors of a single system. For example, when a building has a three-phase, four-wire supply, the neutral and three hot wires, one on each phase, may be installed in a luminaire that has been approved as a raceway. This type of installation is suited to a long continuous row of luminaires, Figure 15-15.

LOADING ALLOWANCE COMPUTATIONS

The branch circuits are usually determined when the luminaire layout is completed. For incandescent luminaires, the VA allowance for each luminaire is based on the wattage rating of the luminaire. *NEC® Section 220-3(b)(4)* stipulates that recessed incandescent lighting fixtures be included at maximum rating. If an incandescent luminaire is rated at 300 watts, it must be included at 300 watts even though a smaller lamp is to be installed. For fluorescent and HID luminaires the VA allowance is based on the rating of the ballast. In the past it was a general practice to estimate this value, usually on the high side. With recent advances in ballast manufacture, and increased interest in reducing energy usage, the practice is to select a specific ballast type and base the allowance on the operation of that ballast. The contractor is required to install an "as good as," or "better than," ballast. Several types of ballasts have been discussed in a previous unit. The design VA and the watts for the luminaires selected for the commercial building are shown in Table 15-1. The following paragraphs discuss the styles of luminaires listed in Table 15-1 and scheduled to be used in the commercial building.

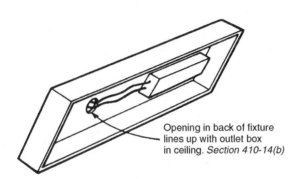

Opening in back of fixture lines up with outlet box in ceiling. *Section 410-14(b)*

Fig. 15-14   Outlet installation for surface-mounted fluorescent fixture.

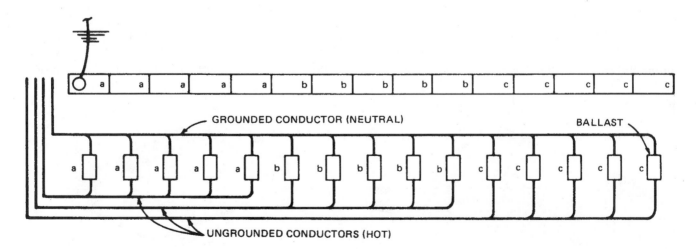

GROUNDED CONDUCTOR (NEUTRAL)

BALLAST

UNGROUNDED CONDUCTORS (HOT)

Fig. 15-15   Multi-wire circuit supplying continuous row of fluorescent fixtures.

Style	Nominal Size	Lamp	Ballast	Lens-Louver	Mounting	Description	VA	Watts
				TABLE 15-1 Luminaire – Lamp Scheduler				
A	18 inches by 4 feet	Two F40/SPEC 30/RS	Energy saving	Wrap-around	Surface		87	75
B	2 by 2 feet	Two FB40/SPEC 30/6	Energy saving	Flat opal	Surface		87	75
C	9 by 51 inches	Two F40/SPEC 30/RS	Energy saving	Clear acrylic	Surface	Enclosed, gasketed	87	75
D	Strip 4 feet long	One F48T12/CW	Standard	Luminous ceiling	Surface		74	64
E	9 inches diameter	Two 26W Quad T4	Compatible	Gold Alzak reflector	Recessed		144	74
F	2 by 4 feet	Four F032/35K	Matching electronic	24 Cell lens-louver	Recessed		132	106
G	1 by 8 feet	Four F032/35K	Matching electronic	24 Cell lens-louver	Recessed		132	106
H	9 inches by 8 feet	Two F40/SPEC 41/RS	Energy saving	Translucent acrylic	Surface		87	75
I	9 inches by 4 feet	Two F40/SPEC 41/RS	Energy saving	Translucent acrylic	Surface		87	75
J	8 inches diameter	One 150W, A21	NA	Fresnel lens	Recessed	IC rated	150	150
K	16 inches square	One 70W, HPS	Standard	Vandal resistant	Surface	With photo control	192	82
L	13 inches by 4 feet	Two F40/CW	Energy saving	None	Surface or hung		87	75
M	7 inches diameter	12V 50W NFL	Transformer	Coilex baffle	Recessed	Adjustable spot	50	50
N	20 inches long	One 60W, 120V	NA	None	Surface	Exposed lamp	60	60
O	4 feet by 20 inches	Three F40/SPEC 41/RS	Two ballasts	Low brightness lens	Recessed		143	129
P	2 feet by 20 inches	Two FB40/SPEC 30/6	One ballast	Low brightness lens	Recessed		87	75
Q	4 feet by 20 inches	Three F40/SPEC 41/RS	Two ballasts	Small cell parabolic	Recessed		143	129

Style A

The Style A luminaire is a popular fluorescent type that features a diffuser extending up the sides of the luminaire, Figure 15-16. These are often called wraparound lenses. This type of luminaire provides good ceiling lighting, which is particularly important for low ceilings. The diffuser is usually available in either an acrylic or polystyrene material. Although a polystyrene diffuser is less expensive, it will yellow as it ages, and quickly becomes unattractive. Diffusers can be specified to be made with an acrylic material that does not yellow with age. The major disadvantage of the Style A luminaire is the difficulty of locating replacements for yellow or broken diffusers. The ballast chosen for this luminaire has an A sound rating, and is of an energy-efficient style that, when operated with F40SPEC30/RS lamps, has a line current of 0.725 ampere at 120 volts (87 VA) and a power rating of 75 watts. This luminaire is used in the beauty salon.

Style B

The Style B luminaire, another popular style, has solid metal sides, Figure 15-17. The bright sides of the Style A luminaire, when used on a low ceiling, may be objectionable to people who must look at them for long periods of time. The Style B reduces this problem. The opal glass of the Style B luminaire provides a soft diffusion of the light, but any flat diffuser may be used. Glass is easily cleaned and does not experience the same aging problems encountered by the plastic materials. This style is used in the sales area of the bakery.

Style C

The Style C luminaire is used where the possible contamination of the area is an important consideration, such as in the bakery, where food is prepared. Style C luminaires are suitable for use in bakeries,

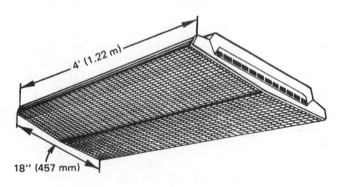

Fig. 15-16   Style A, fluorescent luminaire.

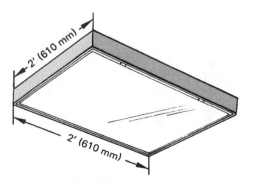

Fig. 15-17   Style B, fluorescent luminaire.

kitchens, slaughterhouses, meat markets, and food-packaging plants. The clear acrylic diffuser of this luminaire protects the area in the event of a broken lamp. At the same time, the interior of the luminaire is kept dry and free from dirt or dust, Figure 15-18. Raceways serving this luminaire may enter from the top or from either end.

Style D

Luminous ceiling systems are used where a high level of diffuse light is required. The Style D system consists of fluorescent light strips (which may be ballasted for rapid-start, high-output, or very-high-output lamps) and a ceiling suspended 18 inches (457 mm) or more below the lamps, Figure 15-19. The ceiling may be of any translucent material, but usually consists of 2 × 2 feet (610 by 610 mm), or 2 × 4 feet (0.6 by 1.22 m) panels that are easily removed for cleaning and lamp replacement. An example of this lighting system is shown in the drugstore.

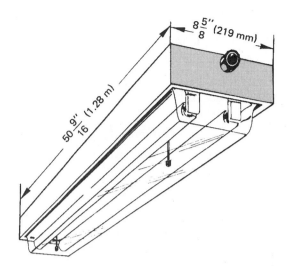

Fig. 15-18   Style C, fluorescent luminaire.

Style E

The Style E luminaire is one of a type commonly referred to as downlights or recessed cans, Figure 15-20. It requires a ceiling opening of under 9 inches in diameter and has a height of 8 inches.

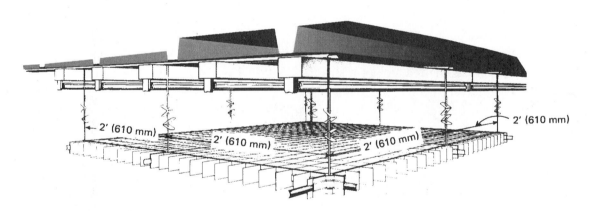

Fig. 15-19   Style D, fluorescent strip lights.

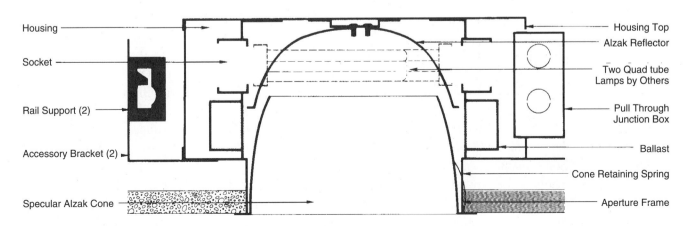

Fig. 15-20   Style E, fluorescent downlight.

This fluorescent version uses two 26-watt quad tube T4 lamps. A single unit will produce about 20 foot-candles on a surface at 9 feet distance. Each lamp has a current rating of 0.6 amperes at 120 volts (72 VA) and uses 37 watts. These are used in several locations in the interior of the commercial building. This lamp is not recommended for exterior applications because of its poor starting characteristics during cold weather.

Fig. 15-21   Style F, fluorescent luminaire.

Style F

The Style F is a 2 × 4 foot recessed fluorescent luminaire, Figure 15-21. It is commonly known as a troffer. This luminaire is equipped with four F032/35K lamps and an electronic ballast. The lamps and the ballast are a matched set, or a system. This system has been developed to maximize the ratio of light output to watts input. Compared to standard F40T12/RS lamps and a magnetic ballast, this combination provides about 160 percent of light per watt. According to the manufacturer's data, each lamp in this luminaire will produce 2900 lumen initially, achieving a lamp-ballast efficacy of 110 lumen per watt.

This type of luminaire is available with many styles of lenses and louvers. For this installation, a lens has been chosen that has the features of a lens but the appearance of a louver. A 2-inch-deep blade arrangement forms square light baffles on the surface of a lens. This gives a strong directionality to the light, concentrating the light downward to the work area. This luminaire and the similar Style G is used in the doctor's office. The lens-louver is also used in the insurance office.

Style G

The Style G luminaire is identical to the Style F luminaire except that it is 1 × 8 feet. A single ballast is used with four F032/35K lamps. This is an identical luminaire to the Style F except for the dimensions.

Style H

The Style H luminaire shown in Figure 15-22 is designed to light corridors, the narrow areas between storage shelves, and other long, narrow spaces. Style H luminaires are available in slightly different forms from various manufacturers and are

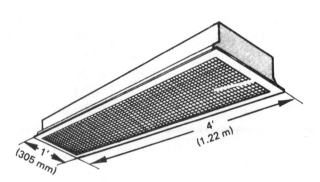

Fig. 15-22   Style H, fluorescent luminaire.

relatively inexpensive. This luminaire has been chosen for use in the second-floor corridor. It is 7 inches wide and 8 feet long. It has two F40 lamps placed in tandem (end to end). A single two-lamp ballast serves 8 feet of luminaire.

Style I

The Style I luminaire is similar to Style H except that, in the former, each four-foot luminaire has two lamps and one ballast. This is used in the drugstore where higher levels of illumination are desirable, Figure 15-23.

Style J

The Style J luminaire, Figure 15-24, is a recessed incandescent downlight luminaire that can be installed in the opening left by removing or omitting a single one-foot-square ceiling tile. This luminaire is a type IC which has been approved to be covered with insulation. It has been specified to be equipped with a fresnel lens for wide distribution of the light and a 150-watt A21 lamp. This is used in exterior locations and in the doctor's office.

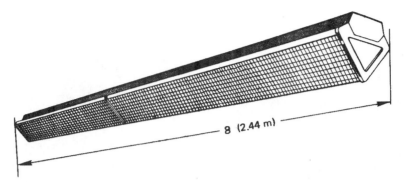

Fig. 15-23   Style I, fluorescent luminaire.

Fig. 15-24   Style J, incandescent downlight.

Style K

The Style K luminaire is used on vertical exterior walls, where it provides a light pattern that covers a large area. The Style K luminaire uses a high-intensity discharge (HID) source such as metal halide or low-pressure sodium, Figure 15-25. This luminaire provides reliable security lighting around a building. The Style K luminaire is completely weatherproof and is equipped with a photoelectric cell to turn the light off during the day.

Style L

The Style L luminaire is of open construction, as shown in Figure 15-26. This type is generally used in storage areas and other locations where it is not necessary to shield the lamps. It is often called an industrial fixture. The Style L luminaires are to be suspended from the ceiling on chains. In this type of installation, a rubber cord runs from the luminaire to a receptacle outlet on the ceiling. The

basement storage areas of both the bakery and the drugstore are good examples of the use of this type of luminaire.

Style M

The Style M luminaire is used to focus light on a specific object, Figure 15-27. This luminaire has a diameter of 7 inches and can be swiveled through 358 degrees laterally and 40 degrees from the zenith. This luminaire utilizes an MR16 lamp in sizes up to 50 watts and in a variety of light distribution patterns. The transformer is provided with the luminaire. At a distance of 4 feet, the 50-watt narrow flood lamp produces 26 footcandles. A dimmer switch, such as specified in the beauty salon, used to operate these luminaires, must be approved for use with low-voltage illumination systems. The dimmer is connected in the supply to the transformers. An ordinary dimmer will, at best, provide sporadic operation.

Fig. 15-25   Style K, HID luminaire.

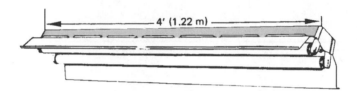

Fig. 15-26   Style L, fluorescent luminaire.

Fig. 15-27   Style M, MR-16 luminaire.

Style N

The Style N luminaire is especially designed for installations where the lamp will be exposed to viewing, Figure 15-28. This luminaire is installed on both sides of a mirror to provide excellent illumination of the face but avoiding the glare that can be a problem when conventional incandescent lamps are installed in a similar fashion. The lamp is a 20-inch linear incandescent with a 60-watt rating.

Style O

The Style O luminaire is 4 feet long and 20 inches wide. It is equipped with three F40T12/RS

Fig. 15-28   Style N, linear incandescent luminaire.

lamps and two ballasts. One of the ballasts supplies two lamps and the other one lamp. This arrangement provides for three levels of illumination in the room by using the single lamp in each luminaire, by using the pairs of lamps in each luminaire, or by using all three lamps. In each of the arrangements, the illumination is uniformly distributed throughout the room. The luminaire is fitted with a white louver and a lens. The Blade,™ which was also used in the doctor's office, provides a high-quality light in addition to having a very pleasing effect on the appearance of the room. This luminaire is used in the staff and the reception areas of the insurance office.

Style P

The Style P luminaire is similar to the Style O luminaire except that the former is only 2 feet long and has a single ballast supplying two FB40/SPEC30/6 lamps. The FB40 is a lamp that was 4 feet long but has been bent into a U-shape for use in luminaires that are only 2 feet long. This arrangement allows the use of the standard rapid-start two-lamp ballast. Two of these

luminaires are used in the staff area of the insurance office to fill in areas of the room where the Style O was too large.

Style Q

The Style Q luminaire is similar to the Style O except for the louvers. The lens specified in this room is a low-brightness lens especially designed for use in computer rooms. A high percentage of the light is directed downward to the horizontal surfaces. A minimum amount of light is produced on the computer monitor's vertical surface.

LOCATION OF LUMINAIRES IN CLOTHES CLOSETS

Clothing, boxes, and other material normally stored in clothes closets are a potential fire hazard.

These items may ignite on contact with the hot surface of an exposed light bulb. The bulb, in turn, may shatter and spray hot sparks and hot glass onto other combustible materials.

NEC® Section 410-8 addresses the special requirements for installing lighting fixtures in clothes closets. It is significant to note that these rules cover ALL clothes closets . . . residential . . . commercial . . . and industrial. In Figures 15-29 and 15-30 "A" represents the width of the storage space above the rod or 12 inches) whichever is greater. "B" represent the depth of the storage space below the rod which is 24 inches in depth and 6 feet, or to the highest rod, in height. See *NEC® Section 410-8(a)*.

WATTS PER SQUARE FOOT CALCULATIONS

In some localities it is necessary that the lighting of a building comply with an energy code. These codes usually evaluate the lighting on the basis of the average watts per square foot of lighting load. Incandescent lamps are rated in watts, so it is just a matter of counting each type of luminaire, then multiplying each by the appropriate lamp watts and adding the products. Fluorescent and HID luminaires are rated by the volt-amperes required for circuit design and the watts required to operate

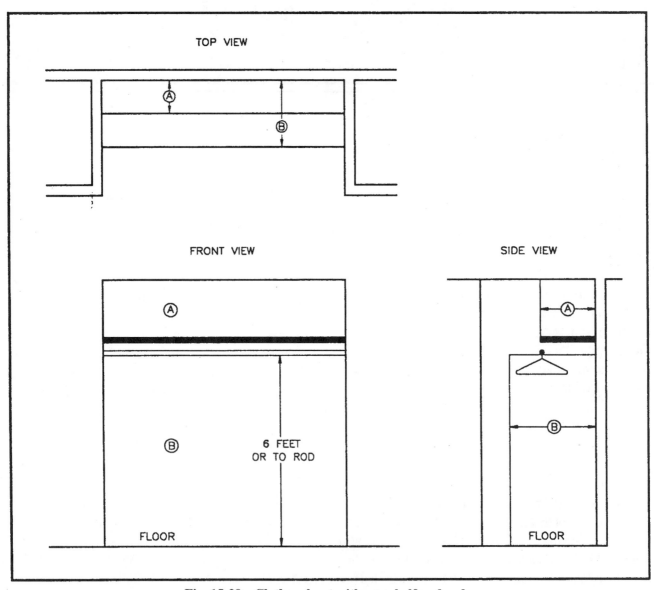

Fig. 15-29   Clothes closet with one shelf and rod.

the luminaire. The watt figure is always lower than the volt-ampere and should be used to determine the watts per square foot for the building. The difference between watts and volt-amperes is determined by the quality of the ballast, the efficiency of the luminaire, and the efficacy of the lamp. Where the quality, the efficiency, and the efficacy are high,

more light will be produced per watt and thus fewer luminaires will be required to achieve the desired lighting level at a given electrical load. After determining the total watts of lighting load, this value is divided by the area to find the watts per square foot (Figure 15-31, Figure 15-32, and Figure 15-33).

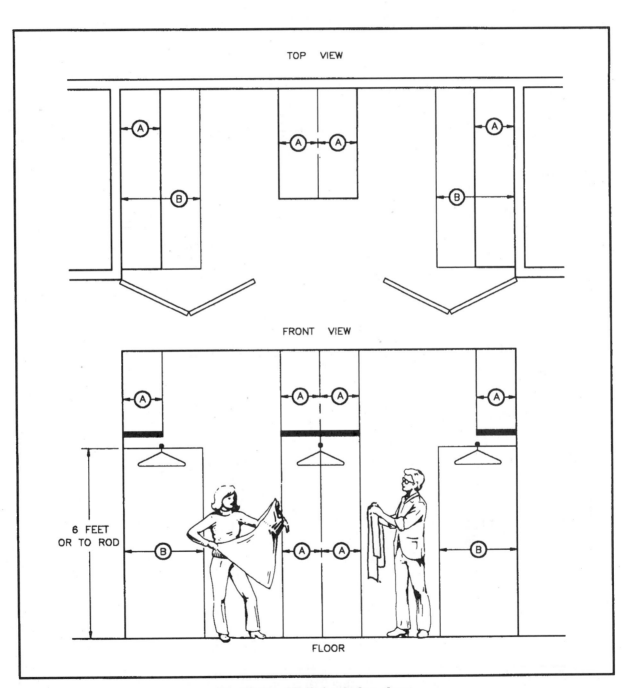

Fig. 15-30   Walk-in clothes closet.

Fig. 15-31   Recessed closet fixture with pull-chain.

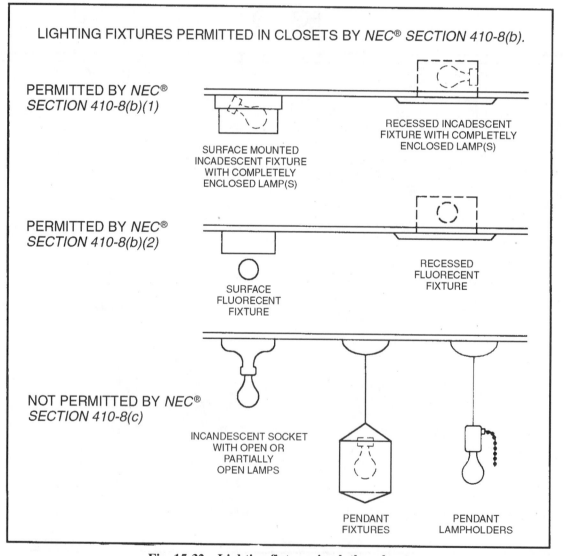

Fig. 15-32   Lighting fixtures in clothes closets.

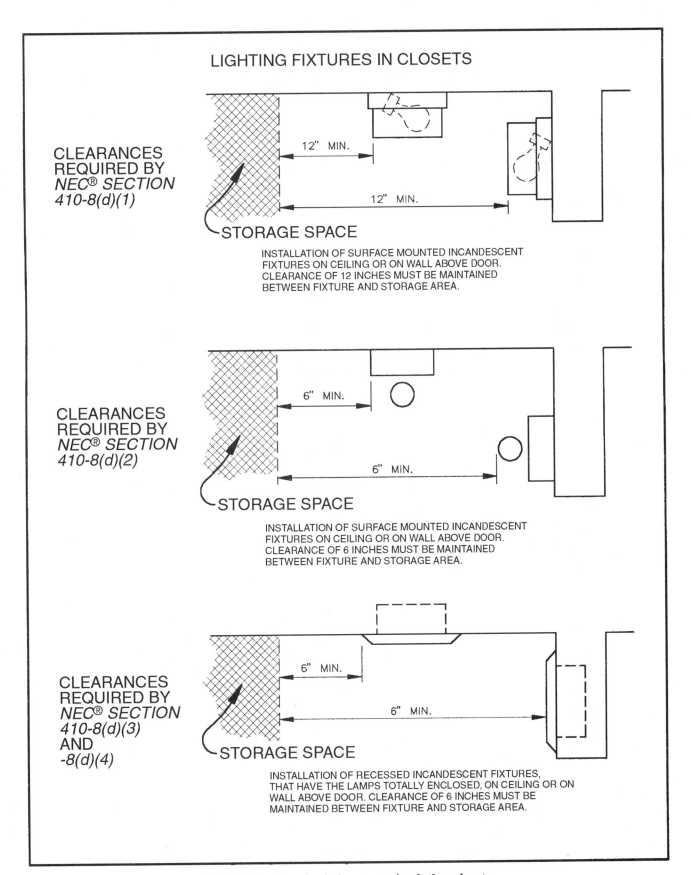

Fig. 15-33   Required clearances in clothes closets.

REVIEW QUESTIONS

Refer to the *National Electrical Code®* or the working drawings when necessary. Where applicable, responses should be written in complete sentences. Write units using unit names, use no abbreviations or symbols (1 foot, not 1' or 1 ft).

Six luminaires, similar to Style J used in the commercial building, are to be installed in a room that is 12 feet by 16 feet with a 9 feet floor to ceiling height. The spacing ratio for the luminaire is 1.0.

1. The maximum distance (in feet) that the luminaires can be separated and achieve uniform illuminance is _____.

2. Three luminaires are to be installed in the row along the long side of the room. Draw a sketch indicating center to center and center to wall distances.

Two luminaires, an 8-foot and a 4-foot luminaire, with dimensions as shown in figure 15-6, are to be installed in tandem (end to end). The end of the 8-foot luminaire is to be 2 feet from the wall.

3. The center of the outlet box should be roughed in at _____ from the wall.

4. The first support should be installed at _____ from the wall.

5. The second support should be installed at _____ from the wall.

6. The final support should be installed at _____ from the wall.

The following questions concern the lighting in the beauty salon.

7. The loading allowance for the beauty salon lighting is _____

8. The watts per square foot lighting loading for the beauty salon is_____

The following question pertains to lighting in general.

9. Of the luminaires selected for the commercial building, the style _____ would be preferred for the illumination of shelving.

The remaining questions pertain to the installation of lighting in a clothes closet. Indicate compliance or violation and give reason. In each case the lighting fixture is located on the ceiling with 10 inches clearance from the storage area.

10. A porcelain socket with a PL fluorescent lamp.   Compliance   Violation

11. A totally enclosed incandescent luminaire.   Compliance   Violation

12. A fluorescent strip luminaire (bare lamp).   Compliance   Violation

UNIT 16

Emergency And Legally Required Standby Power Systems

OBJECTIVES

After studying this unit, the student will be able to

- select and size an emergency power system.
- install an emergency power system.

Many state and local codes require that equipment be installed in public buildings to insure that electric power is provided automatically if the normal power source fails. The electrician should be aware of the special installation requirements of these emergency systems. *NEC® Articles 700* and *701* set forth the requirements for emergency systems.

SOURCES OF POWER

When the need for emergency power is confined to a definite area, such as a stairway, and the power demand in this area is low, then self-contained battery powered units are a convenient and efficient means of providing power. See *NEC® Section 700-12(e)*. In general, these units are wall-mounted and are connected to the normal source by permanent wiring methods. Under normal conditions, this regular source powers a regulated battery charger to keep the battery at full power. When the normal power fails, the circuit is completed automatically to one or more lamps that provide enough light to the area to permit its use, such as lighting a stairway sufficiently to allow people to leave the building. Battery powered units are commonly used in stairwells, hallways, shopping centers, supermarkets, and other public structures.

If the power demand is high (excluding the operation of large motors), central battery power systems are available. These systems usually operate at 32 volts and can service a large number of lamps.

SPECIAL SERVICE ARRANGEMENTS

NEC® Section 700-9(b) states that the wiring from an emergency source wiring shall be kept entirely independent of all other wiring and equipment and shall not enter the same raceway, cable, box, or cabinet with other wiring. Installing a separate service such as illustrated in Figure 16-1 meets the intent of this section.

NEC® Section 701-10 allows the wiring of a legally required standby system to occupy the same raceways, cables, boxes, and cabinets with other general wiring. Furthermore, *NEC® Section 701-11(e)* permits the connection for this system to be made ahead of, *but not within*, the main service disconnect. Figure 16-2 illustrates this arrangement. The connections could be made in a junction box, or to the service conductors outside of the building. Because the size of the taps conductors would be much smaller than the service conductors, cable limiters are installed at the point of the tap. These will limit the available fault-current to withstand the rating of the tap conductors and to the interrupting rating of the disconnect device.

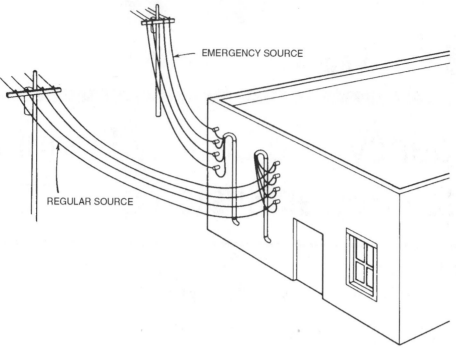

Fig. 16-1   Separate services.

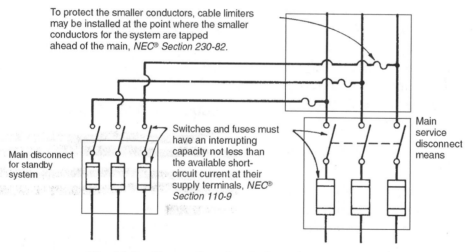

To protect the smaller conductors, cable limiters may be installed at the point where the smaller conductors for the system are tapped ahead of the main, *NEC® Section 230-82.*

Main disconnect for standby system

Switches and fuses must have an interrupting capacity not less than the available short-circuit current at their supply terminals, *NEC® Section 110-9*

Main service disconnect means

Fig. 16-2   Connection ahead of service disconnect means.

EMERGENCY GENERATOR SOURCE

Generator sources may be used to supply emergency power. See *NEC® Section 700-12(b)*. The selection of such a source for a specific installation involves a consideration of the following factors:

- the engine type.
- the generator capacity.
- the load transfer controls.

A typical generator for emergency use is shown in Figure 16-3.

Engine Types and Fuels

The type of fuel to be used in the driving engine of a generator is an important consideration in the installation of the system. Fuels which may be used are LP gas, natural gas, gasoline, or diesel fuel. Factors affecting the selection of the fuel to be used include the availability of the fuel and local regulations governing the storage of the fuel. Natural gas and gasoline engines differ only in the method of supplying the fuel; therefore, in some installations, one of these fuels may be used as a standby alternative for the other fuel, Figure 16-4.

Fig. 16-3 Generator for emergency power supply.

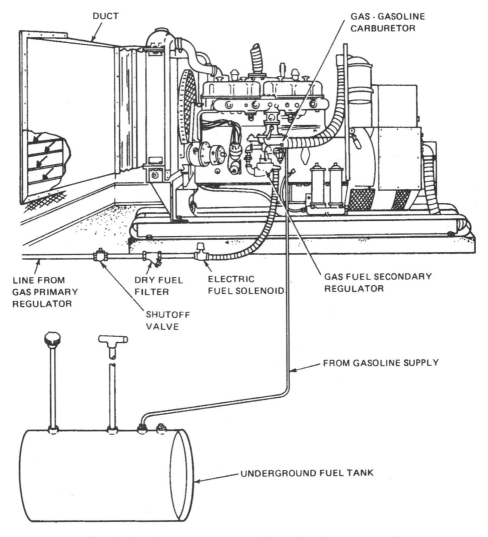

Fig. 16-4 Generator powered by natural gas or gasoline.

An emergency power source that uses gasoline and/or natural gas usually has lower installation and operating costs than a diesel-powered source. However, the problems of fuel storage can be a deciding factor in the selection of the type of emergency generator. Gasoline is not only dangerous, but becomes stale after a relatively short period of time and thus cannot be stored for long periods. If natural gas is used, the Btu content must be greater than 1100 Btu per cubic foot. Diesel-powered generators require less maintenance and have a longer life. This type of diesel system is usually selected for installations having large power requirements since the initial costs closely approach the costs of systems using other fuel types.

Cooling

Smaller generator units are available with either air or liquid cooling systems. Units having a capacity greater than 15 kW generally use liquid cooling. For air-cooled units, it is recommended that the heated air be exhausted to the outside. In addition, a provision should be made to bring in fresh air so that the room where the generator is installed can be kept from becoming excessively hot. Typical installations of air-cooled emergency generator systems are shown in Figures 16-5 and 16-6.

Generator Voltage Characteristics

Generators having any required voltage characteristic are available. A critical factor in the selection of a generator for a particular application is that it must have the same voltage output as the normal building supply system. For the commercial building covered in this text, the generator selected provides 208Y/120-volt, three-phase, 60-Hz power.

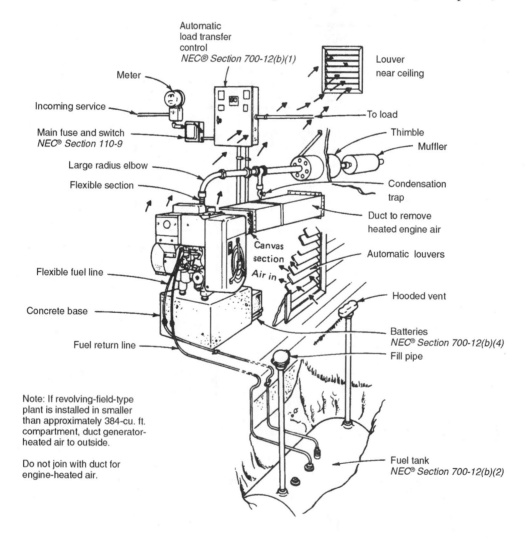

Fig. 16-5   Small generator installation.

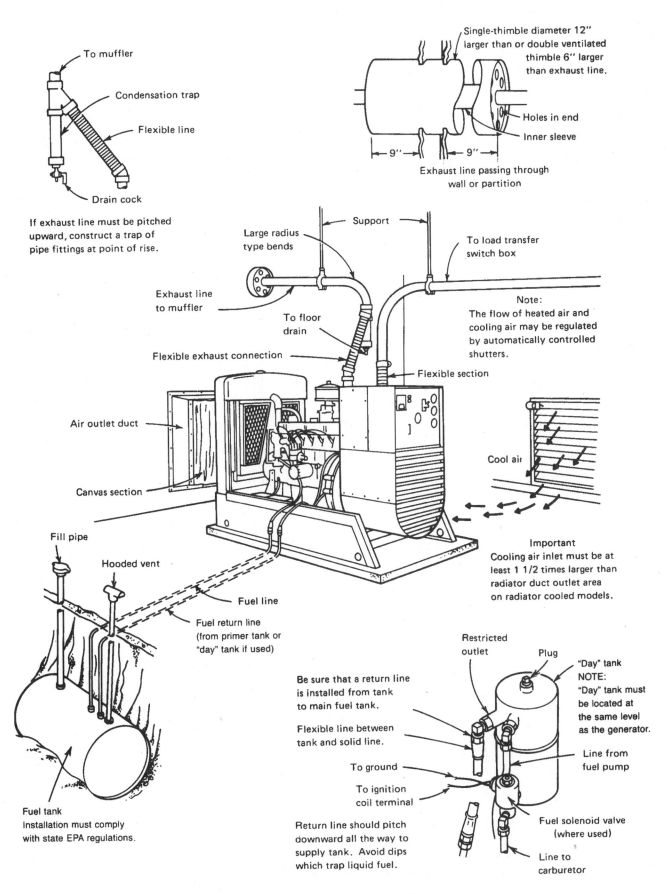

To muffler

Condensation trap

Flexible line

Drain cock

If exhaust line must be pitched upward, construct a trap of pipe fittings at point of rise.

Single-thimble diameter 12" larger than or double ventilated thimble 6" larger than exhaust line.

Holes in end

Inner sleeve

9"   9"

Exhaust line passing through wall or partition

Support

Large radius type bends

To load transfer switch box

Exhaust line to muffler

To floor drain

Note:
The flow of heated air and cooling air may be regulated by automatically controlled shutters.

Flexible exhaust connection

Flexible section

Air outlet duct

Canvas section

Cool air

Important
Cooling air inlet must be at least 1 1/2 times larger than radiator duct outlet area on radiator cooled models.

Fill pipe

Hooded vent

Fuel line

Fuel return line (from primer tank or "day" tank if used)

Fuel tank
Installation must comply with state EPA regulations.

Be sure that a return line is installed from tank to main fuel tank.

Flexible line between tank and solid line.

To ground

To ignition coil terminal

Return line should pitch downward all the way to supply tank. Avoid dips which trap liquid fuel.

Restricted outlet

Plug

"Day" tank
NOTE:
"Day" tank must be located at the same level as the generator.

Line from fuel pump

Fuel solenoid valve (where used)

Line to carburetor

Fig. 16-6   Large generator installation.

Capacity

It is an involved, but extremely important, task to determine the correct size of the engine-driven power system so that it has the minimum capacity necessary to supply the selected equipment. If the system is oversized, additional costs are involved in the installation, operation, and maintenance of the system. However, an undersized system may fail at the critical period when it is being relied upon to provide emergency power. To size the emergency system, it is necessary to determine initially all of the equipment to be supplied with emergency power. If motors are to be included in the emergency system, then it must be determined if all of the motors can be started at the same time. This information is essential to insure that the system has the capacity to provide the total starting inrush kVA required.

The starting kVA for a motor is equivalent to the locked rotor kVA of the motor. This value is determined by selecting the appropriate value from *NEC® Table 430-7(b)* and then multiplying this value by the horsepower. The locked rotor kVA value is independent of the voltage characteristics of the motor. Thus, a 5-hp, code E motor requires a generator capacity of:

$$5 \text{ hp} \times 4.99 \text{ kVA/hp} = 24.95 \text{ kVA}$$

If two motors are to be started at the same time, the emergency power system must have the capacity to provide the sum of the starting kVA values for the two motors.

For a single-phase motor rated at less than ½ hp, Table 16-1 lists the approximate kVA values that may be used if exact information is not available. The power system for the commercial building must supply the following maximum kVA:

Five ⅙ hp C-type motors	
5 × 1.85 kVA	= 9.25 kVA
One ½ hp Code L motor	
½ × 9.99 kVA	= 4.99 kVA
Lighting and receptacle load from emergency panel schedule	= 4.51 kVA
Total	18.75 kVA

Thus, the generator unit selected must be able to supply this maximum kVA load as well as the continuous kVA requirement for the commercial building.

Continuous kVA requirement:

Lighting and receptacle load	4.510 kVA
Motor load	3.763 kVA
Total	8.273 kVA

A check of manufacturers' data shows that a 12-kVA unit is available having a 20-kVA maximum

Table 430-7(b). Locked-Rotor Indicating Code Letters

Code Letter	Kilovolt-Amperes per Horsepower with Locked Rotor	Code Letter	Kilovolt-Amperes per Horsepower with Locked Rotor
A	0–3.14	L	9.0–9.99
B	3.15–3.54	M	10.0–11.19
C	3.55–3.99	N	11.2–12.49
D	4.0–4.49	P	12.5–13.99
E	4.5–4.99	R	14.0–15.99
F	5.0–5.59	S	16.0–17.00
G	5.6–6.29	T	18.0–19.99
H	6.3–7.09	U	29.0–22.39
J	7.1–7.99	V	22.4 and up
K	8.0–8.99		

TABLE 16-1 Approximate kVA Values		
HP	Type	Locked Rotor kVA
1/6	C	1.85
1/6	S	2.15
1/4	C	2.50
1/4	S	2.55
1/3	C	3.0
1/3	S	3.25

C = Capacitor-start motor
S = Split-phase motor

rating for motor starting purposes and a 12-kVA continuous rating. A smaller generator may be installed if provisions, such as time delays on the contactors, can be made to prevent the motors from starting at the same time.

Derangement Signals

According to *NEC® Section 700-7*, a *derangement signal*, a signal device having both audible and visual alarms, shall be installed outside the generator room in a location where it can be readily and regularly observed. The purposes of a device such as the one shown in Figure 16-7 are to indicate: any malfunction in the generator unit, any load on the system, or the correct operation of a battery charger.

Automatic Transfer Equipment

If the main power source fails, equipment must be provided to start the engine of the emergency generator and transfer the supply connection from the regular source to the emergency source, see *NEC® Section 700-6*. These operations can be accomplished by a control panel such as the one shown in Figure 16-8. A voltage-sensitive relay is connected to the main power source. This relay (transfer switch) activates the control cycle when the main source voltage fails, Figure 16-9. The generator motor is started when the control cycle is activated. As soon as the motor reaches the correct speed, a set of contactors is energized to disconnect the load from its normal source and connect it to the generator output.

Wiring

The branch-circuit wiring of emergency systems must be separated from the standard system except for the special conditions noted. Figure 16-10 shows a typical branch-circuit installation for an emergency system. Key-operated switches, Figure 16-11, are installed to prevent unauthorized personnel from operating the lights.

Under certain conditions, emergency circuits are permitted to enter the same junction box as normal circuits, *NEC® Section 700-9(b)*. Figure 16-12 shows an exit light that contains two lamps: one lamp connected to the normal circuit, and the other connected to the emergency circuit.

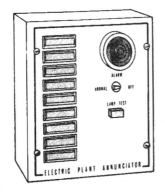

Fig. 16-7   Audible and visual signal alarm annunciator.

Fig. 16-8   Automatic transfer control panel.

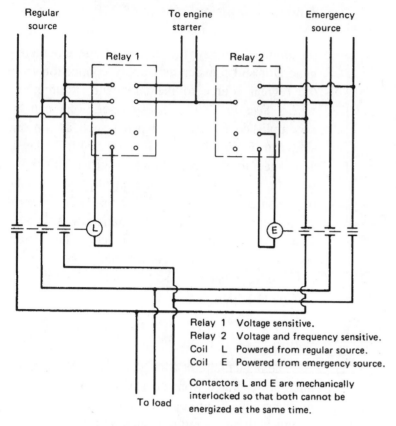

Relay 1 Voltage sensitive.
Relay 2 Voltage and frequency sensitive.
Coil L Powered from regular source.
Coil E Powered from emergency source.

Contactors L and E are mechanically
interlocked so that both cannot be
energized at the same time.

Fig. 16-9   Transfer switch circuitry.

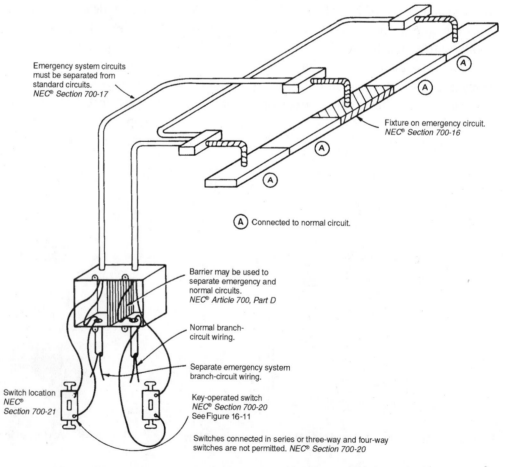

Fig. 16-10   Branch-circuit wiring for emergency and other systems.

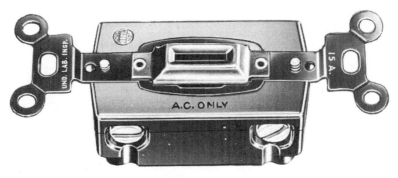

Fig. 16-11   Key-operated switch for emergency lighting control.

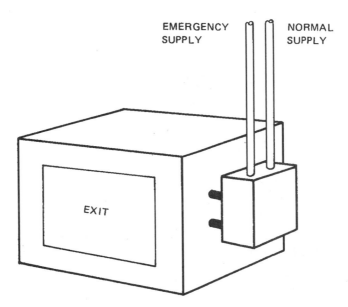

Fig. 16-12   Wiring for emergency and other systems.

REVIEW QUESTIONS

Refer to the *National Electrical Code®* or the working drawings when necessary. Where applicable, responses should be written in complete sentences. Write units using unit names, use no abbreviations or symbols (1 foot, not 1' or 1 ft).

There are three options for providing emergency electrical power to building systems. The options are listed below. For each discuss giving the advantages and disadvantages.

1. A local source of power such as a battery, or a motor-generator.

2. A separate service.

_____

_____

_____

_____

_____

3. A connection ahead of the service main.

_____

_____

_____

_____

_____

Make the following calculations and cite *NEC*® references.

4. The minimum starting kVA for 7½-horsepower, 3-phase, 230-volt, code F motor is:

5. The maximum starting kVA for a 7½-horsepower, 3-phase, 230-volt, code F motor is:

6. The minimum kVA of a generator that will be required to start, simultaneously, two 7½-horsepower, 3-phase, 230-volt, code F motors is:

7. The minimum kVA of a generator that will be required to start two, selectively controlled, 7½-horsepower, 3-phase, 230-volt, code F motors is:

Respond as requested.

8. List the special conditions for emergency circuit wiring that an electrician should know and those items that the occupant/owner should know.

_____

_____

_____

_____

_____

UNIT 17

Overcurrent Protection: Fuses and Circuit Breakers

OBJECTIVES

After studying this unit, the student will be able to

- list and identify the types, classes, and ratings of fuses and circuit breakers.
- describe the operation of fuses and circuit breakers.
- develop an understanding of switch sizes, ratings, and requirements.
- define interrupting rating, short-circuit currents, I^2t, I_p, RMS, and current limitation.
- apply the *National Electrical Code®* to the selection and installation of overcurrent protective devices.
- use the time-current characteristics curves and peak let-through charts.

Overcurrent protection is one of the most important components of an electrical system. The overcurrent device opens an electrical circuit whenever an overload or short circuit occurs. Overcurrent devices in an electrical circuit may be compared to pressure relief valves on a boiler. If dangerously high pressures develop within a boiler, the pressure relief valve opens to relieve the high pressure. In a similar manner, the overcurrent device in an electrical system also acts as a "safety valve."

NEC® Article 240 sets forth the requirements for overcurrent protection. *NEC® Section 240-1* states that overcurrent protection for conductors and equipment is provided to open the circuit if the current reaches a value that will cause an excessive or dangerous temperature in conductors or conductor insulation. *NEC® Sections 110-9,* and *-10* sets forth requirements for interrupting rating and protection against fault current.

Two types of overcurrent protective devices are commonly used: fuses and circuit breakers. The Underwriters Laboratories, Inc. (UL) and the National Electrical Manufacturers Association (NEMA) establish standards for the ratings, types, classifications, and testing procedures for fuses and circuit breakers.

As indicated in *NEC® Section 240-6*, the standard ampere ratings for fuses and nonadjustable circuit breakers are 15, 20, 25, 30, 35, 40, 45, 50, 60, 70, 80, 90, 100, 110, 125, 150, 175, 200, 225, 250, 300, 350, 400, 450, 500, 600, 700, 800, 1000, 1200, 1600, 2000, 2500, 3000, 4000, 5000, and 6000. This *NEC®* section lists additional standard ratings for fuses as 1, 3, 6, 10, and 601 amperes.

Why does the *NEC®* list "standard ampere ratings?" There are many Sections in the Code where permission is given to use "the next standard higher ampere rating overcurrent device" . . . or a requirement to use "a lower than standard ampere rating overcurrent device." For example, in *NEC®* Section 240-3(b), we find that if the ampacity of a conductor does not match a standard ampere rating fuse or breaker, then it is allowable to use the "next standard higher." Therefore, for a conductor that has an ampacity of 115 amperes, the Code permits the overcurrent protection to be sized at 125 amperes.

When protecting conductors where the overcurrent protection is above 800 amperes, we must round down to a "lower than standard ampere rating overcurrent device." An example of this would be if three 500 kcmil copper conductors were installed in parallel. Each has an ampacity of 380 amperes. Therefore,

$$380 \times 3 = 1140 \text{ amperes.}$$

It would be a Code violation to protect these conductors with a 1200 ampere fuse. The correct thing to do is to round down to an ampere rating less than 1200 amperes. Fuse manufacturers provide 1100 ampere rated fuses.

DISCONNECT SWITCHES

Fused switches are available in ratings of 30, 60, 100, 200, 400, 600, 800, 1200, 1600, 2000, 2500, 3000, 5000, and 6000 amperes in both 250 and 600 volts. They are for use with copper conductors unless marked to indicate that the terminals are suitable for use with aluminum conductors. The switches rating is based on 60°C wire (No. 14 through 1 AWG) and 75°C for wires No. 1/0 and larger, unless otherwise marked. See *NEC® Section 110-14(c)*. Switches also may be equipped with ground-fault sensing devices. These will be clearly marked as such.

Switches may have labels that indicate their intended application, such as "Continuous Load Current Not To Exceed 80% Of The Rating Of The Fuses Employed In Other Than Motor Circuits."

- Switches intended for isolating use only are marked "For Isolation Use Only — Do Not Open Under Load."

- Switches that are suitable for use as service switches are marked "Suitable For Use As Service Equipment."

- When a switch is marked "Motor-Circuit Switch," it is only for use in motor circuits.

- Switches of higher quality may have markings such as: "Suitable For Use On A Circuit Capable Of Delivering Not More Than 100,000 RMS (root mean squared) Symmetrical Amperes, 600 Volts Maximum: Use Class T Fuses Having An Interrupting Rating Of No Less Than The Maximum Available Short Circuit Current Of The Circuit."

- Enclosed switches with horsepower ratings in addition to ampere ratings are suitable for use in motor circuits as well as for general use.

- Some switches have a dual-horsepower rating. The larger horsepower rating is applicable when using dual-element, time-delay fuses. Dual-element, time-delay fuses are discussed later on in this unit.

- Fusible bolted pressure contact switches are tested for use at 100% of their current rating, at 600 volts ac and are marked for use on systems having available fault currents of 100,000, 150,000, and 200,000 RMS symmetrical amperes.

For more data regarding fused disconnect switches and fused power circuit devices (bolted pressure contact switches) see UL publications referred to in unit 1 and the *National Electrical Code®* under DISCONNECTING MEANS, SWITCHBOARDS, AND SWITCHES.

In the case of externally adjustable trip circuit breakers, the rating is considered to be the breaker's maximum trip setting, *NEC® Section 240-6*. The exceptions to this are (1) if the breaker has a removable and sealable cover over the adjusting screws, or (2) if it is located behind locked doors accessible only to qualified personnel, or (3) if it is located behind bolted equipment enclosure doors. In these cases, the adjustable setting is considered to be the breaker's ampere rating. This is an important consideration when selecting proper size phase and neutral conductors, equipment grounding conductors, overload relays in motor controllers, and other situations where sizing is based upon the rating or setting of the overcurrent protective device.

FUSES AND CIRCUIT BREAKERS

For general applications, the voltage rating, the continuous current rating, the interrupting rating, and the speed of response are factors that must be considered when selecting the proper fuses and circuit breakers.

Voltage Rating

According to *NEC® Section 110-9*, the voltage rating of a fuse or circuit breaker shall be equal to or greater than the voltage of the circuit in which the fuse or circuit breaker is to be used. In the commercial building, the system voltage is 208/120-volt wye connected. Therefore, fuses rated 250 volts or greater may be used. For a 480/277-volt wye-connected system, fuses rated 600 volts would be used. Fuses and circuit breakers will generally work satisfactorily when used at any voltage less than the fuses' or circuit breakers' rating. For example, there would be no problem using a 600-volt fuse on a 208-volt system, although this practice would not be economically sound.

CAUTION: Fuses or circuit breakers that are marked AC should *not* be installed on DC circuits. Fuses and breakers that are suitable for use on DC circuits will be so marked.

Continuous Current Rating

The continuous current rating of a protective device is the amperes that the device can continuously carry without interrupting the circuit. The standard ratings are listed in *NEC® Section 240-6*. When applied to a circuit, the selection of the rating is usually based on the ampacity of the circuit conductors, although there are notable exceptions to this rule.

For example, referencing *NEC® Table 310-16*, we find that a No. 8 AWG, Type THHN conductor has an ampacity of 55 amperes. The proper overcurrent protection for this conductor would be 40-amperes if the terminals of the connected equipment are suitable for 60°C conductors, or 50-amperes if the terminals of the connected equipment are suitable for 75°C conductors.

NEC® Section 240-3 lists several situations that permit the overcurrent protective device rating to exceed the ampacity of the conductor.

For example, the rating of a branch-circuit short-circuit and ground-fault protection fuse or circuit breaker on a motor circuit may be greater than the

current rating of the conductors that supply the electric motor. This is permitted because properly sized motor overload protection (i.e., 125% of the motors' full-load current rating) will match the conductors' ampacity that are also sized at 125% of the motors' full-load current rating.

Protection of Conductors

- *When the overcurrent device is rated 800-amperes or less.*

  When the ampacity of a conductor does not match the ampere rating of a standard fuse, or does not match the ampere rating of a standard circuit breaker that does not have an overload trip adjustment above its rating, the Code permits the use of the next higher standard ampere-rated fuse or circuit breaker. Adjustable trip circuit breakers were discussed previously. See *NEC® Section 240-3* and Figure 17-1.

- *When the overcurrent device is rated above 800-amperes.*

  When the ampere rating of a fuse or of a circuit breaker that does not have an overload trip adjustment exceeds 800 amperes, the conductor ampacity must be equal to or greater than the rating of the fuse or circuit breaker. The Code allows the use of fuses and circuit breakers that have ratings less than the standard sizes as listed in *NEC® Section 240-6*. Adjustable trip circuit breakers were discussed previously. See *Section 240-3*. Refer to Figure 17-2.

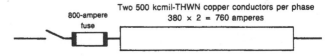

Fig. 17-1 *NEC® Section 240-3(b)* **allows the use of an 800-ampere fuse or circuit breaker.**

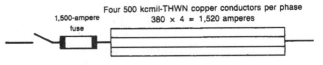

Fig. 17-2 *NEC® Section 240-3(b)* **requires the use of a 1,500-ampere fuse or circuit breaker.**

Interrupting Rating

Interrupting rating is defined in the *NEC®* as "The highest current at rated voltage that a device is intended to interrupt under standard test conditions."

Section 110-9 states that "Equipment intended to interrupt current at fault levels shall have an interrupting rating sufficient for the nominal circuit voltage and the current that is available at the line terminals of the equipment."

Interrupting rating is a maximum rating, and is a measure of a fuse or circuit breaker's ability to safely open an electrical circuit under fault conditions, such as an overload, short-circuit, or ground-fault.

Interrupting rating requirements are found in *Sections 240-60, 240-83,* and *240-86.*

Short-Circuit Current Rating

Electrical equipment may also be marked with a *short-circuit current rating.* Short-circuit current rating is established by the manufacturer of the equipment in conformance to specific UL Standards. Electrical equipment must not be connected to systems capable of delivering more fault current than the equipment's short-circuit current rating. Short-circuit current rating is referenced in *Section 110-10.*

The short-circuit current rating of all electrical equipment is based upon HOW MUCH CURRENT WILL FLOW and HOW LONG THE CURRENT WILL FLOW. Units 17 and 18 discuss the important issues of "Current Limitation" and "Peak Let-Through Current."

Overload currents have the following characteristics:

- They are greater than the normal current flow.

- They are contained within the normal conducting current path.

- If allowed to continue, they will cause overheating of the equipment, conductors, and the insulation of the conductors.

Short-circuit and ground-fault currents have the following characteristics:

- They flow "outside" of the normal current path.

- They may be greater than the normal current flow.

- They may be less than the normal current flow.

Short-circuit and ground-fault currents, which flow outside of the normal current paths, can cause conductors to overheat. In addition, mechanical damage to equipment can occur as a result of the magnetic forces of the large current flow and arcing. Some short-circuit and ground-fault currents may be no larger than the normal load current, or they may be thousands of times larger than the normal load current.

The terms *interrupting rating* and *interrupting capacity* are used interchangeably in the electrical industry, but there is a subtle difference. Interrupting rating is the value of the test circuit capability. Interrupting capacity is the actual current that the contacts of a circuit breaker "see" when opening under fault conditions. You will also hear the term A.I.C. which means *amps interrupting capacity.* In unit 18 you will learn how to calculate available short-circuit currents (fault currents) so you will be able to properly apply breakers, fuses, and other electrical equipment for the available short-circuit current to which they might be subjected.

For example, the test circuit in Figure 17-3 is calibrated to deliver 14,000 amperes of fault current. The standard test circuit allows four feet of conductor to connect between the test bus and the breaker's terminals. The standard test further allows ten inches of conductor per pole to be used as the shorting wire. Therefore, when the impedance of the connecting conductors is taken into consideration, the actual fault current that the contacts of the breaker "see" as they open is approximately 9900 amperes. The label on this circuit breaker is marked "14,000 amperes interrupting rating," yet its true interrupting capacity is 9900 amperes.

That is why it is important to adhere to the requirement of *NEC® Section 110-3(b),* which states:

Installation and Use. Listed or labeled equipment shall be installed and used in accordance with any instructions included in the listing or labeling.

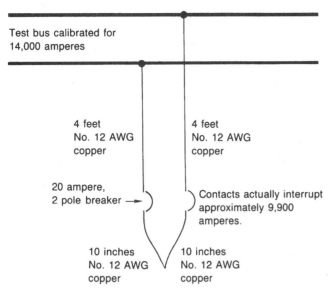

Test bus calibrated for 14,000 amperes

4 feet No. 12 AWG copper

4 feet No. 12 AWG copper

20 ampere, 2 pole breaker →

Contacts actually interrupt approximately 9,900 amperes.

10 inches No. 12 AWG copper

10 inches No. 12 AWG copper

The label on this breaker is marked 14,000 amperes interruping rating. Its actual interrupting capacity is approximately 9,900 amperes. The connecting conductors have an ampacity that is the same as the breaker's ampere rating.

Figure 17-3 Typical laboratory test circuit for a molded case circuit breaker.

Speed of Response

The time required for the fusible element of a fuse to open varies inversely with the magnitude of the current that flows through the fuse. In other words, as the current increases, the time required for the fuse to open decreases. The time-current characteristic of a fuse depends upon its rating and type. A circuit breaker also has a time-current characteristic. For the circuit breaker, however, there is a point at which the opening time cannot be reduced further due to the inertia of the moving parts within the breaker. The time-current characteristic of a fuse or circuit breaker should be selected to match the connected load of the circuit to be protected. Time-current characteristic curves are available from the manufacturers of fuses and circuit breakers.

TYPES OF FUSES

Dual-Element, Time-Delay Fuse

The dual-element, time-delay fuse, Figure 17-4, provides a time delay in the low-overload range to eliminate unnecessary opening of the circuit because of harmless overloads. However, this type

of fuse is extremely responsive in opening on short circuits. This fuse type has two fusible elements connected in series. Depending upon the magnitude of the current flow, one element or the other will open. The thermal cutout element is designed to open when the current reaches a value of approximately 500% of the fuse rating. The short-circuit element opens when a short circuit or heavy overload occurs. That is, the element opens at current values of approximately 500% or more of the fuse rating.

The thermal element is also designed to open at approximately 140°C (284°F), as well as on damaging overloads. In addition, the thermal element will open whenever a loose connection or a poor contact in the fuseholder causes heat to develop. As a result, a true dual-element fuse also offers thermal protection to the equipment in which it is installed.

Dual-element fuses are suitable for use on motor circuits and other circuits having high-inrush characteristics. This type of fuse can be used as well for mains, feeders, subfeeders, and branch circuits. Dual-element fuses may be used to provide back-up protection for circuit breakers, bus duct, and other circuit components that lack an adequate interrupting rating, bracing, or withstand rating (covered later in this unit).

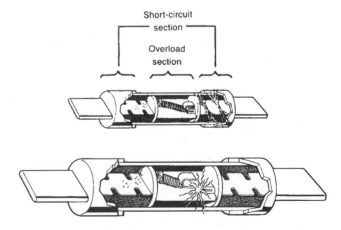

Fig. 17-4 Cutaway view of a dual-element, time-delay fuse. On overloads, the spring-loaded trigger assembly opens. On short circuits or heavy ground faults, the fuse element in the short-circuit section opens. The fuse elements are generally made of copper. (*Courtesy* of Cooper Bussmann, Inc.)

Using Fuses for Motor Overload Protection

Dual-element, time-delay fuses used on single-motor branch circuits are generally sized not to exceed 125% of the full-load running current of the motor in accordance with *NEC® Section 430-32* for motor overload protection. Sizing these fuses slightly higher than the motor's overload setting will provide "back-up" overload protection for the motor. The motor now has double overload protection.

Experience has shown that sizing dual-element, time-delay fuses in the range of 115% to 125% of the motor's full-load current rating meets the *NEC®* requirements for branch-circuit, short-circuit, and ground-fault protection, as well as providing "back-up" motor overload protection.

NEC® Section 430-32 sets forth motor overload protection requirements. There are many questions to ask when searching for the proper level of motor overload protection. These include: what is the motor's starting characteristics; is the motor manually or automatically started; is the motor continuous duty or intermittent duty; does the motor have integral built-in protection; is the motor impedance protected; is the motor larger than one horsepower or is the motor one horsepower or less; is the motor fed by a general-purpose branch circuit; is the motor cord- and plug-connected; will the motor automatically restart if the overload trips? When these questions are answered, then look at *NEC® Section 430-32* for the level of overload protection needed.

The basic, "bare bones" *NEC®* procedure for selecting the overload protection is shown in Table 17-1. The motor type is selected in the first column. The motor nameplate full-load current rating

is increased according to the value in the second column. If this overload protection will not carry the load, or will not allow the motor to start, then overload protection may be sized according to the third column and increased to the next higher size.

Example:

What is the ampere rating of a dual-element time-delay fuse that is to be installed to provide branch-circuit protection as well as overload protection for a 1.15 S.F. motor with a full-load current rating of 16 amperes?

$$16 \times 1.25 = 20 \text{ amperes}$$

Be Careful When Applying Fuses and Breakers on Motor Circuits

▶ In recent years *premium efficiency* motors have entered the scene. Premium efficiency motors have higher starting current characteristics than older style motors. These motors are referred to as Design E or high efficiency Design B motors. Their high starting currents can cause nuisance opening of fuses and nuisance tripping of circuit breakers. This is verified by the fact that the percentages and current values given in *NEC® Section 430-52, Tables 430-150, -151B,* and *-152* were increased since the 1996 edition. Different manufacturers' *same size* premium efficiency motors will have different starting current characteristics. Therefore, be careful when applying some of the newer smaller dimension Class CC, Class J, and Class T fuses on motor circuits. Because these fuses do not have the ability to handle the high inrush currents associated with premium efficiency motors, they cannot be sized at 125% of a motor's full-load ampere rating. They may have to be sized at 150%, 175%, 225%, or as high as 300% to 400%. Instant trip circuit breakers will also experience this problem. Always check the time-current curves of fuses and breakers to make sure that they will handle the momentary motor starting inrush currents without nuisance opening or tripping. Additional information is found in unit 7 of this text. ◀

TABLE 17-1 Overload Protection		
	Maximum overload size NEC® Section 430-32(a)(1) & (c)(1)	Not to Exceed NEC® Section 430-34
Service factor not less than 1.15 Marked not more than 40°C	125%	140%
All other motors	115%	130%

Using Fuses for Motor Branch-Circuit, Short-Circuit, and Ground-Fault Protection

NEC® Table 430-152 shows that for a typical motor branch-circuit, short-circuit, and ground-fault protection, the maximum size permitted for dual-element fuses is based on a maximum of 175% of the full-load current of the motor. An exception to the 175% value is given in *NEC® Section 430-52*, where permission is given to go as high as 225% of the motor full-load current if the lower value is not sufficient to allow the motor to start.

▶ Example:

16 × 1.75 = 28 amperes (maximum)

The next higher standard rating is 30 amperes.

If for some reason the 30-ampere fuse cannot handle the starting current, *NEC® Section 430-52(c) Exception No. 2(b)* allows the selection of a dual-element time-delay fuse not to exceed:

16 × 2.25 = 36 amperes (maximum)

The next lower standard rating is 35 amperes. ◀

Dual-Element, Time-Delay, Current-Limiting Fuses

The dual-element, time-delay, current-limiting fuse, Figure 17-5, operates in the same manner as the standard dual-element, time-delay fuse, Figure 17-4. The only difference between the fuses is that this fuse has a faster response in the short-circuit range and thus is more current limiting. Sizing this type of fuse for motor circuits, where time-delay is necessary to starting the motor, is done according to the procedure discussed in the previous paragraph. The short-circuit element in the current-limiting fuse can be silver or copper surrounded by a quartz sand arc-quenching filler. Silver-link fuses are more current-limiting than copper-link fuses.

Fast Acting Current-Limiting Fuses (Non-Time Delay)

The straight current-limiting fuse, Figure 17-6, has an extremely fast response in both the low-overload and short-circuit ranges. When compared to other types of fuses, this type of fuse has the lowest energy let-through values. Current-limiting fuses are used to provide better protection to mains, feeders and subfeeders, circuit breakers, bus duct, switchboards, and other circuit components that lack an adequate interrupting rating, bracing, or withstand rating.

Current-limiting fuse elements can be made of silver or copper surrounded by a quartz sand arc-quenching filler.

A fast-acting current-limiting fuse does not have the spring-loaded or "loaded link" overload assembly found in dual-element fuses. To apply straight current-limiting fuses for motor circuits, refer to *NEC® Table 430-152* under the column "NON-TIME DELAY FUSE."

To be classified as "current limiting," *NEC® Section 240-11* states that when a fuse or circuit breaker is subjected to heavy (high magnitude) fault currents, the fuse or breaker must reduce the fault current flowing into the circuit to a value less than the fault current that could have flowed into the circuit had there been no fuse or breaker in the circuit.

When used on motor circuits or other circuits having high current-inrush characteristics, the current-limiting non-time delay fuses must be sized at a much higher rating than the actual load. That is, for a motor with a full-load current rating of 10 amperes, a 30- or 40-ampere fast-acting current-limiting fuse may be required to start the motor. In this case, the fuse is considered to be the motor branch-circuit short-circuit protection as required in *NEC® Table 430-152*.

The definition of a current-limiting overcurrent protective device is given in *NEC® Section 240-11*.

Fig. 17-5   Cutaway view of a dual-element, time-delay, current-limiting fuse. On overloads, the spring-loaded trigger assembly opens. On short circuits or heavy ground faults, the fuse elements in the short-circuit section open. The fuse elements are generally made of silver. (*Courtesy* of Bussmann, Cooper Industries.)

Fig. 17-6   Cutaway view of a current-limiting, fast-acting, single-element fuse. (*Courtesy* of Cooper Bussmann, Inc.)

Table 430-152. Maximum Rating or Setting of Motor Branch-Circuit Short-Circuit and Ground-Fault Protective Devices

	Percentage of Full-Load Current			
Type of Motor	**Nontime Delay Fuse[1]**	**Dual Element (Time-Delay) Fuse[1]**	**Instantaneous Trip Breaker**	**Inverse Time Breaker[2]**
Single-phase motors	300	175	800	250
AC polyphase motors other than wound-rotor Squirrel cage—				
Other than Design E	300	175	800	250
Design E	300	175	1100	250
Synchronous[3]	300	175	800	250
Wound rotor	150	150	800	150
Direct current (constant voltage)	150	150	250	150

Note: For certain exceptions to the values specified, see Sections 430-52 through 430-54.

[1]The values in the Nontime Delay Fuse column apply to Time-Delay Class CC fuses.

[2]The values given in the last column also cover the ratings of nonadjustable inverse time types of circuit breakers that may be modified as in Section 430-52.

[3]Synchronous motors of the low-torque, low-speed type (usually 450 rpm or lower), such as are used to drive reciprocating compressors, pumps, etc., that start unloaded, do not require a fuse rating or circuit-breaker setting in excess of 200 percent of full-load current.

Reprinted with permission from NFPA 70-1999.

CARTRIDGE FUSES

The requirements governing cartridge fuses are contained in *NEC® Article 240, Part F*. According to the Code, all cartridge fuses must be marked to show:

- ampere rating.
- voltage rating.
- interrupting rating when over 10,000 amperes.
- current-limiting type, if applicable.
- trade name or name of manufacturer.

The UL and CSA require that the fuse class be indicated as Class G, H, J, K, L, RK1, RK5, T, or CC.

All fuses carrying the UL/CA Class listings (Class G, H, J, K, L, RK1, RK5, T, and CC) and plug fuses are tested on alternating-current circuits and are marked for ac use. When fuses are to be used on direct-current systems, the electrician should consult the fuse manufacturer since it may be necessary to reduce the fuse voltage rating and the interrupting rating to insure safe operation.

The variables in the physical appearance of fuses include length, ferrule diameter, and blade length-width-thickness, as well as other distinctive features. For these reasons, it is difficult to insert a fuse of a given ampere rating into a fuseholder rated for less amperage than the fuse. The differences in fuse construction also make it difficult to insert a fuse of a given voltage into a fuseholder with a higher voltage rating, *NEC® Section 240-60(b)*. Figure 17-7 indicates several examples of the method of insuring that fuses and fuseholders are not mismatched.

NEC® Section 240-60(b) also specifies that fuseholders for current-limiting fuses shall be designed so that they cannot accept fuses that are non-current limiting, Figure 17-8.

In general, low voltage fuses may be used at system voltages that are less than the voltage rating of the fuse. For example, a 600-volt fuse combined with a 600-volt switch may be used on 575-volt, 480/277-volt, 208/120-volt, 240-volt, 50-volt, and 32-volt systems.

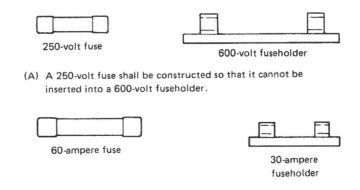

(A) A 250-volt fuse shall be constructed so that it cannot be inserted into a 600-volt fuseholder.

(B) A 60-ampere fuse is constructed so that it cannot be inserted into a 30-ampere fuseholder.

Fig. 17-7   Examples of *NEC® Section 240-60(b)* requirements.

(A)

Fuse A is a Class H, noncurrent-limiting fuse. This fuse does not have the notch required to match the rejection pin in the fuse clip of the Class R fuseblock (C).

(B)

Fuse B is a Class R fuse (either Class RK1 or RK5), which is a current-limiting fuse. This fuse has the required notch on one blade to match the rejection pin in the fuse clip of the Class R fuseblock (C).

(C)   Class R fuseblock

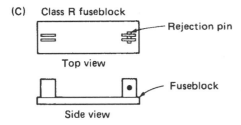

Fig. 17-8   Examples of *NEC® Section 240-60(b)* requirements.

Fig. 17-9   Class H cartridge fuse. Illustration shows renewable-type fuse in which the blown link may be replaced. (*Courtesy* of Cooper Bussmann, Inc.)

Class H. Class H fuses, Figure 17-9, formerly were called *NEC®* or Code fuses. Most low-cost, common, standard nonrenewable one-time fuses are Class H fuses. Renewable-type fuses also come under the Class H classification. Neither the interrupting rating nor the notation *Class H* appears on the label of a Class H fuse. This type of fuse is tested by the Underwriters Laboratories on circuits that deliver 10,000 amperes ac. Class H fuses are available with ratings ranging from 1 ampere to 600 amperes in both 250-volt ac and 600-volt ac types. Class H fuses are not current limiting.

A higher quality nonrenewable one-time fuse is also available, called a Class K5 fuse, which has a 50,000-ampere interrupting rating.

Class K. Class K fuses are grouped into three categories: K1, K5, and K9, Figure 17-10 A through D. These fuses may be UL listed with interrupting ratings in RMS symmetrical amperes in values of 50,000, 100,000, or 200,000 amperes. For each K rating, UL has assigned a maximum level of peak let-through current (I_p) and energy as given by I^2t. Class K fuses have varying degrees of current-limiting ability, depending upon the K rating. Class K1 fuses have the greatest current-limiting ability and Class K9 fuses the least current-limiting ability. A check of various fuse manufacturers' literature reveals that Class K9 fuses are no longer being manufactured. Class K fuses may be listed as time-delay fuses as well. In this case, UL requires that the fuses have a minimum time delay of 10 seconds at 500% of the rated current. Class K fuses are available in ratings ranging from $^1/_{10}$ ampere to 600 amperes at 250- or 600-volts ac. Class K fuses have the same dimensions as Class H fuses.

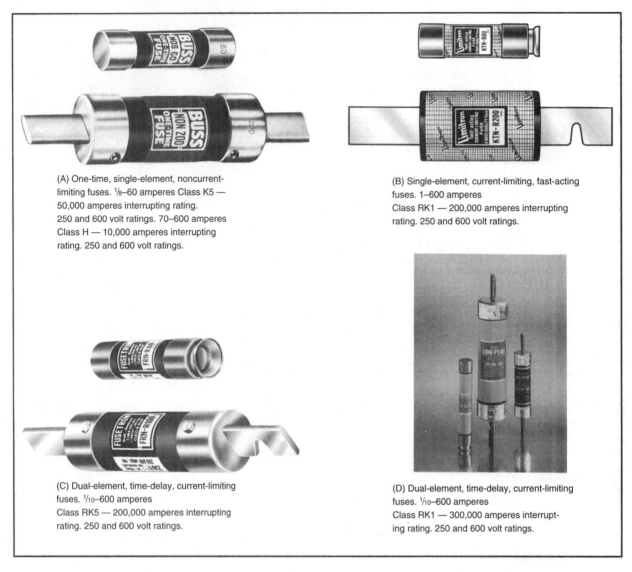

(A) One-time, single-element, noncurrent-limiting fuses. 1/8–60 amperes Class K5 — 50,000 amperes interrupting rating. 250 and 600 volt ratings. 70–600 amperes Class H — 10,000 amperes interrupting rating. 250 and 600 volt ratings.

(B) Single-element, current-limiting, fast-acting fuses. 1–600 amperes Class RK1 — 200,000 amperes interrupting rating. 250 and 600 volt ratings.

(C) Dual-element, time-delay, current-limiting fuses. 1/10–600 amperes Class RK5 — 200,000 amperes interrupting rating. 250 and 600 volt ratings.

(D) Dual-element, time-delay, current-limiting fuses. 1/10–600 amperes Class RK1 — 300,000 amperes interrupting rating. 250 and 600 volt ratings.

Fig. 17-10   Class H, K5, RK1, and RK5 fuses. (*Courtesy* Cooper Bussmann, Inc.)

Class J.   Class J fuses are current limiting and are so marked, Figure 17-11 A and B. They are listed by UL with a minimum interrupting rating of 200,000 RMS symmetrical amperes. Some have a special listing identified by the letters "SP," and have an interrupting rating of 300,000 RMS symmetrical amperes. Certain Class J fuses are also considered to be dual-element, time-delay fuses, and are marked "time-delay." Class J fuses are physically smaller than Class H fuses. Therefore, when a fuseholder is installed to accept a Class J fuse, it will be impossible to install a Class H fuse in the fuseholder, *NEC® Section 240-60(b).*

The Underwriters Laboratories has assigned maximum values of I^2t and I_p that are slightly less than those for Class K1 fuses. Both fastacting, current-limiting Class J fuses and time-delay, current-limiting Class J fuses are available in ratings ranging from 1 ampere to 600 amperes at 600 volts ac.

Class L.   Class L fuses, Figure 17-12 A, B, and C, are listed by UL in sizes ranging from 601 amperes to 6000 amperes at 600 volts ac. These fuses have specified maximum values of I^2t and I_p. They are current-limiting fuses and have a minimum interrupting rating of 200,000 RMS symmetrical amperes. These bolt-type fuses are used in bolted pressure contact switches. Class L fuses are available in both a fast-acting, current-limiting type and a time-delay, current-limiting type. Both types of Class L fuses meet UL requirements.

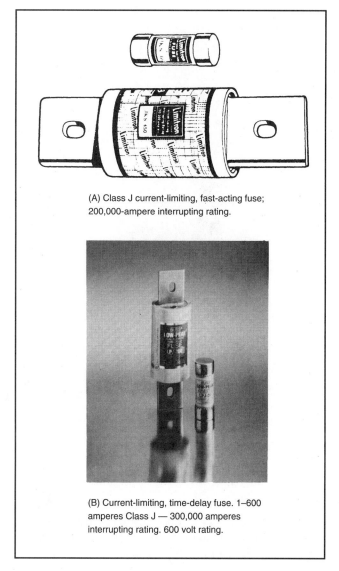

(A) Class J current-limiting, fast-acting fuse; 200,000-ampere interrupting rating.

(B) Current-limiting, time-delay fuse. 1–600 amperes Class J — 300,000 amperes interrupting rating. 600 volt rating.

Fig. 17-11   Class J current-limiting fuses. (*Courtesy Cooper Bussmann, Inc.*)

Some Class L fuses have a special interrupting rating of 300,000 RMS symmetrical amperes. The fuse's label will indicate the part number followed by the letters "SP."

Class T. Class T fuses, Figure 17-13, are current-limiting fuses and are so marked. These fuses are UL listed with an interrupting capacity of 200,000 RMS symmetrical amperes. Class T fuses are physically smaller than Class H or Class J fuses. The configuration of this type of fuse limits its use to fuseholders and switches that will reject all other types of fuses.

Class T fuses rated 600 volts have electrical characteristics similar to those of Class J fuses and are tested in a similar manner by Underwriters Laboratories. Class T fuses rated at 300 volts have lower peak let-through currents and I^2t values than comparable Class J fuses. Many *series-rated* panelboards are listed by Underwriters Laboratories with Class T mains. Because Class T fuses do not have a lot of time-delay (the ability to hold momentary inrush currents such as occurs when a motor is started) they are sized according to the non-time-delay fuse column in *NEC® Table 430-152*.

UL presently lists the Class T fuses in sizes from 1 ampere to 1,200 amperes. Common applications for Class T fuses are for mains, feeders, and branch circuits.

Class T 300-volt fuses may be used on 120/240-volt single-phase, 208/120-volt three-phase four-wire wye, and 240-volt three-phase three-wire delta systems (ungrounded or corner grounded). The *NEC®* permits 300-volt Class T fuses to be installed in single-phase line-to-neutral circuits supplied from three-phase four-wire solidly grounded neutral systems where the line-to-neutral voltage does not exceed 300 volts. The *NEC®* does not permit the use of 300-volt Class T fuses for line-to-line or line-to-line-to-line applications on 480/277-volt three-phase four-wire wye systems. Class T 600-volt fuses may be used on 480/277-volt three-phase four-wire wye, 480-volt three-phase three-wire, and any of the systems where Class T 300-volt fuses are permitted. See *NEC® Section 240-60(a)*.

Class G.   Class G fuses, Figure 17-14, are 480 volt cartridge fuses with small physical dimensions. They are used on circuits of 480 volts or less (for example, 480/277-volt systems). Class G fuses are available in sizes ranging from 0 ampere to 60 amperes and are UL listed at an interrupting capacity of 100,000 RMS symmetrical amperes. To prevent overfusing, Class G fuses are size-limiting within the four categories assigned to their ampere ratings. Therefore, a fuseholder designed to accept a 15-ampere Type SC fuse will not accept a 20-ampere Type SC fuse; and a fuseholder designed to accept a 20-ampere Type SC fuse will not accept a 30-ampere Type SC fuse; and so on for the four categories.

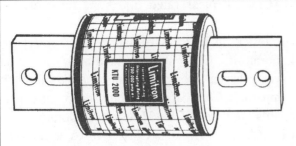

(A) Current-limiting, fast-acting Class L fuse; 200,000-ampere interrupting rating. Links are made of silver. Has very little time delay. Good for protection of circuit breakers and on circuits that *do not* have high inrush loads (such as motors, transformers). Size at 300% on motor circuits and other high inrush loads.

(B) Current-limiting, time-delay, multi-element fuse. 601–6000 amperes Class L — 300,000 amperes interrupting rating. 600 volt rating. Links made of silver. Will hold 500% of rated current for minimum of 4 seconds. Good for use on high inrush circuits (motors, transformers, and other inductive loads). The best choice for overall protection. Generally sized at 150–225% for motors and other high inrush loads.

(C) Current-limiting, time-delay Class L fuse; 200,000-ampere interrupting rating. Links are generally made of copper. Will hold 500% of rated current for a minimum of 10 seconds. Good for use on high inrush circuits (such as motors, transformers), but is the least current-limiting of Class L fuses.

Fig. 17-12   Class L fuses. All Class L fuses are rated 600 volts. Listed is 601 to 6000 ampere ratings. The smallest switch for Class L fuses is 800 amperes. Class L fuses that have elements rated at 600-ampere and less are available. These special ampere-rated fuses are physically the same size as the 800-ampere size. (*Courtesy* Cooper Bussmann, Inc.)

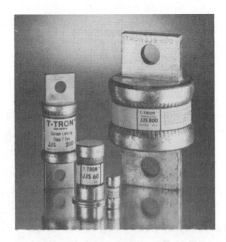

Fig. 17-13   Class T current-limiting, fast-acting fuse; 200,000-ampere interrupting rating. Links are made of silver. Has very little time delay. Good for the protection of circuit breakers and on circuits that *do not* have high inrush loads (such as motors, transformers). Size at 300-400% for motors and other high inrush loads.

Ampere Rating	Dimensions
0-15	13/32" × 1 5/16"
16-20	13/32" × 1 13/32"
21-30	13/32" × 1 5/8"
31-60	13/32" × 2 1/4"

Fig. 17-14   Class G fuses. (*Courtesy* Cooper Bussmann, Inc.)

Class G fuses are current limiting. They may be used for the protection of ballasts, electric heat, and similar loads. They are UL listed for branch-circuit protection.

Class R. This fuse is a nonrenewable cartridge type and has a minimum interrupting rating of 200,000 RMS symmetrical amperes. The peak let-through current (I_p) and the total clearing energy (I^2t) values are specified for the individual case sizes. The values of I^2t and I_p are specified by UL based on short-circuit tests at 50,000, 100,000, and 200,000 amperes.

Class R fuses are divided into two subclasses: Class RK1 and Class RK5. The Class RK1 fuse has characteristics similar to those of the Class K1 fuse. The Class RK5 fuse has characteristics similar to those of the Class K5 fuse. These fuses must be marked either Class RK1 or RK5. In addition, they are marked to be current limiting.

Some Class RK1 fuses have a special interrupting rating of 300,000 RMS symmetrical amperes. The fuse's label will indicate the part number followed by the letters "SP."

The ferrule-type Class R fuse has a rating range of $^1/_{10}$ ampere to 60 amperes and can be distinguished by the annular ring on one end of the case, Figure 17-15A. The knife blade-type Class R fuse has a rating range of 61 amperes to 600 amperes and has a slot in the blade on one end, Figure 17-15B. When a fuseholder is designed to accept a Class R fuse, it will be impossible to install a standard Class H or Class K fuse (these fuses do not have the annular ring or slot of the Class R fuse). The requirements for noninterchangeable cartridge fuses and fuseholders are covered in *NEC® Section 240-60(b)*. However, the Class R fuse can be installed in older style fuse clips on existing installations. As a result, the Class R fuse may be called a *one-way rejection fuse*.

Electrical equipment manufacturers will provide the necessary rejection-type fuseholders in their equipment, which is then tested with a Class R fuse at short-circuit current values such as 50,000, 100,000, or 200,000 amperes. Each piece of equipment will be marked accordingly.

Class CC. Class CC fuses are primarily used for control circuit protection, for the protection of motor control circuits, ballasts, small transformers, and so on. They are UL listed as branch-circuit fuses. Class CC fuses are rated at 600 volts or less and have a 200,000-ampere interrupting rating in sizes from $^1/_{10}$ ampere through 30 amperes. These fuses measure $1^1/_2" \times {}^{13}/_{32}"$ and can be recognized by a "button" on one end of the fuse, Figure 17-16. This "button" is unique to Class CC fuses. When a fuseblock or fuseholder that has the matching Class CC rejection feature is installed, it is impossible to insert any other $1^1/_2" \times {}^{13}/_{32}"$ fuses. Only a Class CC fuse will fit into these special fuseblocks and fuseholders. A Class CC fuse can be installed in a standard fuseholder.

Plug Fuses

NEC® Article 240, Part E enumerates the requirements for plug fuses. The electrician will be most concerned with the following requirements for plug fuses, fuseholders, and adapters:

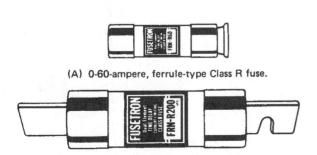

(A) 0-60-ampere, ferrule-type Class R fuse.

Fig. 17-15   Class R cartridge fuses. (*Courtesy* Cooper Bussmann, Inc.)

Fig. 17-16   Class CC fuses with rejection feature. (*Courtesy* Cooper Bussmann, Inc.)

- They shall not be used in circuits exceeding 125 volts between conductors, except on systems having a grounded neutral with no conductor having more than 150 volts to ground. This situation is found in the 120/208-volt system in the commercial building covered in this text, or in the case of a 120/240-volt, single-phase system.

- They shall have ampere ratings of 0 to 30 amperes.

- They shall have a hexagonal configuration for ratings of 15 amperes and below.

- The screw shell must be connected to the load side of the circuit.

- Edison-base plug fuses may be used only as replacements in existing installations where there is no evidence of overfusing or tampering.

- All new installations shall use fuseholders requiring Type S plug fuses or fuseholders with a Type S adapter inserted to accept Type S fuses only.

- Type S plug fuses are classified 0 to 15 amperes; 16 to 20 amperes; and 21 to 30 amperes.

Prior to the installation of fusible equipment, the electrician must determine the ampere rating of the various circuits. An adapter of the proper size is then inserted into the Edison-base fuseholder; finally, the proper size of Type S fuse can be inserted into the fuseholder. The adapter makes the fuseholder nontamperable and noninterchangeable. For example, after a 15-ampere adapter is inserted

into a fuseholder for a 15-ampere circuit, it is impossible to insert a Type S or an Edison-base fuse having a larger ampere rating into this fuseholder without removing the 15-ampere adapter. Adapters are designed so that they are extremely difficult to remove.

Type S fuses and suitable adapters, Figure 17-17, are available in a large selection of ampere ratings ranging from 3/10 ampere to 30 amperes. When a Type S fuse has dual elements and a time-delay feature, it does not open unnecessarily under momentary overloads, such as the current surge caused by the startup of an electric motor. On heavy overloads or short circuits, this type of fuse opens very rapidly.

Summary of Fuse Applications

Table 17-2 provides rating and application information for the fuses supplied by one manufacturer. The table indicates the Class, voltage rating, current range, interrupting ratings, and typical applications for the various fuses. A table of this type may be used to select the proper fuse to meet the needs of a given situation.

TESTING FUSES

As mentioned at the beginning of this text, the Occupational Safety and Health Act (OSHA) clearly states that electrical equipment must not be worked on when it is energized. There have been too many injuries to those intentionally working on equipment "hot," or thinking the power is off, only to find that it is still energized. If equipment is to be worked on "hot," then proper training, fire resistant clothing,

(A) Dual-element type S fuse

(B) Adapter for type S fuses

(C) Ordinary plug fuse, nontime-delay, Edison-base type

(D) Fusetron dual-element, time delay plug fuse, Edison-base type

Fig. 17-17   Plug fuse types W, T, and S with adapter. (*Courtesy* Cooper Bussmann, Inc.)

TABLE 17-2   Buss Fuses: Their Ratings, Class, and Application**

FUSE & AMPERE RATING RANGE	Voltage Rating	Symbol	Industry Class	Interrupting Rating in Sym. RMS Amperes	Application All Types Recommended For Protection of General Purpose Circuits & Components
LOW-PEAK time-delay FUSE 601–6000A	600 V	KRP-C (SP)	L	300,000 A	Will hold 500% of its rating for at least 4 seconds. High Capacity Main, Feeder, Br. Ckts. & Large Motors Circuits. Has more time-delay than KTU. For motors, can be sized 150–225% of motor FLA
LIMITRON fast-acting FUSE 601–6000A	600 V	KTU	L	200,000 A	High Capacity Main, Feeder, Br. Ckts. & Circuit Breaker Protection. For motors, size 300% of motor FLA
LIMITRON time-delay FUSE 601–4000A	600 V	KLU	L	200,000 A	High Capacity Mains, Feeders, and Circuits for large motors. Has good time-delay characteristics. Will hold 500% of its rating for 10 seconds. Copper links, thus not as current-limiting as KTU or KRPC type.
FUSETRON dual-element FUSE ¹/₁₀–600A	250 V / 600V	FRN-R / FRS-R	RK5 / RK5	200,000 A	Main, Feeder, Br. Ckts. & circuit breaker protection. Especially recommended for Motors, Welders, & Transformers and any loads having high inrush characteristics. Will hold 500% for 10 seconds. Size 115%–175% of motor FLA
LOW-PEAK dual-element FUSE ¹/₁₀–600A	250 V / 600 V	LPN-RK (SP) / LPS-RK (SP)	RK1 / RK1	300,000 A	Main, Feeder, Br. Ckts. & circuit breaker protection. Especially recommended for Motors, Welders, & Transformers. (More current-limiting than Fusetron dual-element Fuse.) Will hold 500% for 10 seconds. Size at 115%–175% of motor FLA
LIMITRON fast-acting FUSE 1–600A	250 V / 600 V	KTN-R / KTS-R	RK1 / RK1	200,000 A	Main, Feeder, Br. Ckts. & circuit breaker protection. Especially recommended for Circuit Breaker Protection. (High Degree of current-limitation.) Because these fuses are fast-acting, size at 300% or more for motors.
TRON fast-acting FUSE 1–1200A (300 V) 1–800A (600 V)	300 V / 600 V	JJN / JJS	T	200,000 A	Main, Feeder, Br. Ckts. & circuit breaker protection. Circuit Breaker Protection Small Physical Dimensions. Smaller than Class J. (High Degree of current limitation.) Because these fuses are fast-acting, size at 300% or more for motors.
LOW-PEAK time-delay FUSE 1–600A	600 V	LPJ (SP)	J	300,000 A	Main, Feeder, Br. Ckts., & circuit breaker protection. Motor & Transformer Ckts. Size at 150%–225% for motors.
LIMITRON fast-acting FUSE 1–600A	600 V	JKS	J	200,000 A	Main, Feeder, & Br. Ckts. Circuit Breaker Protection. Because these fuses are fast acting, size at 300% or more for motors.
ONE-TIME FUSE ¹/₈–600A	250 V / 600 V	NON / NOS	1–60 A, K5 / 65–600 A, H / 1–60 A, K5 / 65–600 A, H	50,000 A / 10,000 A / 50,000 A / 10,000 A	General Purpose — where fault currents are low. For motors, size at 300% or more.
SUPERLAG renewable (replaceable) link fuse 1–600A	250 V / 600 V	REN / RES	H	10,000 A	Offers feature of being able to replace links. Use only where fault currents are low. Follow manufacturer's instructions for replacing links. Is not current limiting. Generally sized at 300% for motors.
SC FUSE 1–60A	480 V	SC	G	100,000 A	General Purpose Branch Circuits. For motors, size at 300% or more.
FUSETRON dual-element PLUG FUSE ³/₁₀–30A, Edison Base	125 V	T	***	10,000 A	General Purpose Sizes ³/₁₀ A thru 14A excellent for motors. Size at 115%–125% for motors. Branch circuit sizes 15-20-25-30A.
FUSTAT dual-element type S PLUG FUSE ³/₁₀–30A	125 V	S	S****	10,000 A	General Purpose, Sizes ³/₁₀ A thru 14A excellent for motors. Size at 115%–125% for motors. Branch circuit sizes 15-20-25-30A.
ORDINARY PLUG FUSE Edison Base, ¹/₂–30A	125 V	W	***	10,000 A	General Purpose. Less Time Delay than Types T & S. Not recommended for motor circuits unless sized at

NOTE: For motor circuits, refer to *NEC® Article 430*, especially *Table 430-152.*

** Table 17-1 gives general application recommendations from one manufacturer's technical data. For critical applications, consult electrical equipment and specific fuse manufacturers' technical data.

*** Listed as Edison Base Plug Fuse.

**** Difficult to change ampere ratings because of different thread on screw base for different ampere ratings.

and protective gear (rubber blankets, insulated tools, goggles, rubber gloves, etc.) need to be used. A second person should be present when working electrical equipment "hot." OSHA has specific "lock-out" and "tag-out" rules for working on energized electrical equipment.

WHEN POWER IS TURNED ON. On "live" circuits, extreme caution must be exercised when checking fuses. There are many different voltage readings that can be taken, such as line-to-line, line-to-ground, line-to-neutral, etc.

Using a voltmeter, the first step is to be sure to set the scale to its highest voltage setting, then change to a lower scale after you are sure you are within the range of the voltmeter. For example, when testing what you believe to be a 120-volt circuit, it is wise to first use the 600-volt scale, then try the 300-volt scale, and then use the 150-volt scale — JUST TO BE SURE!

Taking a voltage reading across the bottom (load side) of fuses — either fuse-to-fuse, fuse-to-neutral, or fuse-to-ground — can show voltage readings because even though a fuse might have opened, there can be "feed-back" through the load. You could come to a wrong conclusion. Taking a voltage reading from the line side of a fuse to the load side of a fuse will show "open-circuit voltage" if the fuse has blown and the load is still connected. This also can result in a wrong conclusion.

Reading from line-to-load side of a good fuse should show zero voltage or else an extremely small voltage across the fuse.

Always read carefully the instructions furnished with electrical test equipment such as voltmeters, ohmmeters, etc.

WHEN POWER IS TURNED OFF. This is the safest way to test fuses. Remove the fuse from the switch, then take a resistance reading across the

fuse using an ohmmeter. A good fuse will show zero to very minimal resistance. An open (blown) fuse will generally show a very high resistance reading.

Cable Limiters

Cable limiters are used quite often in commercial and industrial installations where parallel cables are used on service entrances and feeders.

Cable limiters differ in purpose from fuses which are used for overload protection. Cable limiters are short-circuit devices that can *isolate* a faulted cable, rather than having the fault open the entire phase. They are selected on the basis of conductor size; for example, a 500 kcmil cable limiter would be used on a 500 kcmil conductor.

Cable limiters are available for cable-to-cable or cable-to-bus installation, for either aluminum or copper conductors. The type of cable limiter illustrated in Figure 17-18 is for use with a 500 kcmil copper conductor, and is rated at 600 volts with an interrupting rating of 200,000 amperes.

Cable limiters are also used where taps are made ahead of the mains, such as they are in the commercial building where the emergency system is tapped ahead of the main fuses (as shown in Figure 16-2).

Figure 17-19 is an example of how cable limiters may be installed on the service of the commercial building. A cable limiter is installed at each end of each 500 kcmil conductor. Three conductors per phase terminate in the main switchboard where the service is then split into two 800-ampere bolted pressure switches.

Figure 17-20 shows how cable limiters are used where more than one customer is connected to one transformer.

Cable limiters are permitted, *NEC® Section 230-82.*

Fig. 17-18   Cable limiter for 500 kcmil copper conductor. (Courtesy Cooper Bussmann, Inc.)

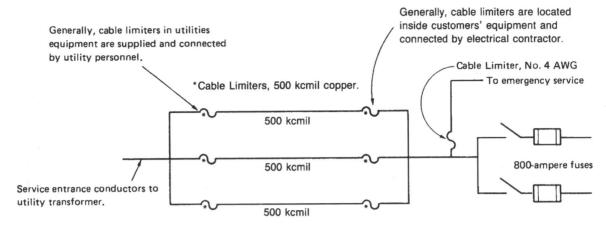

Generally, cable limiters in utilities equipment are supplied and connected by utility personnel.

Generally, cable limiters are located inside customers' equipment and connected by electrical contractor.

Cable Limiter, No. 4 AWG
To emergency service

*Cable Limiters, 500 kcmil copper.

500 kcmil

500 kcmil

800-ampere fuses

Service entrance conductors to utility transformer.

500 kcmil

Fig. 17-19   Use of cable limiters in a service entrance.

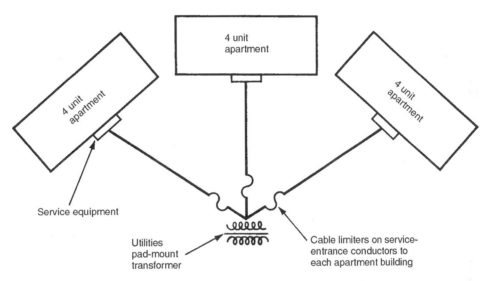

4 unit apartment

4 unit apartment

4 unit apartment

Service equipment

Utilities pad-mount transformer

Cable limiters on service-entrance conductors to each apartment building

Fig. 17-20   Some utilities install cable limiters where more than one customer is connected to one transformer. Thus, if one customer has problems in the main service equipment, the cable limiter on that service will open, isolating the problem from other customers on the same transformer. This is common for multiunit apartments, condos, and shopping centers.

Cable limiters are installed in the "hot" phase conductors only. They are not installed in a grounded conductor.

DELTA, THREE-PHASE, GROUNDED "B" PHASE SYSTEM

See Unit 13 for a discussion of delta-connected three-phase systems.

Fuses shall be installed in series with ungrounded conductors for overcurrent protection, *NEC® Section 240-20*. This connection is sometimes called a "corner ground." The Code prohibits overcurrent devices from being connected in series with any conductor that is intentionally grounded, *NEC® Section 240-22*.

However, there are certain instances where a three-pole, three-phase switch may be installed, where it is permitted to install fuses in two poles only. This would be in the case of a delta, three-phase, corner grounded "B" phase system, Figure 17-21. A device called a *solid neutral* can be installed in the "B" phase, as shown in the figure. Note that the switch has two fuses and one solid neutral installed.

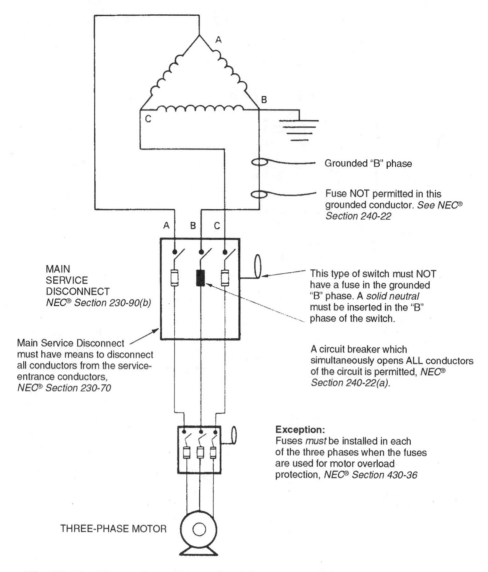

Grounded "B" phase

Fuse NOT permitted in this grounded conductor. *See NEC® Section 240-22*

MAIN
SERVICE
DISCONNECT
NEC® Section 230-90(b)

This type of switch must NOT have a fuse in the grounded "B" phase. A *solid neutral* must be inserted in the "B" phase of the switch.

Main Service Disconnect must have means to disconnect all conductors from the service-entrance conductors, *NEC® Section 230-70*

A circuit breaker which simultaneously opens ALL conductors of the circuit is permitted, *NEC® Section 240-22(a).*

Exception:
Fuses *must* be installed in each of the three phases when the fuses are used for motor overload protection, *NEC® Section 430-36*

THREE-PHASE MOTOR

Fig. 17-21   Three-phase, three-wire delta system with grounded "B" phase.

Figure 17-21 shows a three-pole service-entrance switch. *NEC® Section 230-75* permits a bus bar to be used as a disconnecting means for the grounded conductor. Thus, a two-pole service-entrance switch is permitted if the grounded conductor can be "disconnected" from the service-entrance conductors, Figure 17-22.

Solid Neutrals

A solid neutral, Figure 17-23, is made of copper bar that has exactly the same dimensions as a fuse for a given ampere rating and voltage rating. For

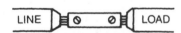

Fig. 17-22   *NEC® Section 230-75* allows this for disconnecting the grounded conductor at the service disconnect.

Fig. 17-23   Examples of a solid neutral.

example, a solid neutral rated at 100 amperes, 600 volts would be installed in a switch rated at 100 amperes, 600 volts.

"Solid neutrals" are generally used in retrofit situations.

Solid neutrals are installed in situations, for example, where a 3-pole, 3-fuse disconnect switch is used on a grounded "B" phase system, Figure 17-21. *NEC® Section 240-20* requires that "a fuse shall be connected in series with each ungrounded conductor." The solid neutral would be inserted into the grounded "B" phase leg instead of a fuse. Other examples of where a solid neutral might be used is where a 3-pole, 3-fuse disconnect switch is used on a 3-wire, 240/120-volt, single-phase system, or where a fusible disconnect is used only as a disconnecting means, and the user does not want any fuses to be installed in that particular switch.

TIME-CURRENT CHARACTERISTIC CURVES AND PEAK LET-THROUGH CHARTS

The electrician must have a basic amount of information concerning fuses and their application to be able to make the correct choices and decisions for everyday situations that arise on an installation.

The electrician must be able to use the following types of fuse data:

- time-current characteristic curves, including total clearing and minimum melting curves.

- peak let-through charts.

Fuse manufacturers furnish this information for each of the fuse types they produce.

The Use of Time-Current Characteristic Curves

The use of the time-current characteristic curves shown in Figures 17-24 and 17-25 can be demonstrated by considering a typical problem. Assume that an electrician must select a fuse to protect a motor that is governed by *NEC® Section 430-32(a)*. This section indicates that the running overcurrent protection shall be based on not over 125% of the full-load running current of the motor.

The motor in this example is assumed to have a full-load current of 24 amperes. Therefore, the size of the required protective fuse is determined as follows:

$$24 \times 1.25 = 30 \text{ amperes}$$

Now the question must be asked: is this 30-ampere fuse capable of holding the inrush current of the motor (which is approximately four to five times the running current) for a sufficient length of time for the motor to reach its normal speed? For this problem, the inrush current of the motor is assumed to be 100 amperes.

Refer to Figure 17-24. At the 100-ampere line, draw a horizontal line until it intersects with the 30-ampere fuse line. Then draw a vertical line down to the base line of the chart. At this point, it can be seen that the 30-ampere fuse will hold 100 amperes for approximately 40 seconds. In this amount of time, the motor can be started. In addition, the fuse will provide running overload protection as required by *NEC® Section 430-32(a)*.

If the time-current curve for thermal overload relays in a motor controller is checked, it will show that the overload element will open the same current in much less than 40 seconds. Therefore, in the event of an overload, the thermal overload elements will operate before the fuse opens. If the overload elements do not open for any reason, or if the contacts of the controller weld together, then the properly sized dual-element fuse will open to reduce the possibility of motor burnout. The above method is a simple way to obtain backup or double motor protection.

Referring again to Figure 17-24, assume that a short circuit of approximately 500 amperes occurs. Find the 500-ampere line at the left of the chart and then, as before, draw a horizontal line until it intersects with the 30-ampere fuse line. From this intersection point, drop a vertical line to the base line of the chart. This value indicates that the fuse will open the fault in slightly over $^2/_{100}$ (0.02) seconds. On a 60-hertz system (60 cycles per second), one cycle equals 0.016 seconds. Therefore, a 30-ampere fuse will clear a 500-ampere fault in just over one cycle.

AVERAGE MELTING TIME-CURRENT CHARACTERISTIC CURVES

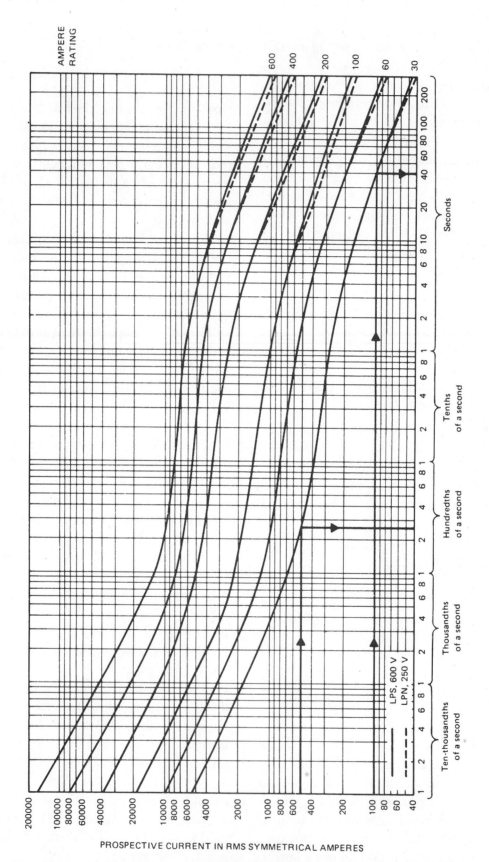

PROSPECTIVE CURRENT IN RMS SYMMETRICAL AMPERES

Fig. 17-24   Average melting time-current characteristic curves for one family of current-limiting, dual-element, time-delay fuses. Time-current curves such as this are available from the manufacturers of fuses and circuit breakers.

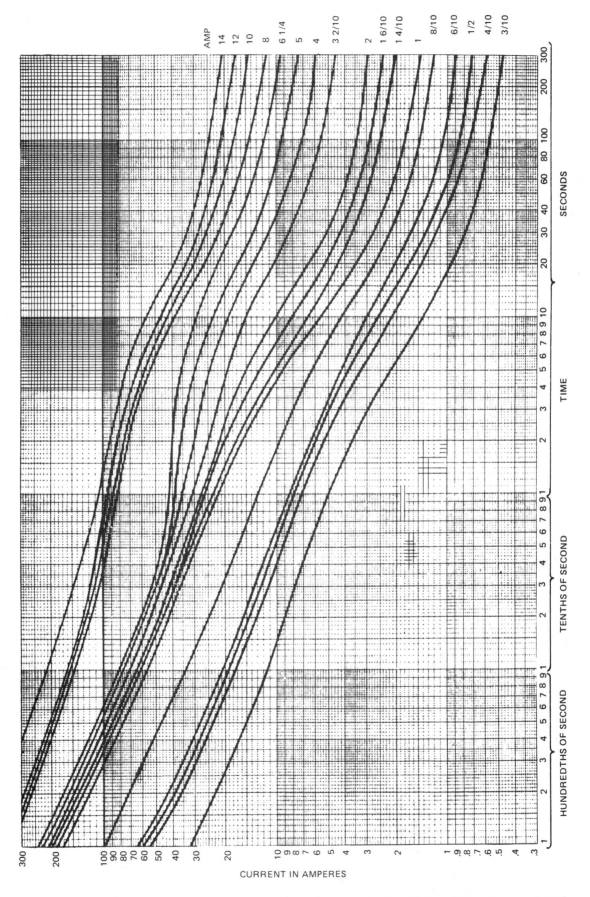

Fig. 17-25 Average melting time-current curve for one family of time-delay fuses rated ³/₁₀ through 14 amperes.

The Use of Peak Let-Through Charts

It is important to understand the use of *peak let-through charts* if one is to properly match the short-circuit current rating of electrical equipment with the let-through current values of the overcurrent protective devices (fuses or circuit breakers).

The short-circuit current ratings of electrical equipment such as panelboards, bus duct, switchboards, controllers, and similar electrical equipment is established by UL Standards. Testing the equipment is done at a specified nominal voltage and at a specific value of fault current. The equipment must be capable of safely carrying the fault current at rated voltage for a given length of time as required by the UL Standards.

Short-circuit current ratings marked on electrical equipment are usually given in root mean square (RMS) values. For example, a panelboard might be marked "22,000 Amperes RMS Symmetrical" ... a switchboard might be marked "50,000 Amperes Symmetrical."

Although not marked on electrical equipment, UL Standards also refer to peak current (I_p) values that are associated with a given RMS current value for a specific power factor.

The short-circuit current rating (withstand rating) for condutors is found in the Insulated Cable Engineering Association (ICEA) standards.

Peak let-through charts give us a good indication of the current-limiting effects of a current-limiting fuse or circuit breaker under "bolted fault" conditions. Current-limitation is one of the principles used to obtain a series-rating listing for electrical equipment. Another principle used to obtain a series-rating is referred to as "dynamic impedance," where the available fault-current is further reduced because of the added impedance of the arc between the contacts of a circuit breaker as it is opening. Series-rating of panels is discussed later in this unit.

To use charts such as Figure 17-26, find the 40,000 ampere point on the base line of the chart, draw a vertical line upward to Line A-B, then move

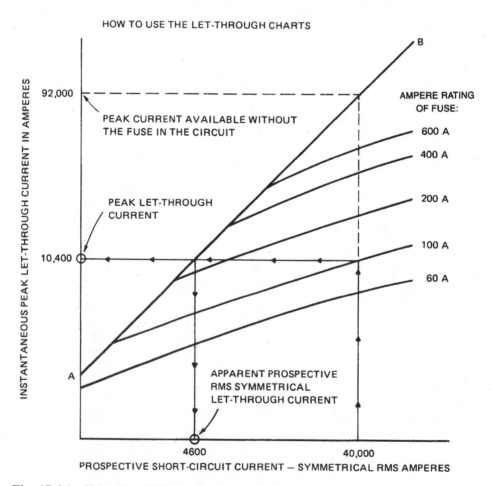

Fig. 17-26   Using the let-through charts to determine peak let-through current and apparent prospective RMS symmetrical let-through current.

horizontally to the left margin where the value of 92,000 amperes is found. This is the peak current, I_p, at the first half cycle.

To find the peak let-through current of the 100-ampere fuse, start again at the 40,000 ampere point on the base line, go vertically upward to Line A-B, then horizontally to the left margin, where the value of 10,400 is found. This is the peak let-through current, I_p, of this particular 100-ampere fuse.

To find the apparent RMS let-through current of this fuse, start again at the 40,000 ampere point on the base line, go vertically upward to Line A-B, then horizontally to the left until Line A-B is reached, then vertically downward to the base line. The value

at this point on the base line is 4,600 amperes. This is the apparent RMS amperes let-through current of this particular 100-ampere fuse.

Figure 17-27 shows how a current limiting main overcurrent device can be applied to "Listed" series-rated panels, discussed in more detail a little later on in this unit.

Figure 17-28 shows a peak let-through chart for a family of dual-element fuses. Fuse and circuit breaker manufacturers provide charts of this type for various sizes and types of their fuses and circuit breakers.

What are these charts used for? Let's take an example of a 400 ampere busway that is marked as having 22,000 ampere bracing. The available fault current of the system has been determined to be approximately 50,000 amperes. In Figure 17-27, draw a vertical line upward from 50,000 amperes on the base line up to the 400 ampere fuse line. From this point, go horizontally to the left to Line A-B. From this point, draw a line vertically downward to the base line and read an apparent RMS current value of slightly more than 10,000 amperes. Therefore, the 400 ampere busway with a bracing of 22,000 amperes is adequately protected by the 400 ampere fast-acting, current-limiting fuse.

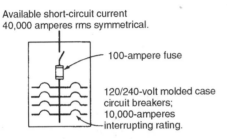

Fig. 17-27   Example of a fuse used to protect circuit breakers. This installation meets the requirements of *NEC® Section 110-9* **when panel is listed as a Series-Rated panel.**

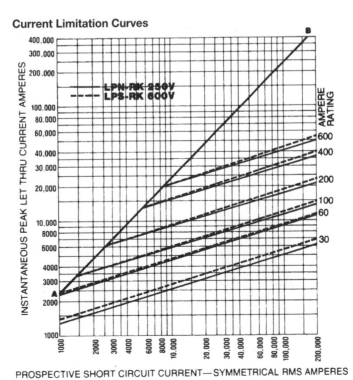

Fig. 17-28   Current-limiting effect of a dual-element, time-delay Class RK1 fuse.

Branch Breaker			Main Fuse		Interrupting Rating		
Frame	Poles	Max. Amp.	Type	Max. Amp.	VAC	KA	Phase
ABC	2	15–100	J,T	200	120/240	200	1
DEF	2,3	15–100	R	100	120/240	200	1
GHI	2,3	70–225	J,T	400	600	100	1,3
JKL	1	15–100	J,T	200	277	200	1
MNO	2,3	15–150	R	200	600	200	1,3
PQR	2,3	70–400	L	1200	480	100	1,3
STU	2,3	400	J,T,R	600	480	100	1,3
VWX	2,3	400	J,T,R	400	480	200	1,3
SYX	1,2	30	RK5	200	240	100	3

TABLE 17-3   Typical Table Showing Sizes and Types of Fuses

Typical table published by manufacturers showing the maximum sizes and types of fuses that may be used to protect circuit breakers against fault currents that exceed the breakers' maximum interrupting rating. This table was modeled from various manufacturers' "series-connected" technical information. Always refer to the specific manufacturer's data.

Table 17-3 shows how a manufacturer might indicate the maximum size and type of fuse that will protect given circuit breakers against fault currents that exceed the breakers' interrupting rating. This information is found in the UL *Recognized Components Directory* under "Circuit Breakers – Series Connected." Circuit breakers are tested according to the requirements of UL Standard 489, panelboards are tested to UL Standard 67, switchboards are tested to UL Standard 891, and motor controllers are tested to UL 508.

CIRCUIT BREAKERS

In *NEC® Article 100* a circuit breaker is defined as "*a device that is designed to open and close a circuit by nonautomatic means and to open the circuit automatically on a predetermined overcurrent without being damaged itself when properly applied within its rating.*"

Molded case circuit breakers are the most common type in use today, Figure 17-29. The tripping mechanism of this type of breaker is enclosed in a molded plastic case. The *thermal-magnetic* type of

Fig. 17-29   Molded case circuit breaker.

circuit breaker is covered in this unit. Another type is the power circuit breaker, which is larger, of heavier construction, and more costly. Power circuit breakers are used in large industrial applications. However, these breakers perform the same electrical functions as the smaller molded case breakers.

Circuit breakers are also available with solid-state, adjustable trip ratings and settings. When compared to the standard molded case circuit breaker, these breakers are very costly.

Molded case circuit breakers are listed by the Underwriters Laboratories and are covered by NEMA Standards. *NEC® Sections 240-80* through *240-83* state the basic requirements for circuit breakers and are summarized as follows:

- Breakers shall be trip free so that if the handle is held in the *On* position, the internal mechanism trips the breaker to the *Off* position.

- The breaker shall clearly indicate if it is in the *On* or *Off* position.

- The breaker shall be nontamperable; that is, it cannot be readjusted (to change its trip point or time required for operation) without dismantling the breaker or breaking the seal.

- The rating shall be durably marked on the breaker. For the smaller breakers with ratings of 100 amperes or less and 600 volts or less, this marking must be molded, stamped, or etched on the handle or other portion of the breaker that will be visible after the cover of the panel is installed.

- Every breaker having an interrupting rating other than 5000 amperes shall have its interrupting rating shown on the breaker.

- The *NEC® Section 240-83(d)*, and the Underwriters Laboratories require that when a circuit breaker is to be used as a toggle switch, the breaker must be so tested by the Laboratories. A typical use of a circuit breaker as a toggle switch is in a panel when it is desired to control 120- or 277-volt lighting circuits (turn them on and off) instead of installing separate toggle switches. Breakers listed for use as toggle switches will bear the letters *SWD* on the label. Breakers not marked in this manner must not be used as switches.

- A circuit breaker should not be loaded to more than 80% of its current rating for loads that are likely to be on for three hours or more, unless the breaker is marked otherwise, such as those breakers tested by the Underwriters Laboratories at 100% loading.

- If the voltage rating on a circuit breaker is marked with a single voltage (example: 240 volts), the breaker may be used in grounded or ungrounded systems where the voltage between any two conductors does not exceed the breaker's marked voltage rating. **CAUTION:** Do not use a two-pole single voltage marked breaker to protect a three-phase, three-wire corner-grounded circuit unless the breaker is marked with both one-phase and three-phase marking.

- If the voltage rating on a circuit breaker is marked with a slash voltage (example: 480/277 volts or 120/240 volts), the breaker may be used where the voltage to ground does not exceed the breaker's lower voltage marking, and the voltage between any two conductors does not exceed the breaker's higher voltage marking.

Thermal-Magnetic Circuit Breakers

A thermal-magnetic circuit breaker contains a bimetallic element. On a continuous overload, the bimetallic element moves until it unlatches the inner tripping mechanism of the breaker. Harmless momentary overloads do not cause the tripping of the bimetallic element. If the overload is heavy, or if a short circuit occurs, then the mechanism within the circuit breaker causes the breaker to interrupt the circuit instantly. The time required for the breaker to open the circuit completely depends upon the magnitude of the fault current and the mechanical condition of the circuit breaker. This time may range from approximately one-half cycle to several cycles.

Circuit breaker manufacturers calibrate and set the tripping characteristic for most molded case breakers. Breakers are designed so that it is difficult to alter the set tripping point, *Section 240-82*. For certain types of breakers, however, the trip coil can be changed physically to a different rating. Adjustment provisions are made on some breakers to permit the magnetic trip range to be changed. For example, a breaker rated at 100 amperes may have an external adjustment screw with positions marked HI-MED-LO. The manufacturer's application data for this breaker indicates that the magnetic tripping occurs at 1500 amperes, 1000 amperes, or 500 amperes respectively for the indicated positions. These settings usually have a tolerance of ±10%.

The *ambient-compensated* type of circuit breaker is designed so that its tripping point is not affected by an increase in the surrounding temperature. An ambient-compensated breaker has two elements: the first element heats due to the current passing through it and because of the heat of the surrounding air; the second element is affected only by the ambient temperature. These elements act in opposition to each other. In other words, as the tripping element tends to lower its tripping point because of a high ambient temperature, the second element exerts an opposing force that stabilizes the tripping point. Therefore, current through the tripping element is the only factor that causes the element to open the circuit. Most breakers are calibrated for use in up to 40°C (104°F) ambient. This is marked on the breaker as 40°C.

Factors that can affect the proper operation of a circuit breaker include moisture, dust, vibration, corrosive fumes and vapors, and excessive tripping and switching. As a result, care must be taken when locating and installing circuit breakers and all other electrical equipment.

The interrupting rating of a circuit breaker is marked on the breaker label. The electrician should check the breaker carefully for the interrupting rating since the breaker may have several voltage ratings with a different interrupting rating for each. For example, assume that it is necessary to select a breaker having an interrupting rating at 240 volts of at least 50,000 amperes. A close inspection of the breaker may reveal the following data:

Voltage	Interrupting Rating
240 volts	65,000 amperes
480 volts	25,000 amperes
600 volts	18,000 amperes

Recall that for a fuse, the interrupting rating is marked on the fuse label. This rating is the same for any voltage up to and including the maximum voltage rating of the fuse.

The standard full-load ampere ratings of non-adjustable circuit breakers are the same as those for fuses according to *NEC® Section 240-6*. As previously mentioned, additional standard ratings of fuses are 1, 3, 6, 10, and 601 amperes.

The time-current characteristics curves for circuit breakers are similar to those for fuses. A typical circuit breaker time-current curve is shown in Figure 17-30. Note that the current is indicated in percentage values of the breaker trip unit rating. Therefore, according to this graph, the 100-ampere breaker being considered will:

1. carry its 100-ampere (100%) rating indefinitely.

2. carry 300 amperes (300%) for a minimum of 25 seconds and a maximum of 70 seconds.

3. unlatch its tripping mechanism in 0.0032 seconds (approximately one-quarter cycle) with a current of 5000 amperes.

4. interrupt the circuit in a maximum time of 0.016 seconds (one cycle) with a current of 5000 amperes (5000%).

The same time-current curve can be used to determine that a 200-ampere circuit breaker will:

1. carry its 200-ampere (100%) rating indefinitely.

2. carry 600 amperes (300%) for a minimum of 25 seconds and a maximum of 70 seconds.

3. unlatch its tripping mechanism in 0.0032 seconds (approximately one-quarter cycle) with a current of 5000 amperes.

4. interrupt the circuit in a maximum time of 0.016 seconds (one cycle) with a current of 5000 amperes (2500%).

This example shows that if a short circuit occurs in the magnitude of 5000 amperes, then both the 100-ampere breaker and the 200-ampere breaker installed in the same circuit will open together because they have the same unlatching times. In many instances, this action is the reason for otherwise unexplainable power outages (unit 18). This is rather common when heavy (high value) fault currents occur on circuits protected with molded case circuit breakers.

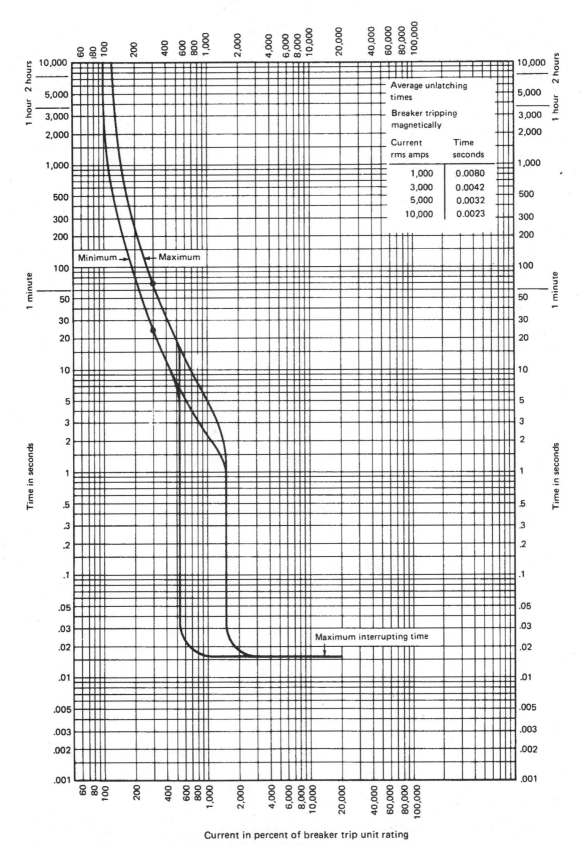

Fig. 17-30 Typical molded case circuit-breaker, time-current curve.

Common Misapplication

A common violation of *NEC® Sections 110-9* and *110-10* is the installation of a main circuit breaker (such as a 100-ampere breaker) that has a high interrupting rating (such as 50,000 amperes) while making the assumption that the branch-circuit breakers (with interrupting ratings of 10,000 amperes) are protected adequately against the 40,000-ampere short circuit, Figure 17-31.

Standard circuit breakers with high interrupting ratings cannot protect a standard circuit breaker with a lower interrupting rating. This is confirmed in the UL *Electrical Construction Equipment Directory* in the circuit breaker section:

> An interrupting rating on a circuit breaker included in a piece of equipment does not automatically qualify the equipment in which the circuit breaker is installed for use on circuits with higher available currents than the rating of the equipment itself.

SERIES-RATED EQUIPMENT

Where high available fault currents indicate the need for high interrupting breakers or fuses, a "fully rated" system is generally used. In a "fully rated" system, all breakers and fuses are rated for the fault current available at the point of application. Another less costly way to safely match a main circuit breaker or main fuses ahead of branch circuit breakers is to use listed *series-rated* equipment. Series-rated equipment is accepted by most electrical inspectors across the country. Series-rated equipment is also referred to as *series-connected*.

Underwriters Laboratories lists series-rated panelboards with breaker main/breaker branches and fused main/breaker branches. Figure 17-32 shows a series-rated panel that may be used where the available fault current does not exceed 22,000 amperes. The panel has a 22,000 AIC main breaker and branch circuit breakers rated 10,000 AIC. This series arrangement of breakers results in a cushion-

AVAILABLE SHORT-CIRCUIT CURRENT
AT PANEL 40,000 AMPERES.

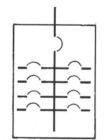

PANEL IS NOT "SERIES-RATED."

*100-AMPERE MAIN BREAKER WITH INTER-
RUPTING RATING OF 50,000 AMPERES.

MOLDED CASE CIRCUIT BREAKERS; INTER-
RUPTING RATING OF 10,000 AMPERES.

*NOTE: UNLESS "SERIES-RATED," THE MAIN BREAKER CANNOT
PROTECT THE BRANCH BREAKERS AGAINST A SHORT CIRCUIT OF
40,000 AMPERES; BRANCH BREAKERS ARE CAPABLE OF INTERRUPTING
10,000 AMPERES.

Fig. 17-31   Violation of *NEC® Section 110-9*.

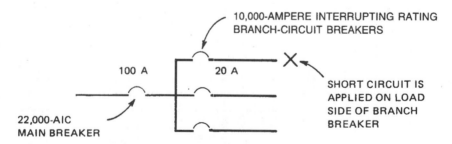

10,000-AMPERE INTERRUPTING RATING
BRANCH-CIRCUIT BREAKERS

100 A   20 A

SHORT CIRCUIT IS
APPLIED ON LOAD
SIDE OF BRANCH
BREAKER

22,000-AIC
MAIN BREAKER

Fig. 17-32   Series-rated circuit breakers. In this example, both the 20-ampere breaker and the 100-ampere main breaker trip off under high-level fault conditions.

ing effect when the main breaker contacts and the branch breaker contacts simultaneously open under fault conditions that exceed the interrupting rating of the branch circuit breaker. The impedance of the two arcs in series reduces the fault current that the branch breakers actually "see." The impedance of the arc is called "dynamic impedance."

One disadvantage of series-rated systems is that should a heavy fault current occur on any of the branch circuits, the main breaker will also trip off, causing a complete loss of power to the panel. Likewise, a series-rated main fuse/breaker branches panel would also experience a power outage, but the possibility exists that only one fuse would open, leaving one half (for a single-phase panel) or two-thirds (for a three-phase, four-wire panel) of the power on.

For normal overload conditions on a branch circuit, the branch breaker only will trip off, and should not cause the main breaker to trip or main fuses to open. Refer to unit 18 for more data on the subject of selectivity and non-selectivity.

CAUTION: Be extremely careful when installing series-rated breakers. Do *not* "mix" different manufacturer's breakers. Do *not* use any breakers that have not been UL listed as being suitable for use in combination with one another.

The UL *Recognized Component Directory* (Yellow Book) has this to say about series-connected breakers:

> The devices covered under this category are incomplete in certain constructional features or restricted in performance capabilities and are intended for use as components of complete equipment submitted for investigation rather than for direct separate installation in the field. The final acceptance of the component is dependent upon its installation and use in complete equipment submitted to Underwriters Laboratories Inc.

NEC® Section 240-86(a) requires that the manufacturer of a series-rated panel must legibly mark the equipment. The label will show the many ampere ratings and catalog numbers of those breakers that are suitable for use with that particular piece of equipment:

NEC® Section 110-22 extends *Section 240-86(a)* by requiring that there be "field marking" of series-related equipment. Most electrical inspectors expect the electrical contractor to attach this labeling to the equipment. For example, the main distribution equipment of the Commercial building is located remote from the individual tenants' panelboards. The electrical contractor will affix legible and durable labels to each disconnect switch in the main distribution equipment, identifying the size and type of overcurrent that has been installed for the proper protection of the downstream panelboards, Figure 17-33.

This CAUTION — SERIES RATED SYSTEM labeling will alert future users of the equipment that they should not indiscriminately replace existing overcurrent devices or install additional overcurrent devices with sizes and types of fuses or circuit breakers not compatible with the series-rated system as originally designed and installed. Otherwise the integrity of the series-rated system might be sacrificed, creating a hazard to life and property.

SERIES-RATED SYSTEMS WHERE ELECTRIC MOTORS ARE CONNECTED

Series-rated systems have available fault current higher than the interrupting rating of the load side breakers and have an acceptable overcurrent device of higher interrupting rating ahead of the load side breakers. The series combinations' interrupting rating is marked on the end use equipment, *NEC® Section 240-86(a)*.

Because electric motors contribute short-circuit current under fault conditions on an electrical system, *NEC® Section 240-86(b)* sets forth two requirements when series-rated systems are applied where electric motors are involved.

CAUTION — SERIES COMBINATION
SYSTEM RATED _____ AMPERES
IDENTIFIED REPLACEMENT
COMPONENTS REQUIRED

Figure 17-33   Typical label.

1. Do not connect electric motors between the load side of the higher-rated overcurrent device and on the line side of the lower-rated overcurrent device.

2. The sum of the connected motor full-load currents shall not exceed 1% of the interrupting rating of the lower-rated circuit breaker. For example, if the interrupting rating of the lower-rated breakers in a series-rated panel is 10,000 amperes, then the 1% motor load limitation is 100 amperes. Therefore, the full-load currents of all of the motors connected to this series-rated panel must not exceed 100 amperes.

CURRENT-LIMITING BREAKERS

A current-limiting circuit breaker will limit the let-through energy (I^2t) to something less than the I^2t of a one-half cycle wave.

The label on this type of breaker will show the words "CURRENT-LIMITING." The label will also indicate the breaker's let-through characteristics or will indicate where to obtain the let-through characteristic data. This let-through data is necessary to assure adequate protection of downstream components (wire, breakers, controllers, bus bar bracing, and so on). Therefore, it is important when installing circuit breakers to insure that not only the circuit breakers have the proper interrupting rating (capacity), but also that all of the components connected to the circuit downstream from the breakers are capable of withstanding the let-through current of the breaker. To use the previous problem as an example, this means that the branch-circuit conductors must be capable of withstanding 40,000 amperes for approximately one cycle (the opening time of the breaker).

COST CONSIDERATIONS

As previously discussed, there are a number of different types of circuit breakers to choose from. The selection of the type to use depends upon a number of factors, including the interrupting rating, selectivity, space, and cost. When an installation requires something other than the standard molded-case-type circuit breaker, the use of high interrupting types or current-limiting types may be necessary. The electrician and/or design engineer

Table 17-4 List Price Cost Comparisons		
Type	Interrupting Rating	List Price
Standard, plug-in	10,000	$165.00
Standard, bolt-on	22,000	322.00
Standard, bolt-on	100,000	390.00
Current-limiting	200,000	1163.00
Electronic solid-state having adjustable trips and ground fault protection, 225 frame size	65,000	3323.00

must complete a "short-circuit study" to assure that the overcurrent protective devices provide proper protection against short-circuit conditions per *National Electrical Code®* requirements.

The list price cost comparisons in Table 17-4 are for various 100-ampere, 240-volt, panelboard-type circuit breakers.

Molded case circuit breakers with an integral current-limiting fuse are available. The thermal element in this type of breaker is used for low overloads, the magnetic element is used for low-level short circuits, and the integral fuse is used for short circuits of high magnitude. The interrupting capacity of this breaker/fuse combination (sometimes called a *limiter*) is higher than that of the same breaker without the fuse.

The selection of the ampacity of circuit breakers for branch circuits is governed by the requirements of *NEC® Article 240*.

MOTOR CIRCUITS

▶ *NEC® Table 430-152* shows that on motor branch circuits, the maximum setting of an inverse-time circuit breaker must not exceed 250% of the full-load current of the motor. For an instantaneous trip circuit breaker, permitted only in "listed" combination controllers, the maximum setting is 800% of the full-load current of Design B, C, and D motors. For Design E energy efficient motors, the maximum setting is 1100%.

NEC® Section 430-52(c), Exception No. 1, recognizes that the 800% and 1100% settings listed in *NEC® Table 430-152* might not allow the motor to start, particularly in the case of energy efficient motors designated as Design E motors. Design E motors have a high starting inrush current that can cause an instantaneous circuit breaker to nuisance trip or a fuse to open. If an "engineering evaluation" can demonstrate the need to exceed the percentages shown in *NEC® Table 430-152*, then:

- the 800% setting may be increased to a maximum of 1300%.

- the 1100% setting may be increased to a maximum of 1700%. ◄

Additional discussion regarding motor circuit design is found in Unit 7.

HACR CIRCUIT BREAKERS

Breakers suitable for the protection of heating, air-conditioning, and refrigeration equipment where multimotor or combination loads are involved must be marked "LISTED HACR TYPE," see Figures 17-34, 17-35, and 17-36. Note that "H" means "heating," "AC" means "air conditioning," and "R" means "refrigeration." The equipment to be protected will also bear the HACR reference on the equipment nameplate. Refer to the UL Green Book.

CAUTION: Carefully read the label on the equipment to be connected. When the label states "FUSES," fuses must be used for the overcurrent protection. Circuit breakers are not permitted in this case.

Some labels might read "maximum size overcurrent device," in which case fuses or circuit breakers could be used for branch-circuit overcurrent protection.

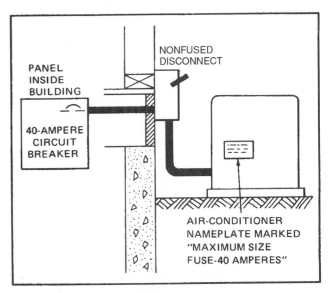

Fig. 17-35   This installation violates *NEC® Section 110-3(b)*. Although the disconnect switch is within sight of the air conditioner, it does not contain fuses. Note that the branch-circuit protection is provided by the 40-ampere circuit breaker inside the building. Note also that the nameplate requires that the branch-circuit protection be 40-ampere fuses maximum. If fused branch-circuit protection were provided at the panel inside the building, the installation would meet the Code requirements.

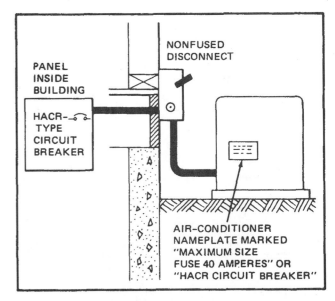

Fig. 17-36   Type HACR circuit breaker (two-pole, 15 through 60-ampere, 120/240-volt) are UL listed for group motor application such as found on heating (H), air conditioning (AC), and refrigeration (R). The label on these breakers will have the letters HACR. The equipment must also be tested and marked "Suitable for protection by a Type HACR breaker." If the HACR marking is not found on both the circuit breaker and the equipment, the installation does not meet the Code, unless fuses are installed for the branch-circuit overcurrent protection.

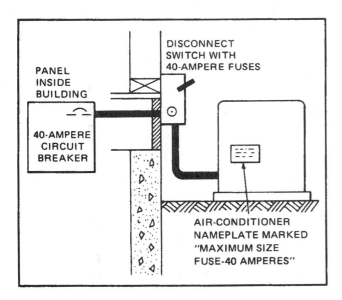

Fig. 17-34   This installation complies with *NEC® Section 440-14*. The disconnect switch is within sight of the unit and contains the 40-ampere fuses called for on the air-conditioner nameplate as the branch-circuit protection.

REVIEW QUESTIONS

Refer to the *National Electrical Code®* or the working drawings when necessary. Where applicable, responses should be written in complete sentences. Write units using unit names, use no abbreviations or symbols (1 foot, not 1' or 1 ft).

1. Why must overcurrent protection be provided?

    _____

    _____

    _____

    _____

Several factors should be understood when selecting overcurrent protective devices. In your own words, define each of the following as they apply to overcurrent devices.

2. Voltage rating:

    _____

    _____

    _____

    _____

3. Continuous current rating:

    _____

    _____

    _____

    _____

4. Interrupting rating:

    _____

    _____

    _____

    _____

5. Speed of response:

    _____

    _____

    _____

    _____

6. Current-limitation:

    _____

    _____

    _____

    _____

7. Peak let-through:

_____

_____

_____

_____

NEC® Table 430-152 gives four options for selecting the *maximum* size motor branch-circuit, short-circuit, and ground-fault protective devices. These options are listed below. For each option, determine the correct rating or setting for a 25-hp, three-phase, 230-volt motor. The motor is marked "Design E."

8. Non-time-delay fuses:

_____

_____

9. Dual-element, time-delay fuses:

_____

_____

10. Instantaneous trip breakers:

_____

_____

11. Inverse time breakers:

_____

_____

12. Define the following terms:
    a. I^2t _____
    b. I_p _____
    c. RMS _____

13. Class J fuses (will fit) (will not fit) into standard fuse clips. (Circle the correct answer.)

Using the chart from Figure 17-24, find the opening time for a 60-ampere fuse under the following conditions:

14. A 300-ampere load will cause the fuse to open in approximately _____ seconds.

15. A 400-ampere load will cause the fuse to open in approximately _____ seconds.

16. An 800-ampere short circuit will cause the fuse to open in approximately _____ seconds.

17. A 5000-ampere short circuit will cause the fuse to open in approximately _____ seconds.

Using the chart from Figure 17-28, determine the approximate current values for a 60-ampere, 250-volt fuse when the prospective short-circuit current is 40,000 amperes.

18. The instantaneous peak let-through current is approximately _____ amperes.

19. The apparent RMS current is approximately _____ amperes.

20. A section of plug-in busway nameplate indicates that the busway is braced for 14,000 amperes. The available fault current is 30,000 amperes. Using Figure 17-28, determine if a 200-ampere, 250-volt, dual-element fuse will limit the current sufficiently to protect the busway against a possible fault of 30,000 amperes. Explain.

_____

_____

_____

21. In a thermal-magnetic circuit breaker, overloads are sensed by the _____ element and short circuits are sensed by the _____ element.

22. A cable limiter is: (Circle the correct answer.)
    a. a short-circuit device only.   T   F
    b. not to be used for overload protection of conductors.   T   F
    c. generally connected to both ends of large paralleled conductors so
       that if a fault occurs on one of the conductors, that faulted cable is
       isolated from the system.   T . F

23. Listed series-rated equipment allows the branch-circuit breakers to have a lower interrupting rating than the main. What two sections of the *National Electrical Code*® cover the marking of series-rated systems with the words **CAUTION — SERIES RATED SYSTEM**? *Sections* _____

Using the chart in Figure 17-30, provide the unlatching time and interrupting time (also referred to as *opening time*) for a 50-ampere circuit breaker under the conditions stated below. Unlatching time and interrupting time is critical when designing a selectively coordinated electrical system.

24. For a load of 300 amperes, the minimum interrupting time is _____ seconds.

25. For a load of 300 amperes, the maximum interrupting time is _____ seconds.

26. For a 1000-ampere fault, the average unlatching time is _____ seconds.

27. For a 5000-ampere fault, the average unlatching time is _____ seconds.

28. The ampere range of Class L fuses is from _____ to _____ amperes.

29. The voltage rating of a plug fuse is _____ volts.

30. The interrupting rating of a Class J fuse is _____ amperes.

31. When the nameplate of HVAC equipment is marked "Maximum Size Fuse 50-Amperes," is it permitted to connect the equipment to a 50-ampere circuit breaker?

_____

UNIT 18

Short-Circuit Calculations and Coordination of Overcurrent Protective Devices

OBJECTIVES

After studying this unit, the student will be able to

- perform short-circuit calculations using the point-to-point method.
- calculate short-circuit currents using the appropriate tables and charts.
- define the terms *coordination, selective systems*, and *nonselective systems*.
- define the term interrupting rating and explain its significance.
- use time-current curves.

The student must understand the intent of *NEC®* *Sections 110-9* and *110-10*. That is, to insure that the fuses and/or circuit breakers selected for an installation are capable of interrupting the current at the rated voltage that may flow under any condition (overload, short circuit, or ground fault) with complete safety to personnel and without damage to the panel, load center, switch, or electrical equipment in which the protective devices are installed.

An overloaded condition resulting from a miscalculation of load currents will cause a fuse to blow or a circuit breaker to trip in a normal manner. However, a miscalculation, a guess, or ignorance of the magnitude of the available short-circuit currents may result in the installation of breakers or fuses having inadequate interrupting ratings. Such a situation can occur even though the load currents in the circuit are checked carefully. **Breakers or fuses having inadequate interrupting ratings need only be subjected to a short circuit to cause them to explode, resulting in injury to personnel and**

serious damage to the electrical equipment. *The interrupting rating of an overcurrent device is its maximum rating and must not be exceeded.*

In any electrical installation, individual branch circuits are calculated as has been discussed previously in this text. After the quantity, size, and type of branch circuits are determined, these branch-circuit loads are then combined to determine the size of the feeder conductors to the respective panelboards. Most consulting engineers will specify that a certain number of spare branch-circuit breakers be installed in the panelboard, plus a quantity of spaces that can be used in the future.

For example, a certain computation may require a minimum of sixteen 20-ampere branch circuits. The specification might call for a twenty-four circuit panelboard with sixteen active circuits, four spares, and four spaces.

The next step is to determine the interrupting rating requirements of the fuses or circuit breakers to be installed in the panel. *NEC® Section 110-9* is an all-encompassing section that covers the interrupting rating requirements for services, mains, feeders, subfeeders, and branch-circuit overcurrent of devices. For various types of equipment, normal currents can be determined by checking the equipment nameplate current, voltage, and wattage ratings. In addition, an ammeter can be used to check for normal and overloaded circuit conditions.

A standard ammeter must not be used to read short-circuit current, as this practice will result in damage to the ammeter and possible injury to personnel.

SHORT-CIRCUIT CALCULATIONS

The following sections will cover several of the basic methods of determining available short-circuit currents. As the short-circuit values given in the arious tables are compared with the actual calculations, it will be noted that there are slight variances in the results. These differences are due largely to (1) the rounding off of the numbers in the calculations and (2) variations in the resistance and reactance data used to prepare the tables and charts. For example, the value of the square root of three (1.732) is used frequently in three-phase calcula-

Determining the Short-Circuit Current at the Terminals of a Transformer Using the Impedance Formula

PROBLEM 1:

Assume that the three-phase transformer installed by the utility company for the commercial building has a rating of 300 kVA at 208Y/120 volts with an impedance of 2% (from the transformer nameplate). The available short-circuit current at the secondary terminals of the transformer must be determined. To simplify the calculations, it is also assumed that the utility can deliver unlimited short-circuit current to the primary of the transformer. In this case, the transformer primary is known as an *infinite bus* or an *infinite primary*.

The first step is to determine the normal full-load current rating of the transformer:

$$I \text{ (at the secondary)} = \frac{\text{kVA} \times 1,000}{E \times 1.73} = \frac{300 \times 1,000}{208 \times 1.73}$$

$$= 834 \text{ amperes normal full load}$$

Using the impedance value given on the nameplate of the transformer, the next step is to find a multiplier that can be used to determine the short-circuit current available at the secondary terminals of the transformer.

The factor of 0.9 shown in the calculations below reflects the fact that the transformer's actual impedance might be 10% less than marked on the nameplate and would be a worst case condition. In electrical circuits, the lower the impedance, the higher the current.

If the transformer is marked 2% impedance:

Then, multiplier $= \dfrac{100}{2 \times 0.9} = 55.556$

and, $834 \times 55.556 = 46,334$ amperes of short-circuit current.

If the transformer is marked 1% impedance:

Then, multiplier $= \dfrac{100}{1 \times 0.9} = 111.111$

and, $834 \times 111.111 = 92,667$ amperes of short-circuit current.

If the transformer is marked 4% impedance:

Then, multiplier $= \dfrac{100}{4 \times 0.9} = 27.778$

and, $834 \times 27.778 = 23,167$ amperes of short-circuit current.

tions. Depending on the accuracy required, values of 1.7, 1.73, or 1.732 can be used in the calculations.

In actual practice, the available short-circuit current at the load side of a transformer is less than the values shown in Problem 1. However, this simplified method of finding the available short-circuit currents will result in values that are conservative.

The actual impedance value on a UL listed 25 kVA or larger transformer can vary plus or minus 10% from the transformer's marked impedance. The actual impedance could be as low as 1.8% and as high as 2.2%. This will affect the available fault current calculations.

For example, in Problem 1, the *marked* impedance is reduced by 10% to reflect the transformer's possible *actual* impedance. The calculations show this "worst case" scenario. All short-circuit examples in this text have the marked transformer impedance values reduced by 10%.

Another factor that affects fault current calculations is voltage. Utility companies are allowed to vary voltage to their customers within a certain range. This might be plus or minus 10% for power services, and 5% for lighting services. The higher voltage will result in a larger magnitude of fault current.

Another source of short-circuit current comes from electric motors that are running at the time the fault occurs. This is covered later on in this unit.

Thus, it can be seen that no matter how much data we plug into our fault current calculations, there are many variables that are out of our control. What we hope for is to arrive at a result that is reasonably accurate so that our electrical equipment is reasonably safe insofar as interrupting ratings and withstand ratings are concerned.

In addition to the methods of determining available short-circuit currents that are provided below, there are computer software programs that do the calculations. These programs are fast, par-

ticularly when there are many points in a system to be calculated.

Determining the Short-Circuit Current at the Terminals of a Transformer Using Tables

Table 18-1 is a table of the short-circuit currents for a typical transformer. NEMA and transformer manufacturers publish short-circuit tables for many sizes of transformers having various impedance values. Table 18-1 provides data for a 300-kVA, three-phase transformer with an impedance of 2%. According to the table, the symmetrical short-circuit current is 42,090 amperes at the secondary terminals of a 208Y/120-volt transformer (refer to the zero-foot row of the table). This value is on the low side because the manufacturer that developed the table did not allow for the ± impedance variation allowed by the UL standard. The table shown in Table 18-2 indicates that the available short-circuit current at the secondary of the transformer is 46,334 amperes.

Determining the Short-Circuit Current at Various Distances from a Transformer Using the Table in Figure 18-1

The amount of available short-circuit current decreases as the distance from the transformer increases, as indicated in Table 18-1. See Problem 2.

Determining Short-Circuit Currents at Various Distances from Transformers, Switchboards, Panelboards, and Load Centers Using the Point-to-Point Method

A simple method of determining the available short-circuit currents (also referred to as fault current) at various distances from a given location is the *point-to-point method*. Reasonable accuracy is obtained when this method is used with three-phase and single-phase systems.

PROBLEM 2:

For a 300-kVA transformer with a secondary voltage of 208 volts, find the available short-circuit current at a main switch that is located 25 feet from the transformer. The main switch is supplied by four No. 750 kcmil copper conductors per phase in steel conduit.

Refer to Table 18-1 and read the value of 39,090 amperes in the column on the right-hand side of the table for a distance of 25 feet.

TABLE 18-1   Symmetrical Short-Circuit Currents at Various Distances from a Liquid Filled Transformer

WIRE-SIZE (COPPER)

VOLTS	DIST. (FT.)	#14	#12	#10	#8	#6	#4	#1	0	00	000	2-000	0000	250 kcmil	2-250 kcmil	3-300 kcmil	350 kcmil	2-350 kcmil	3-350 kcmil	3-400 kcmil	500 kcmil	2-500 kcmil	750 kcmil	4-750 kcmil
208	0	42090	42090	42090	42090	42090	42090	42090	42090	42090	42090	42090	42090	42090	42090	42090	42090	42090	42090	42090	42090	42090	42090	(42090)
	5	6910	10290	14730	19970	25240	29840	34690	35770	36640	37340	39610	37930	38270	40100	40870	38840	40410	40960	41030	39300	40650	39650	41460
	10	3640	5610	8460	12350	17090	22230	29030	30760	32210	33410	37340	34420	35030	38270	39710	36040	38840	39870	40010	36850	39300	37480	40840
	25	1500	2360	3670	5650	8430	12150	18930	21170	23240	25090	31710	26750	27780	33590	36560	29550	34780	36930	37230	31020	35730	32190	(39090)
	50	760	1200	1890	2950	4530	6810	11740	13670	15610	17510	25090	19320	20520	27780	32250	22660	29550	32850	33340	24520	31020	26050	36480
	100	380	600	960	1510	2350	3610	6610	7920	9320	10810	17510	12320	13380	20520	26010	15400	22660	26850	27530	17250	24520	18860	32190
	200	190	300	480	760	1190	1860	3510	4280	5140	6090	10810	7860	7860	13380	18660	9360	15400	19590	20370	10820	17250	12150	26050
	500	80	120	190	310	480	760	1460	1800	2180	2630	4990	3130	3500	6510	10030	4290	7820	10770	11400	5100	9120	5870	16670
	1000	40	60	100	150	240	380	740	910	1110	1350	2630	1620	1820	3500	5650	2250	4290	6140	6560	2710	5100	3160	10310
	5000	10	10	20	30	50	80	150	180	230	280	550	330	380	740	1260	470	930	1380	1490	570	1130	670	2560
240	0	37820	37820	37820	37820	37820	37820	37820	37820	37820	37820	37820	37820	37820	37820	37820	37820	37820	37820	37820	37820	37820	37820	37820
	5	7750	11330	15810	20720	25260	28940	32560	33340	33960	34460	36080	34870	35120	36420	36960	35520	36640	37020	37070	35840	36800	36090	37370
	10	4140	6320	9400	13430	18040	22670	28230	29560	30660	31550	34460	32290	32730	35120	36140	33470	35520	36260	36350	34060	35840	34510	36930
	25	1720	2700	4180	6360	9360	13190	19640	21620	23380	24920	30240	26260	27090	31650	33860	28480	32530	34130	34340	29610	33230	30510	35680
	50	870	1380	2160	3360	5130	7620	12730	14630	16480	18230	24920	19850	20900	27090	30610	22740	28480	31060	31430	24300	29610	25570	33780
	100	440	700	1100	1730	2680	4100	7380	8770	10220	11720	18230	13220	14240	20900	25600	16150	22740	26280	26630	17860	24300	19310	30510
	200	220	350	550	880	1370	2130	3990	4830	5770	6790	11720	7880	8650	14240	19200	10190	16150	20030	20710	11650	17860	12960	25570
	500	90	140	220	350	560	870	1670	2050	2490	2990	5610	3540	3960	7230	10890	4820	8590	11620	12250	5700	9920	6520	17200
	1000	40	70	110	180	280	440	850	1050	1280	1550	2990	1850	2080	3960	6300	2560	4820	6820	7270	3070	5700	3570	11130
	5000	10	10	20	40	60	90	170	210	260	320	630	380	430	860	1440	540	1070	1580	1710	660	1290	770	2910
480	0	18910	18910	18910	18910	18910	18910	18910	18910	18910	18910	18910	18910	18910	18910	18910	18910	18910	18910	18910	18910	18910	18910	18910
	5	10450	12820	14750	16150	17080	17690	18200	18310	18400	18470	18690	18520	18550	18730	18800	18610	18760	18810	18810	18650	18780	18690	18850
	10	6750	9170	11630	13780	15400	16530	17540	17740	17910	18040	18470	18150	18210	18550	18690	18320	18610	18710	18720	18400	18650	18470	18800
	25	3180	4740	6770	9150	11520	13570	15690	16160	16540	16840	17830	17100	17250	18040	18380	17490	18180	18410	18440	17690	18280	17840	18630
	50	1680	2590	3900	5680	7840	10170	13190	13960	14600	15120	16840	15560	15820	17250	17870	16260	17490	17940	18000	16610	17690	16890	18360
	100	860	1350	2090	3180	4680	6600	9820	10810	11690	12460	15120	13130	13540	15820	16930	14240	16260	17060	17170	14810	16610	15260	17840
	200	440	690	1080	1680	2560	3810	6370	7320	8240	9110	12460	9930	10010	13540	15300	11370	14240	15530	15710	12150	14810	12780	16890
	500	180	280	440	700	1080	1670	3040	3640	4290	4960	8010	5560	6140	9360	11820	7050	10320	12190	12500	7880	11150	8600	14550
	1000	90	140	220	350	550	860	1620	1970	2370	2800	4960	3270	3610	6140	8520	4300	7050	8940	9290	4960	7880	5560	11830
	5000	20	30	40	70	110	180	340	420	510	620	1210	750	840	1610	2600	1040	1980	2820	3020	1250	2350	1450	4730
600	0	15130	15130	15130	15130	15130	15130	15130	15130	15130	15130	15130	15130	15130	15130	15130	15130	15130	15130	15130	15130	15130	15130	15130
	5	10210	11790	12920	13690	14180	14500	14770	14820	14870	14900	15010	14930	14940	15040	15070	14970	15050	15080	15080	15000	15060	15010	15100
	10	7270	9270	11010	12350	13280	13890	14410	14520	14610	14680	14900	14730	14770	14940	15020	14820	14970	15020	15030	14870	15000	14900	15070
	25	3740	5370	7280	9230	10920	12200	13410	13670	13870	14040	14570	14170	14250	14680	14850	14380	14750	14870	14890	14490	14800	14570	14980
	50	2040	3080	4500	6270	8170	9950	11940	12400	12770	13060	14040	13310	13460	14250	14590	13700	14380	14620	14650	13900	14490	14050	14840
	100	1060	1650	2510	3730	5290	7080	9650	10350	10930	11420	13060	11840	12090	13460	14080	12510	13700	14150	14210	12850	13900	13120	14570
	200	540	850	1330	2040	3040	4390	6840	7640	8390	9050	11420	9640	10010	12090	13160	10640	12510	13290	13390	11160	12850	11580	14050
	500	220	350	550	860	1350	2010	3550	4180	4830	5480	8180	6110	6530	9210	10960	7310	9900	11210	11400	7990	10470	8560	12700
	1000	110	170	280	440	680	1050	1950	2360	2800	3270	5480	3760	4110	6530	8540	4780	7310	8860	9120	5410	7990	5970	10940
	5000	20	40	60	90	140	220	420	520	640	770	1470	910	1030	1930	3030	1260	2340	3270	3480	1510	2750	1740	5180

TABLE 18-2 Short-Circuit Currents Available from Various Size Transformers				
Voltage and Phase	KVA	Full Load Amps	% Impedance†† (Nameplate)	Short Circuit Amps†
120/240 1 ph.*	25	104	1.58	11,574
	37½	156	1.56	17,351
	50	209	1.54	23,122
	75	313	1.6	32,637
	100	417	1.6	42,478
	167	695	1.8	60,255
120/208 3 ph.**	25	69	1.6	4,791
	50	139	1.6	9,652
	75	208	1.11	20,821
	100	278	1.11	27,828
	150	416	1.07	43,198
	225	625	1.12	62,004
	300	833	1.11	83,383
	500	1388	1.24	124,373
	750	2082	3.5	66,095
	1000	2776	3.5	88.167
	1500	4164	3.5	132,190
	2000	5552	5.0	123,377
	2500	6950	5.0	154,444
277/480 3 ph.	112½	135	1.0	15,000
	150	181	1.2	16,759
	225	271	1.2	25,082
	300	361	1.2	33,426
	500	601	1.3	51,362
	750	902	3.5	28,410
	1000	1203	3.5	38,180
	1500	1804	3.5	57,261
	2000	2406	5.0	53,461
	2500	3007	5.0	66,822

* Single-phase values are L-N values at transformer terminals. These figures are based on change in turns ratio between primary and secondary, 100,000 KVA primary, zero feet from terminals of transformer, 1.2 (%×) and 1.5 (%R) multipliers for L-N vs. L-L reactance and resistance values and transformer ×/R ratio = 3.

** Three-phase short-circuit currents based on "infinite" primary.

†† U.L. listed transformers 25 KVA or greater have a ±10% impedance tolerance. Short-circuit amps reflect a "worst case" condition.

† Fluctuations in system voltage will affect the available short-circuit current. For example, a 10% increase in system voltage will result in a 10% increase in the available short-circuit currents shown in the table.

The following procedure demonstrates the use of the point-to-point method:

Step 1.   Determine the full-load rating of the transformer in amperes from the transformer nameplate, tables, or the following formulas:

a. For three-phase transformers:

$$I_{FLA} = \frac{kVA \times 1,000}{E_{L\text{-}L} \times 1.73}$$

where $E_{L\text{-}L}$ = Line-to-line voltage

b. For single-phase transformers:

$$I_{FLA} = \frac{kVA \times 1,000}{E_{L\text{-}L}}$$

Step 2.   Find the percent impedance (Z) on nameplate of transformer.

Step 3.   Find the transformer multiplier "M_1":

$$M_1 = \frac{100}{\text{transformer \% impedance (Z)} \times 0.9}$$

Note: Because the marked transformer impedance can vary ±10% per the UL standard, the 0.9 factor above takes this into consideration to show "worst case" conditions.

Step 4. Determine the transformer let-through short-circuit current at the secondary terminals of transformer. Use tables or the following formula:

a. For 3-phase transformers (L-L-L):

$$I_{SCA} = transformer_{FLA} \times multiplier \text{ "}M_1\text{"}$$

b. For 1-phase transformers (L-L):

$$I_{SCA} = transformer_{FLA} \times multiplier \text{ "}M_1\text{"}$$

c. For 1-phase transformers (L-N):

$$I_{SCA} = transformer_{FLA} \times multiplier \text{ "}M_1\text{"} \times 1.5$$

Note: The 1.5 factor is explained in Step 5 in the paragraph marked "X."

Step 5. Determine the "f" factor:

a. For 3-phase faults:

$$f = \frac{1.73 \times L \times I}{N \times C \times E_{L\text{-}L}}$$

b. For 1-phase, line-to-line (L-L) faults on 1-phase, center tapped transformers:

$$f = \frac{2 \times L \times I}{N \times C \times E_{L\text{-}L}}$$

c. For 1-phase, line-to-neutral (L-N) faults on 1-phase, center tapped transformers:

$$f = \frac{2 \times L \times I}{N \times C \times E_{L\text{-}N}}$$

where . . .

L = the length of the circuit to the fault, in feet.

I = the available fault current in amperes at the beginning of the circuit.

C = the constant derived from Table 18-3 for the specific type of conductors and wiring method.

E = the voltage, line-to-line or line-to-neutral. See Step 4a, b, and c to decide which voltage to use.

N = the number of conductors in parallel.

X = the fault current in amperes at the transformer terminals. At the secondary terminals of a single-phase, center-tapped transformer, the L-N fault current is higher than the L-L fault current. At some distance from the terminals, depending upon the wire size and type, the L-N fault current is lower than the L-L fault

current. This can vary from 1.33 to 1.67 times. These figures are based on the change in the turns ratio between the primary and secondary, infinite source impedance, a distance of zero feet from the terminals of the transformer, and 1.2% reactance *(X)* and 1.5% resistance *(R)*.for the L-N vs L-L resistance and reactance values. For simplicity, in Step 4c we used an approximate multiplier of 1.5. First, do the L-N calculation at the transformer secondary terminals, Step 4c, then proceed with the point-to-point method. See Figure 18-2 for example.

Step 6. After finding the "f" factor, refer to Table 18-4 and locate in Chart M the appropriate value of the multiplier "M_2" for the specific "f" value. Or, calculate as follows:

$$M_2 = \frac{1}{1+f}$$

Step 7. Multiply the available fault current at the beginning of the circuit by the multiplier "M_2" to determine the available symmetrical fault current at the fault.

$$I_{SCA} \text{ at fault} = I_{SCA} \text{ at beginning of circuit} \times \text{"}M_2\text{"}$$

Motor Contribution. All motors running at the instant a short circuit occurs contribute to the short-circuit current. The amount of current from the motors is equal approximately to the starting (locked rotor) current for each motor. This current value depends upon the type of motor, its characteristics, and its code letter. Refer to *NEC® Table 430-7(b)*. It is a common practice to multiply the full-load ampere rating of the motor by 4 or 5 to obtain a close approximation of the locked rotor current and provide a margin of safety. For energy-efficient motors, multiply the motor's full-load current rating by 6 to 8 times for a reasonable approximation of fault-current contribution. The current contributed by running motors at the instant a short circuit occurs is added to the value of the short-circuit current at the main switchboard prior to the start of the point-to-point calculations for the rest of the system. To simplify the following problems, motor contributions have not been added to the short-circuit currents.

TABLE 18-3   "C" Values to be Used in Step 5 of the Method of Calculating Available Short-Circuit Currents Discussed in this Unit

AWG or kcmil	Copper Conductors Three Single Conductors						Copper Conductors Three Conductor Cable					
	Steel Conduit			Nonmagnetic Conduit			Steel Conduit			Nonmagnetic Conduit		
	600V	5KV	15KV	600V	5KV	15KV	600V	5KV	15KV	600V	5KV	15KV
14	389	—	—	389	—	—	389	—	—	389	—	—
12	617	—	—	617	—	—	617	—	—	617	—	—
10	981	—	—	981	—	—	981	—	—	981	—	—
8	1557	1551	1557	1556	1555	1558	1559	1557	1559	1559	1558	1559
6	2425	2406	2389	2430	2417	2406	2431	2424	2414	2433	2428	2420
4	4779	3750	3695	3825	3789	3752	3830	3811	3778	3837	3823	3798
3	4760	4760	4760	4802	4802	4802	4760	4790	4760	4802	4802	4802
2	5906	5736	5574	6044	5926	5809	5989	5929	5827	6087	6022	5957
1	7292	7029	6758	7493	7306	7108	7454	7364	7188	7579	7507	7364
1/0	8924	8543	7973	9317	9033	8590	9209	9086	8707	9472	9372	9052
2/0	10755	10061	9389	11423	10877	10318	11244	11045	10500	11703	11528	11052
3/0	12843	11804	11021	13923	13048	12360	13656	13333	12613	14410	14118	13461
4/0	15082	13605	12542	16673	15351	14347	16391	15890	14813	17482	17019	16012
250	16483	14924	13643	18593	17120	15865	18310	17850	16465	19779	19352	18001
300	18176	16292	14768	20867	18975	17408	20617	20051	18318	22524	21938	20163
350	19529	17385	15678	22736	20526	18672	22646	21914	19821	24904	24126	21982
400	20565	18235	16365	24296	21786	19731	24253	23371	21042	26915	26044	23517
500	22185	19172	17492	26706	23277	21329	26980	25449	23125	30028	28712	25916
600	22965	20567	17962	28033	25203	22097	28752	27974	24896	32236	31258	27766
750	24136	21386	18888	28303	25430	22690	31050	30024	26932	32404	31338	28303
1000	25278	22539	19923	31490	28083	24887	33864	32688	29320	37197	35748	31959

AWG or kcmil	Aluminum Conductors Three Single Conductors						Aluminum Conductors Three Conductor Cable					
	Steel Conduit			Nonmagnetic Conduit			Steel Conduit			Nonmagnetic Conduit		
	600V	5KV	15KV	600V	5KV	15KV	600V	5KV	15KV	600V	5KV	15KV
14	236	—	—	236	—	—	236	—	—	236	—	—
12	375	—	—	375	—	—	375	—	—	375	—	—
10	598	—	—	598	—	—	598	—	—	598	—	—
8	951	950	951	951	950	951	951	951	951	951	951	951
6	1480	1476	1472	1481	1478	1476	1481	1480	1478	1482	1481	1479
4	2345	2332	2319	2350	2341	2333	2351	2347	2339	2353	2349	2344
3	2948	2948	2948	2958	2958	2958	2948	2956	2948	2958	2958	2958
2	3713	3669	3626	3729	3701	3672	3733	3719	3693	3739	3724	3709
1	4645	4574	4497	4678	4631	4580	4686	4663	4617	4699	4681	4646
1/0	5777	5669	5493	5838	5766	5645	5852	5820	5717	5875	5851	5771
2/0	7186	6968	6733	7301	7152	6986	7327	7271	7109	7372	7328	7201
3/0	8826	8466	8163	9110	8851	8627	9077	8980	8750	9242	9164	8977
4/0	10740	10167	9700	11174	10749	10386	11184	11021	10642	11408	11277	10968
250	12122	11460	10848	12862	12343	11847	12796	12636	12115	13236	13105	12661
300	13909	13009	12192	14922	14182	13491	14916	14698	13973	15494	15299	14658
350	15484	14280	13288	16812	15857	14954	15413	15490	15540	16812	17351	16500
400	16670	15355	14188	18505	17321	16233	18461	18063	16921	19587	19243	18154
500	18755	16827	15657	21390	19503	18314	21394	20606	19314	22987	22381	20978
600	20093	18427	16484	23451	21718	19635	23633	23195	21348	25750	25243	23294
750	21766	19685	17686	23491	21769	19976	26431	25789	23750	25682	25141	23491
1000	23477	21235	19005	28778	26109	23482	29864	29049	26608	32938	31919	29135

Ampacity	Plug-In Busway		Feeder Busway		High Imped. Busway
	Copper	Aluminum	Copper	Aluminum	Copper
225	28700	23000	18700	12000	—
400	38900	34700	23900	21300	—
600	41000	38300	36500	31300	—
800	46100	57500	49300	44100	—
1000	69400	89300	62900	56200	15600
1200	94300	97100	76900	69900	16100
1350	119000	104200	90100	84000	17500
1600	129900	120500	101000	90900	19200
2000	142900	135100	134200	125000	20400
2500	143800	156300	180500	166700	21700
3000	144900	175400	204100	188700	23800
4000	—	—	277800	256400	—

TABLE 18-4 Chart to Convert "f" Values to "M" Values When Using the Point-to-Point Method.			
f	**M**	**f**	**M**
0.01	0.99	1.20	0.45
0.02	0.98	1.50	0.40
0.03	0.97	2.00	0.33
0.04	0.96	3.00	0.25
0.05	0.95	4.00	0.20
0.06	0.94	5.00	0.17
0.07	0.93	6.00	0.14
0.08	0.93	7.00	0.13
0.09	0.92	8.00	0.11
0.10	0.91	9.00	0.10
0.15	0.87	10.00	0.09
0.20	0.83	15.00	0.06
0.30	0.77	20.00	0.05
0.40	0.71	30.00	0.03
0.50	0.67	40.00	0.02
0.60	0.63	50.00	0.02
0.70	0.59	60.00	0.02
0.80	0.55	70.00	0.01
0.90	0.53	80.00	0.01
1.00	0.50	90.00	0.01
		100.00	0.01

$$M = \frac{1}{1 + f}$$

SHORT-CIRCUIT CURRENT VARIABLES

Phase-to-Phase-to-Phase Fault

Three-phase fault current value determined in Step 6 is the *approximate* current that will flow if the three "hot" phase conductors of a three-phase system are shorted together, in what is commonly referred to as a "bolted fault." This is the worst case condition.

Phase-to-Phase Fault

To obtain the *approximate* short-circuit current values when two "hot" conductors of a three-phase system are shorted together, use 87% of the three-phase current value. In other words, if the three-phase current value is 20,000 amperes when the three "hot" lines are shorted together (L-L-L value), then the short-circuit current to two "hot" lines shorted together (L-L) is approximately:

$$20{,}000 \times 0.87 = 17{,}400 \text{ amperes}$$

Phase-to-Neutral (Ground)

For solidly grounded three-phase systems, such as the 208Y/120 volt system that supplies the com-

mercial building, the phase-to-neutral (ground) bolted short-circuit value can vary from 25% to 125% of the L-L-L bolted short-circuit current value.

It is rare that the L-N (or L-G) fault current would exceed the L-L-L fault current. Therefore, it is common practice to consider the L-N (or L-G) short-circuit current value to be the same as the L-L-L short-circuit current value.

In summary:

L-L-L bolted short-circuit current = 100%
L-L   bolted short-circuit current =   87%
L-N   bolted short-circuit current = 100%

Example:

If the three-phase (L-L-L) fault current has been calculated to be 20,000 amperes, then the L-N fault current is approximately:

$$20{,}000 \times 1.00 = 20{,}000 \text{ amperes}$$

The main concern is to provide proper interrupting rating for the overcurrent protective devices, and adequate withstand rating for the equipment. Therefore, for most three-phase electrical systems, the line-line-line "bolted" fault current value will provide the desired level of safety.

Arcing Fault Values (Approximate)

There are times when one wishes to know the values of arcing faults. These values vary considerably. The following are acceptable approximations.

Type of Fault	480 Volts	208 Volts
L-L-L	0.89	0.12
L-L	0.74	0.02
L-G	0.38*	0

*Some reference books indicate this value to be 0.19.

Example:

What is the approximate line-to-ground arcing fault value on a 480/277-volt system where the line-line-line fault current has been calculated to be 30,000 amperes.

Solution: $30{,}000 \times 0.38 = 11{,}400$ amperes

Fault Current at Main Switchboard

As shown in Problem 3, the fuses or circuit breakers located in the main switchboard of the commercial building must have an interrupting

PROBLEM 3:

It is desired to find the available short-circuit current at the main switchboard of the commercial building. Once this value is known, the electrician can provide overcurrent devices with adequate interrupting ratings and the proper bus bar bracing within the switchboard (see *NEC® Sections 110-9* and *110-10*). Figure 18-1 shows the actual electrical system for the commercial building. As each of the following steps in the point-to-point method is examined, refer to the tables given in Tables 18-1, 18-2, 18-3, and 18-4 to determine the necessary values of C, f, and M.

Step 1. $I_{FLA} = \dfrac{kVA \times 1,000}{E_{L-L} \times 1.73} = \dfrac{300 \times 1,000}{208 \times 1.73} = 834$ amperes

Step 2.   Multiplier $= \dfrac{100}{Trans.\ \%Z \times 0.9} = \dfrac{100}{2 \times 0.9} = 55.556$

Step 3. $I_{SCA} = 834 \times 55.556 = 46,334$

Step 4. $\dfrac{1.73 \times L \times I}{3 \times C \times E_{L-L}} = \dfrac{1.73 \times 25 \times 46,334}{3 \times 22,185 \times 208} = 0.145$

Step 5. $M_2 = \dfrac{1}{1+f} = \dfrac{1}{1 + 0.145} = 0.87$

Step 6.   The short-circuit current available at the line side lugs on the main switchboard is: $46,334 \times 0.87 = 40,311$ RMS symmetrical amperes. (You could also refer to Figure 18-4, then approximate the multiplier "M_2" by noting that "f" value of 0.145 is very close to 0.15, where "M" equals 0.87 in the chart.)

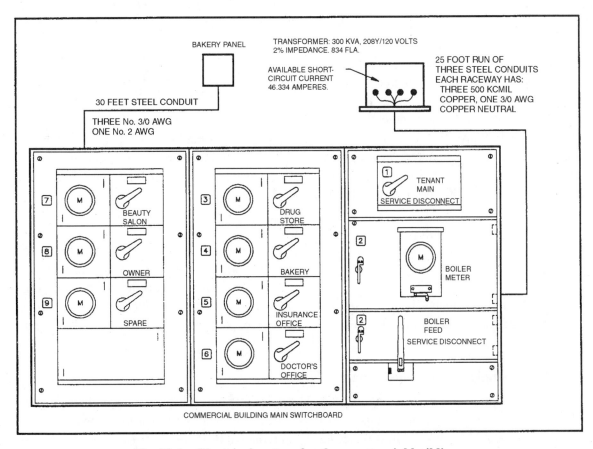

Fig. 18-1   Electrical system for the commercial building.

capacity of at least 40,311 RMS symmetrical amperes. It is good practice to install protective devices having an interrupting rating at least 25% greater than the actual calculated available fault current. This practice generally provides a margin of safety to permit the rounding off of numbers, as well as compensating for a reasonable amount of short-circuit contribution from any electrical motors that may be running at the instant the fault occurs.

The fuses specified for the commercial building have an interrupting rating of 200,000 amperes (see the Specifications). In addition, the switchboard bracing is specified to be 50,000 amperes.

If current-limiting fuses are installed in the main switchboard feeders protecting the various panelboards, breakers having an interrupting rating of 10,000 amperes may be installed in the panelboards. This installation meets the requirements of *NEC® Section 110-9*. If necessary, the student should review the sections covering the application of peak let-through charts in Unit 17.

If non-current-limiting overcurrent devices (standard molded case circuit breakers) are to be installed in the main switchboard, breakers having adequate interrupting ratings must be installed in the panelboards. A short-circuit study must be made for each panelboard location to determine the value of the available short-circuit current.

The cost of circuit breakers increases as the interrupting rating of the breaker increases. The most economical protection system generally results when current-limiting fuses are installed in the main switchboard to protect the breakers in the panelboards. In this case, the breakers in the panelboards will have the standard 10,000 ampere interrupting rating.

See Problem 4 as an example of short-circuit calculations for a Single-Phase Transformer.

Summary

1. To meet the requirements of *NEC® Sections 110-9*, and *110-10*, it is absolutely necessary

to determine the available fault currents at various points on the electrical system. If a short-circuit study is not done, the selection of overcurrent devices may be in error, resulting in a hazard to life and property.

2. To install a *fully-rated system*, use fuses and circuit breakers than have an interrupting rating not less than the available fault current at the point of application.

3. To install a *series-rated system*, use "listed" series-rated electrical equipment where the available fault-current exceeds the short-circuit rating of the downstream lower-rated circuit breakers to be installed in the system. Series-rated electrical equipment is available with fuse/breaker and breaker/breaker combinations. the lower interrupting rated circuit breakers are protected by a higher interrupting rated main breaker. Tested in combination, this arrangement is called "series-rated," Series-rated electrical equipment is marked with its maximum short-circuit rating and the type of overcurrent devices permitted to be used in the specific series-rated combination. See Problem 4.

COORDINATION OF OVERCURRENT PROTECTIVE DEVICES

While this text cannot cover the topic of electrical system coordination (selectivity) in detail, the following material will provide the student with a working knowledge of this important topic. See Problem 4.

What Is Coordination?

Electrical system coordination is defined in *NEC® Section 240-12*. A situation known as *nonselective coordination* occurs when a fault on a branch circuit opens not only the branch-circuit overcurrent device, but also the feeder overcurrent device,

---

PROBLEM 4. Single-Phase Transformer:

　　This problem is illustrated in Figure 18-2. The point-to-point is used to determine the currents for both line-to-line and line-to-neutral faults for a 167-kVA, 2% impedance transformer on a 120/240-volt, single-phase system. Note that the impedance has been reduced by 10%.

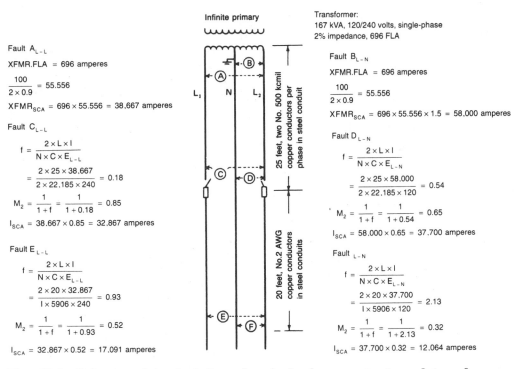

Fault A_{L-L}

XFMR.FLA = 696 amperes

$$\frac{100}{2 \times 0.9} = 55.556$$

$XFMR_{SCA} = 696 \times 55.556 = 38,667$ amperes

Fault C_{L-L}

$$f = \frac{2 \times L \times I}{N \times C \times E_{L-L}}$$

$$= \frac{2 \times 25 \times 38.667}{2 \times 22.185 \times 240} = 0.18$$

$$M_2 = \frac{1}{1 + f} = \frac{1}{1 + 0.18} = 0.85$$

$I_{SCA} = 38.667 \times 0.85 = 32.867$ amperes

Fault E_{L-L}

$$f = \frac{2 \times L \times I}{N \times C \times E_{L-L}}$$

$$= \frac{2 \times 20 \times 32.867}{I \times 5906 \times 240} = 0.93$$

$$M_2 = \frac{1}{1 + f} = \frac{1}{1 + 0.93} = 0.52$$

$I_{SCA} = 32.867 \times 0.52 = 17.091$ amperes

Infinite primary

Transformer:
167 kVA, 120/240 volts, single-phase
2% impedance, 696 FLA

Fault B_{L-N}

XFMR.FLA = 696 amperes

$$\frac{100}{2 \times 0.9} = 55.556$$

$XFMR_{SCA} = 696 \times 55.556 \times 1.5 = 58,000$ amperes

Fault D_{L-N}

$$f = \frac{2 \times L \times I}{N \times C \times E_{L-N}}$$

$$= \frac{2 \times 25 \times 58.000}{2 \times 22.185 \times 120} = 0.54$$

$$M_2 = \frac{1}{1 + f} = \frac{1}{1 + 0.54} = 0.65$$

$I_{SCA} = 58.000 \times 0.65 = 37.700$ amperes

Fault $_{L-N}$

$$f = \frac{2 \times L \times I}{N \times C \times E_{L-N}}$$

$$= \frac{2 \times 20 \times 37.700}{I \times 5906 \times 120} = 2.13$$

$$M_2 = \frac{1}{1 + f} = \frac{1}{1 + 2.13} = 0.32$$

$I_{SCA} = 37.700 \times 0.32 = 12.064$ amperes

25 feet, two No. 500 kcmil copper conductors per phase in steel conduit

20 feet, No. 2 AWG copper conductors in steel conduits

Fig. 18-2   Point-to-point calculations for single-phase, center-tapped transformer. Calculations show L-L and L-N values.

Figure 18-3. Nonselective systems are installed unknowingly and cause needless power outages in portions of an electrical system that should not be affected by a fault.

A *selectively coordinated system*, Figure 18-4, is one in which *only* the overcurrent device immediately upstream from the fault opens. Obviously, the installation of a selective system is much more desirable than a nonselective system.

The importance of selectivity in an electrical system is covered extensively throughout *NEC® Article 517*. This article pertains to health care facilities, where maintaining electrical power is extremely important. The unexpected loss of power in certain areas of hospitals, nursing care centers, and similar health care facilities can be catastrophic.

The importance of selectivity (system coordination) is also emphasized in *NEC® Sections 240-12, 230-95,* and *620-62.* These sections refer to industrial installations where additional hazards would be introduced should a nonorderly shutdown occur. The Code defines coordination, in part, as the proper localizing of a fault condition to restrict outages to the equipment affected.

Some local electrical codes require that all circuits, feeders, and mains in buildings such as schools, shopping centers, assembly halls, nursing homes, retirement homes, churches, restaurants, and any other places of public occupancy be selectively

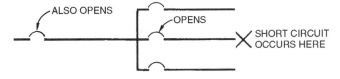

THE FAULT ON THE BRANCH-CIRCUIT TRIPS BOTH THE BRANCH-CIRCUIT BREAKER AND THE FEEDER CIRCUIT BREAKER. AS A RESULT, POWER TO THE PANEL IS CUT OFF AND CIRCUITS THAT SHOULD NOT BE AFFECTED ARE NOW OFF.

Fig. 18-3   Schematic of nonselective system.

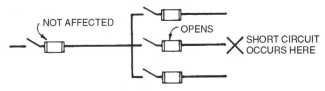

ONLY THE BRANCH-CIRCUIT FUSE OPENS. ALL OTHER CIRCUITS AND THE MAIN FEEDER REMAIN ON.

Fig. 18-4   Schematic of selective system.

coordinated so as to minimize the dangers associated with total power outages.

Nonselectivity of an electrical system is not considered good design practice, and is generally accepted only as a trade-off for a low-cost installation.

It is advisable to check with the authority enforcing the Code before proceeding too far in the selection of overcurrent protective devices for a specific installation.

By knowing how to determine the available short-circuit current and ground-fault current, the electrician can make effective use of the time-current curves and peak let-through charts (Unit 17) to find the length of time required for a fuse to open or a circuit breaker to trip.

What Causes Nonselectivity?

In Figure 18-5, a short circuit in the range of 3000 amperes occurs on the load side of a 20-ampere breaker. The magnetic trip of the breaker is adjusted permanently by the manufacturer to unlatch at a current value equal to 10 times its rating or 200 amperes. The feeder breaker is rated at 100 amperes; the magnetic trip of this breaker is set by the manufacturer to unlatch at a current equal to 10 times its rating or 1000 amperes.

This type of breaker generally cannot be adjusted in the field. Therefore, a current of 200 amperes or more will cause the 20-ampere breaker to trip instantly. In addition, any current of 1000 or more will cause the 100-ampere breaker to trip instantly.

For the breakers shown in Figure 18-5, a momentary fault of 3000 amperes will trip (unlatch) both breakers. Since the flow of current in a series circuit is the same in all parts of the circuit, the 3000-ampere fault will trigger both magnetic trip mechanisms. The time-current curve shown in figure 17-30 indicates that for a 3000-ampere fault, the unlatching time for both breakers is 0.0042 second and the interrupting time for both breakers is 0.016 second.

The term "interrupting time" refers to the time it takes for the circuit breakers' contacts to open, thereby stopping the flow of current in the circuit. Refer to Figures 18-5 and 18-6.

This example of a nonselective system should make apparent to the student the need for a thorough study and complete understanding of time-current curves, fuse selectivity ratios, and unlatching time data for circuit breakers. Otherwise, a blackout may occur, such as the loss of exit and emergency lighting. The student must be able to determine available short-circuit currents (1) to insure the

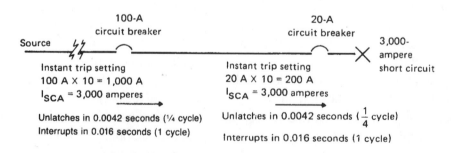

Fig. 18-5   Nonselective system verification.

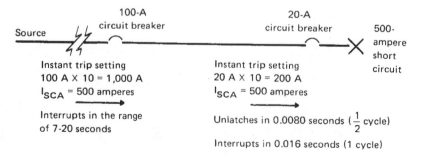

Fig. 18-6   Selective system verification.

proper selection of protective devices with adequate interrupting ratings and (2) to provide the proper coordination as well.

Figure 18-6 shows an example of a selective circuit. In this circuit, a fault current of 500 amperes trips the 20-ampere breaker instantly (the unlatching time of the breaker is approximately 0.0080 second and its interrupting time is 0.016 second). The graph in Figure 17-30 indicates that the 100-ampere breaker interrupts the 500-ampere current in a range from 7 to 20 seconds. This relatively lengthy trip time range is due to the fact that the 500-ampere fault acts upon the current thermal trip element only, and does not affect the magnetic trip element, which operates on a current of 1000 amperes or more.

Selective System Using Fuses

The proper choice of the various classes and types of fuses is necessary if selectivity is to be achieved, Figure 18-7. Indiscriminate mixing of fuses of different classes, time-current characteristics, and even manufacturers may cause a system to become nonselective.

To insure selective operation under low-overload conditions, it is necessary only to check and compare the time-current characteristic curves of fuses. Selectivity occurs when the curves do not cross one another. See Problem 5.

Fuse manufacturers publish *Selectivity Guides,* similar to the one shown in Table 18-5, to be used for short-circuit conditions. When using these guides, selectivity is achieved by maintaining a specific amperage ratio between the various classes and types of fuses. A selectivity chart is based on any fault current up to the maximum interrupting ratings of the fuses listed in the chart.

Selective System Using Circuit Breakers

Circuit-breaker manufacturers publish time-current characteristic curves and unlatching information, see Figure 17-30.

For normal overload situations, a circuit breaker having an ampere rating lower than the ampere rating of an upstream circuit breaker will trip. The upstream breaker will not trip. The system is "selective."

For low-level faults less than the instantaneous trip setting of an upstream circuit breaker, a circuit breaker having an ampere rating lower than the ampere rating of the upstream breaker will trip, and the upstream breaker will not trip. The system is "selective," Figure 18-6.

For fault current levels above the instantaneous trip setting of the upstream circuit breaker, both the branch circuit breaker and the upstream circuit breaker will trip off. The system is "nonselective," Figure 18-5.

There are no ratio selectivity charts for breakers as there are for fuses, Table 18-5.

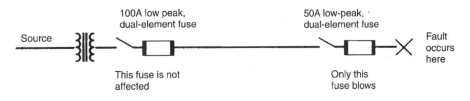

Fig. 18-7   Selective system using fuses.

PROBLEM 5:

   It is desired to install 100-ampere dual-element fuses in a main switch, and 50-ampere dual-element fuses in the feeder switch. Is this combination of fuses "selective"?

   Refer to the Selectivity Guide in Table 18-5 using the *line* and *column* marked Ⓐ. Because 100:50 is a 2:1 ratio, the installation is "selective." In addition, any fuse of the same type having a rating of less than 50 amperes will also be "selective" with the 100-ampere main fuses. Thus, for any short circuit or ground fault on the load side of the 50-ampere fuses, only the 50-ampere fuses will open.

TABLE 18-5   Ratios for Selectivity

LINE SIDE FUSE	LOAD SIDE FUSE									(A)
	KRP-CSP LOW-PEAK time-delay Fuse 601–6000A Class L	KTU LIMITRON fast-acting Fuse 601–6000A Class L	KLU LIMITRON time-delay Fuse 601–4000A Class L	KTN-R, KTS-R LIMITRON fast-acting Fuse 0–600A Class RK1	JJS, JJN TRON fast-acting Fuse 0–1200A Class T	JKS LIMITRON quick-acting Fuse 0–600A Class J	FRN-R, FRS-R FUSETRON dual-element Fuse 0–600A Class RK5	LPN-RK-SP, LPS-RK-SP LOW-PEAK dual-element Fuse 0–600A Class RK1	LPJ-SP LOW-PEAK time-delay Fuse 0–600A Class J	SC Type Fuse 0–60A Class G
KRP-CSP LOW-PEAK time-delay Fuse 601–6000A Class L	2:1	2:1	2.5:1	2:1	2:1	2:1	4:1	2:1	2:1	N/A
KTU LIMITRON fast-acting Fuse 601–6000A Class L	2:1	2:1	2.5:1	2:1	2:1	2:1	6:1	2:1	2:1	N/A
KLU LIMITRON time-delay Fuse 601–4000A Class L	2:1	2:1	2:1	2:1	2:1	2:1	4:1	2:1	2:1	N/A
KTN-R, KTS-R LIMITRON fast-acting Fuse 0–600A Class RK1	N/A	N/A	N/A	3:1	3:1	3:1	8:1	3:1	3:1	4:1
JJN, JJS TRON fast-acting Fuse 0–1200A Class T	N/A	N/A	N/A	3:1	3:1	3:1	8:1	3:1	3:1	4:1
JKS LIMITRON quick-acting Fuse 0–600A Class J	N/A	N/A	N/A	3:1	3:1	3:1	8:1	2:1	2:1	4:1
FRN-R, FRS-R FUSETRON dual-element Fuse 0–600A Class RK5	N/A	N/A	N/A	1.5:1	1.5:1	1.5:1	2:1	1.5:1	1.5:1	1.5:1
LPN-RK-SP, LPS-RK-SP LOW-PEAK dual-element Fuse 0–600A Class RK1 (A)	N/A	N/A	N/A	3:1	3:1	3:1	8:1	2:1	2:1	4:1
LPJ-SP LOW-PEAK time-delay Fuse 0–600A Class J	N/A	N/A	N/A	3:1	3:1	3:1	8:1	2:1	2:1	4:1
SC Type Fuse 0–60A Class G	N/A	N/A	N/A	2:1	2:1	2:1	4:1	3:1	3:1	2:1

N/A = NOT APPLICABLE

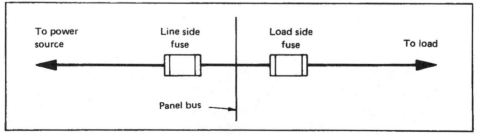

Selectivity Guide. This chart is one manufacturer's selectivity guide. All manufacturers of fuses have similar charts, but the charts may differ in that the ratios for given combinations of fuses may not be the same as in this chart. It is important to use the technical data for the particular type and manufacturer of the fuses being used. Intermixing fuses from various manufacturers may result in a nonselective system.

SINGLE PHASING

The *National Electrical Code®* requires that all three-phase motors be provided with running overcurrent protection in each phase, *NEC® Article 430, Part C*. A line-to-ground fault generally will open one fuse. There will be a resulting increase of 173-200% in the line current in the remaining two connected phases. This increased current will be sensed by the motor fuses and overload relays when such fuses and relays are sized at 125% or less of the full-load current rating of the motor. Thus, the fuses and/or the overload relays will open before the motor windings are damaged. When properly matched, the overload relays will open before the fuses. The fuses offer "backup" protection if the overload relays fail to open for any reason.

A line-to-line fault in a three-phase motor will blow two fuses. In general, the operating coil of the motor controller will drop out, thus providing protection to the motor winding.

To reduce *single-phasing* problems, each three-phase motor must be provided with individual overload protection through the proper sizing of the overload relays and fuses. Phase failure relays are also available. However, can other equipment be affected by a single-phasing condition?

In general, loads that are connected line-to-neutral or line-to-line, such as lighting, receptacles, and electric heating units will not burn out under a single-phasing condition. In other words, if one main fuse blows, then two-thirds of the lighting, receptacles, and electric heat will remain on. If two main fuses blow, then one-third of the lighting remains on, and a portion of the electric heat connected line-to-neutral will stay on.

Nothing can prevent the occurrence of single-phasing. What must be detected is the increase in current that occurs under single-phase conditions.

This is the purpose of motor overload protection as required in *NEC® Article 430, Part C*.

It is essential to maintain some degree of lighting in occupancies such as stores, schools, offices, and health care facilities (such as nursing homes). A total blackout in these public structures has the potential for causing extensive personal injury due to panic. A loss of one or two phases of the system supplying a building should not cause a complete power outage in the building.

REVIEW QUESTIONS

Refer to the *National Electrical Code®* or the working drawings when necessary. Where applicable, responses should be written in complete sentences. Write units using unit names, use no abbreviations or symbols (1 foot, not 1' or 1 ft).

Refer to the table in figure 18-1, which shows the symmetrical short-circuit currents for a 300-kVA transformer with 2% impedance.

1. The current on a 208-volt system at a distance of 50 feet using 1 AWG conductors will be _____ amperes.

2. The current on a 480-volt system at a distance of 100 feet using 500-kcmil conductors will be _____ amperes.

3. In your own words define, and give a hypothetical example of, a selectively coordinated system.

_____

_____

_____

_____

_____

4. Identify some common causes of nonselectivity.

_____

_____

_____

_____

_____

_____

Indicate if the following systems would be selective or nonselective. Support your answer.

5. A KRP-C 2000-ampere fuse is installed on a feeder that serves an LPS-R 600-ampere fuse.

_____

_____

_____

_____

6. A KTS-R 400-ampere fuse is installed on a feeder that serves an FRS-R 200-ampere Fustron fuse.

_____

_____

_____

_____

7. A panelboard is protected by a main 225-ampere circuit breaker and contains several 20-ampere breakers for branch-circuit protection. The breakers are all factory set to unlatch at 10 times their rating. Mark true or false in the space provided.

   a. For a 500-ampere short circuit or ground fault on the load side of a 20-ampere breaker, only the 20-ampere breaker will trip off.   _____

   b. For a 3000-ampere short circuit or ground fault on the load side of the 20-ampere breaker, the 20-ampere breaker and the 225-ampere breaker will trip off.   _____

Refer to the following drawing to make the requested calculations. Use the point-to-point method and show all calculations.

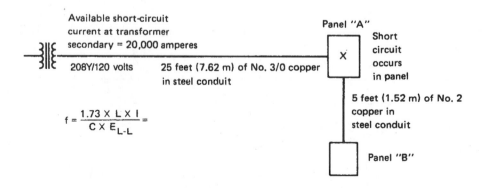

8. Calculate the short-circuit current at panelboard A.

9. Calculate the short-circuit current at panelboard B.

Calculate the following fault current values for a three-phase L-L-L bolted short circuit that has been calculated to be 40,000 amperes.

10. The approximate value of a line-to-line fault will be _____ amperes.

11. The approximate value of a line-to-ground fault will be _____ amperes.

UNIT 19

Equipment and Conductor Short-Circuit Protection

OBJECTIVES

After studying this unit, the student will be able to

- understand that all electrical equipment has a withstand rating.
- discuss withstand rating of conductors.
- understand important *National Electrical Code®* sections that pertain to interrupting rating, available short-circuit current, current-limitation, effective grounding, bonding, and temperature limitation of conductors.
- understand what the term *ampere squared seconds* means.
- perform calculations to determine how much current a copper conductor can safely carry for a specified period of time before being damaged or destroyed.
- refer to charts to determine conductor withstand ratings.
- understand that the two forces present when short circuits or overloads occur are THERMAL and MAGNETIC.
- discuss the 10-foot "tap" conductor maximum overcurrent protection.

All electrical equipment, including switches, motor controllers, conductors, bus duct, panelboards, load centers, switchboards, and so on, have an ability to withstand a certain amount of electrical energy for a given amount of time before damage to the equipment occurs. This gives rise to the term *component short-circuit current rating* that is found in the *NEC®* as well as numerous UL Standards. This term is synonymous with the term *withstand rating*.

Underwriters Laboratories standards specify certain test criteria for the above equipment. For example, switchboards must be capable of withstanding a given amount of fault current for at least three cycles. Note that both the amount of current and the length of time for the test is specified.

Simply stated, component short-circuit current rating is the ability of the equipment to "hold together" for the time it takes the overcurrent protective device (fuse or circuit breaker) to respond to the fault condition.

Acceptable damage levels are well defined in the various Underwriters Standards for electrical equipment. Where the equipment is intended actually to break current, such as a fuse or circuit breaker, the equipment is marked with its interrupting rating.

Electrical equipment manufacturers conduct exhaustive tests to determine the withstand and/or interrupting ratings of their products.

Equipment tested and listed is often marked with the size and type of overcurrent protection required. For example, the label on a motor controller might indicate a maximum size fuse for different sizes of thermal overloads used in that controller. This marking indicates that fuses must be used for the overcurrent protection. A circuit breaker would not be permitted, see *NEC® Section 110-3(b)*.

The ability of a current-limiting overcurrent device to limit the let-through energy to a value less than the amount of energy that the electrical system is capable of delivering means that the equipment can be protected against fault current values of high magnitude. Equipment manufacturers specify current-limiting fuses to minimize the potential damage that might occur in the event of a high-level short circuit or ground fault.

The electrical engineer and/or electrical contractor must perform short-circuit studies, then determine the proper size and type of current-limiting overcurrent protective device that can be used ahead of the electrical equipment that does not have an adequate withstand or interrupting rating for the available fault current to which it will be subjected. The calculation of short-circuit currents is covered in Unit 18.

Study the normal circuit, overloaded circuit, and short-circuit diagrams (Figure 19-1 A-D) and observe how Ohm's law is applied to these circuits. The calculations for the ground fault circuit are not shown, as the impedance of the return ground path can vary considerably.

CONDUCTOR WITHSTAND RATING

Up to this point in the text, we have covered in detail how to compute conductor ampacities for branch circuits, feeders, and service-entrance conductors, with the basis being the connected load and/or volt-amperes per square foot. Then the demand factors and other diversity factors are applied. If we have followed the Code rules for calculating conductor size, the conductors will not be damaged under overloaded conditions because the conductors will have the proper ampacity and will

be protected by the proper size and type of overcurrent protective device.

NEC® Tables such as *Table 310-16*, set forth the ampacity ratings for various sizes and insulation types of conductors. These tables consider normal loading conditions.

But what happens to these conductors under high-level short-circuit and ground-fault conditions? Small conductors often are the weak link in an electrical system.

When short circuits and/or ground faults occur, the connected loads are bypassed, thus the term "short circuit." All that remains in the circuit are the conductors and other electrical devices such as breakers, switches, and motor controllers. The impedance (ac resistance) of these devices is extremely low, so for all practical purposes, the most significant opposition to the flow of current when a fault occurs is the conductor impedance. Fault-current calculations are covered in Unit 18.

The following discussion covers the actual ability of a conductor to maintain its integrity for the time it is called upon to carry fault current, instead of burning off and causing additional electrical problems. It is important that the electrician, electrical contractor, electrical inspector, and consulting engineer have a good understanding of what happens to a conductor under medium- to high-level fault conditions.

The withstand rating of a conductor, such as an equipment grounding conductor, main bonding jumper, or any other current-carrying conductor, reveals that the conductor can withstand a "certain amount of current for a given amount of time." This is the short-time withstand rating of the conductor.

Let us review some key *National Electrical Code®* sections that focus on the importance of equipment and conductor withstand rating, available fault currents, and circuit impedance.

- The interrupting rating is: "the highest current at rated voltage that a device is intended to interrupted under standard test conditions." See *NEC® Article 100*.

- Listed or labeled equipment shall be installed and used in accordance with any instructions included in the listing or labeling. See *NEC® Section 110-3(b)*

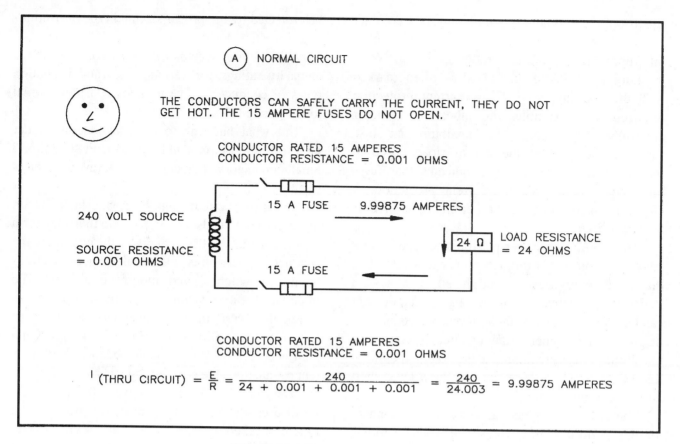

(A) NORMAL CIRCUIT

THE CONDUCTORS CAN SAFELY CARRY THE CURRENT, THEY DO NOT GET HOT. THE 15 AMPERE FUSES DO NOT OPEN.

CONDUCTOR RATED 15 AMPERES
CONDUCTOR RESISTANCE = 0.001 OHMS

240 VOLT SOURCE

SOURCE RESISTANCE = 0.001 OHMS

15 A FUSE   9.99875 AMPERES

24 Ω   LOAD RESISTANCE = 24 OHMS

15 A FUSE

CONDUCTOR RATED 15 AMPERES
CONDUCTOR RESISTANCE = 0.001 OHMS

$$I \text{ (THRU CIRCUIT)} = \frac{E}{R} = \frac{240}{24 + 0.001 + 0.001 + 0.001} = \frac{240}{24.003} = 9.99875 \text{ AMPERES}$$

Fig. 19-1A   Normal circuit current.

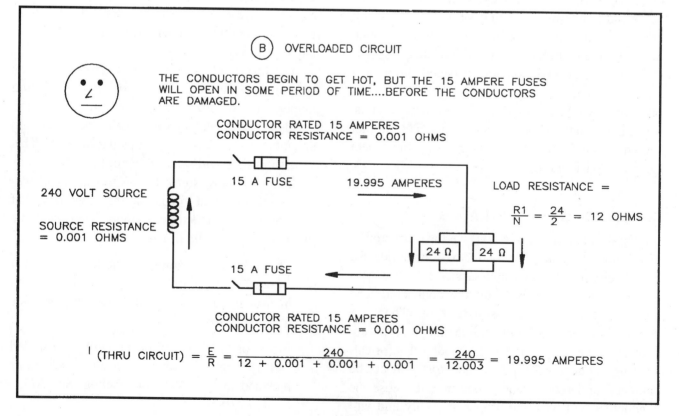

(B) OVERLOADED CIRCUIT

THE CONDUCTORS BEGIN TO GET HOT, BUT THE 15 AMPERE FUSES WILL OPEN IN SOME PERIOD OF TIME....BEFORE THE CONDUCTORS ARE DAMAGED.

CONDUCTOR RATED 15 AMPERES
CONDUCTOR RESISTANCE = 0.001 OHMS

240 VOLT SOURCE

SOURCE RESISTANCE = 0.001 OHMS

15 A FUSE   19.995 AMPERES

LOAD RESISTANCE =

$$\frac{R1}{N} = \frac{24}{2} = 12 \text{ OHMS}$$

24 Ω   24 Ω

15 A FUSE

CONDUCTOR RATED 15 AMPERES
CONDUCTOR RESISTANCE = 0.001 OHMS

$$I \text{ (THRU CIRCUIT)} = \frac{E}{R} = \frac{240}{12 + 0.001 + 0.001 + 0.001} = \frac{240}{12.003} = 19.995 \text{ AMPERES}$$

Fig. 19-1B   Overloaded circuit current.

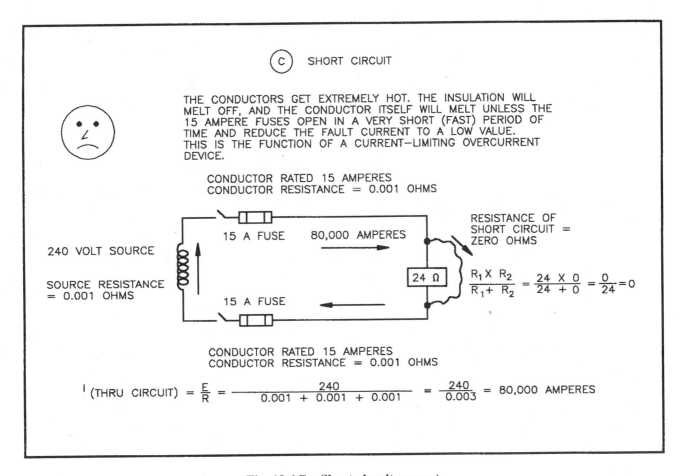

Fig. 19-1C Short circuit current.

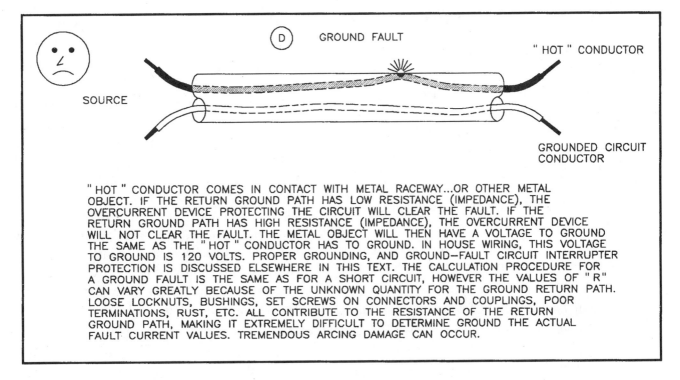

Fig. 19-1D Ground-fault current.

- Equipment intended to break current at fault levels shall have an interrupting rating sufficient for the system voltage and the current that is available at the line terminals of the equipment. See *NEC® Section 110-9.*

▶ Circuit Impedance and Other Characteristics. The over-current protective devices, the total impedance, the component short-circuit current ratings, and other characteristics of the circuit to be protected shall be selected and coordinated to permit the circuit protective devices used to clear a fault to do so without extensive damage to the electrical components of the circuit. This fault shall be assumed to be either between two or more of the circuit conductors, or between any circuit conductor and the grounding conductor or enclosing metal raceway. Listed products applied in accordance with their listing shall be considered to meet the requirements of this section. See *NEC® Section 110-10.* ◀

- Service equipment shall be suitable for the short-circuit current available at its supply terminals. See *NEC® Section 110-9.*

- Overcurrent protection for conductors and equipment is provided to open the circuit if the current reaches a value that will cause an excessive or dangerous temperature in conductors or conductor insulation. See also *NEC® Sections 110-9 and -10* for requirements for interrupting rating and protection against fault currents. See *NEC® Section 240-1.*

- A current-limiting overcurrent protective device is a device that, when interrupting currents in its current-limiting range, will reduce the current flowing in the faulted circuit to a magnitude substantially less than that obtainable in the same circuit if the device were replaced with a solid conductor having comparable impedance. See *NEC® Section 240-11.*

- The required label marking for fuses, one item being the fuse's interrupting rating is set forth in *NEC® Section 240-60(c).*

- The required label marking for circuit breakers, one item being the circuit breaker's interrupting rating, is set forth in *NEC® Section 240-83(c).*

- The requirements for "series-rated" equipment are contained in *NEC® Section 240-86.*

- Ground electrical system so as to limit the voltage imposed by lightning, line surges, or unintentional contact with higher voltage lines so as to stabilize the voltage to earth during normal operation. See *NEC® Section 250-2(a).*

- Ground conductive electrical conductors and equipment to limit the voltage to ground on these materials. See *NEC® Section 250-2(b).*

- Metal piping and other electrically conductive materials that might become energized must be bonded together so as to establish an effective path for fault currents. See *NEC® Section 250-2(c).*

- The fault current path shall:
  1. be permanent and electrically continuous.
  2. be capable of safely carrying the maximum fault likely to be imposed on it.
  3. Shall have sufficiently low impedance to limit the voltage to ground, and to facilitate the operation of the overcurrent devices under fault condition.

  The earth shall not be used as the sole equipment grounding conductor or fault current path. See *NEC® Section 250-2(d).*

- Bonding shall be provided where necessary to assure electrical continuity and the capacity to conduct safely any fault current likely to be imposed. See *NEC® Section 250-90.*

- Metal raceways, cable trays, cable armor, cable sheath, enclosures, frames, fittings, and other metal non-current-carrying parts that are to serve as grounding conductors with or without the use of supplementary equipment grounding conductors shall be effectively bonded where necessary to assure electrical continuity and the capacity to conduct safely any fault current likely to be imposed on them. Any nonconductive paint, enamel, or similar coating shall be removed at threads, contact points, and contact surfaces or be connected by means of fittings so designed as to make such removal unnecessary. See *NEC® Section 250-96(a).*

- The size of the grounding electrode conductor for a grounded or ungrounded system shall not be less than given in *NEC® Table 250-66.* See *NEC® Section 250-66.*

- The size of copper, aluminum, or copper-clad aluminum equipment grounding conductors shall not be less than given in *NEC® Table 250-122*. See *NEC® Section 250-122*.

- Attention should be given to the requirement that equipment grounding conductors may need to be sized larger than shown in *NEC® Table 250-122* because of the need to comply with *NEC® Section 250-2* regarding effective grounding. This would generally be necessary where available fault currents are high, and the possibility of burning off the equipment grounding conductor under fault conditions exists. See *NEC® Section 250-122*.

- *NEC® Section 310-10* is rather lengthy, presenting in detail the requirement that "no conductor shall be used in such a manner that its operating temperature will exceed that designated for the type of insulated conductor involved." This section should be read carefully.

CONDUCTOR HEATING

The value of energy (heat) generated during a fault varies as the square of the rms (root-mean-square) current multiplied by the time (duration of fault) in seconds. This value is expressed as:

$$I^2t$$

The term I^2t is called "ampere squared seconds."

Since watts $= I^2R$, we could say that the damage that might be expected under severe fault conditions can be related to (1) the amount of current flowing during the fault, (2) the time in seconds that the fault current flows, and (3) the resistance of the fault path. This relationship is expressed as:

$$I^2Rt$$

Since the value of resistance under severe fault conditions is generally extremely low, we can simply think in terms of how much current *(I)* is flowing and for how long *(t)* it will flow.

It is safe to say that whenever an electrical system is subjected to a high-level short circuit or ground fault, less damage will occur if the fault current can be limited to a low value, and if the faulted circuit is cleared as fast as possible.

It is important to understand the time-current characteristics of fuses and circuit breakers in order to minimize equipment damage. Time-current characteristic curves were discussed in Unit 17.

CALCULATING AN INSULATED (75°C THERMOPLASTIC) CONDUCTOR'S SHORT TIME WITHSTAND RATING

Copper conductors can withstand:

- one ampere (rms current)
- for five seconds
- for every 42.25 circular mils of cross-sectional area.

Note in the above statement that both current (how much) and time (how long) are included.

Let's take a real-world situation. If we wish to provide an equipment grounding conductor for a circuit protected by a 60-ampere overcurrent device, we find in *NEC® Table 250-122* that a No. 10 AWG copper conductor is the MINIMUM size permitted.

Referring to *Chapter 9, Table 8, NEC®*, we find that the cross-sectional area of a No. 10 AWG conductor is 10,380 circular mils.

This No. 10 copper conductor has a five-second withstand rating of:

$$\frac{10,380 \text{ circular mils}}{42.25 \text{ circular mils}} = 246 \text{ amperes}$$

This means that 246 amperes is the maximum amount of current that a No. 10 copper insulated conductor can carry for five seconds without being damaged. This is the No. 10 conductor's short time "withstand rating" and is the most current that it can safely carry for five seconds without damage to its insulation occurring.

Stating this information using the thermal stress *(heat)* formula, we have:

$$\text{Thermal stress (heat)} = I^2_{\text{rms}}t$$
where I = rms current in amperes and
t = time in seconds

Thus, the No. 10 copper conductor's withstand rating is:

$$I^2t = 246 \times 246 \times 5 = 302,580 \text{ ampere squared seconds}$$

With this basic information, we can easily determine the short time withstand rating of this No. 10 copper insulated conductor for other values of time and/or current. The No. 10 copper conductor's one-second withstand rating is:

I^2t = ampere squared seconds

I^2 = $\dfrac{\text{ampere squared seconds}}{t}$

I = $\sqrt{\dfrac{\text{ampere squared seconds}}{t}}$

I = $\sqrt{\dfrac{302,580}{1}}$

I = 550 amperes

The No. 10 copper conductor's one-cycle withstand rating (given that the approximate opening time of a typical molded case circuit breaker is one cycle, or 1/60th of a second, or 0.0167 second) is:

I = $\sqrt{\dfrac{302,580}{0.0167}}$

I = 4257 amperes

The No. 10 copper conductor's one-quarter-cycle withstand rating (given that the typical clearing time for a current-limiting fuse is approximately one-fourth of a cycle, or 0.004 second) is:

I = $\sqrt{\dfrac{302,580}{0.004}}$

I = 8697 amperes

Therefore, a conductor can be subjected to large values of fault current if the clearing time is kept very short.

When applying current-limiting overcurrent devices, it is important to use peak let-through charts to determine the apparent rms let-through current before applying the thermal stress formula. For example, in the case of a 60-ampere Class RK1 current-limiting fuse, the apparent rms let-through current with an available fault current of 40,000 amperes is approximately 3000 amperes. This fuse will clear in approximately 0.004 second. See Figures 17-26, 19-2, and 19-4.

I^2t let-through of current-limiting fuse = $3000 \times 3000 \times 0.004 = 36,000$ ampere squared seconds

Because a current-limiting overcurrent device is used in this example, the No. 10 AWG copper equip-

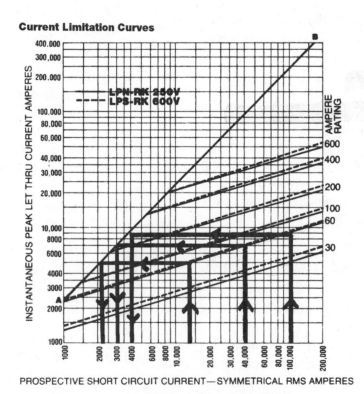

Fig. 19-2   Current-limiting effect of Class RK1 fuses. The technique required to use these charts is covered in Unit 17.

ment grounding conductor could be used where the available fault current is 40,000 amperes.

With an available fault current of 100,000 amperes, the apparent rms let-through current of the 60-ampere Class RK1 current-limiting fuse is approximately 4000 amperes.

I^2t let-through of current-limiting fuse =
$4000 \times 4000 \times 0.004 = 64,000$ ampere squared seconds

Remember that, as previously discussed, the withstand I^2t rating of a No. 10 copper conductor is 302,580 ampere squared seconds.

Example:

A 75° thermoplastic insulated Type THW copper conductor can withstand 4200 amperes for one cycle.

a. What is the I^2t withstand rating of the conductor?

b. What is the I^2t let-through value for a non-current-limiting circuit breaker that takes one cycle to open? The available fault current is 40,000 amperes.

c. What is the I^2t let-through value for a current-limiting fuse that opens in 0.004 second when subjected to a fault of 40,000 amperes? The apparent rms let-through current is approximately 4600 amperes. Refer to Figure 17-26.

d. Which overcurrent device (b or c) will properly protect the conductor under the 40,000-ampere available fault current?

Answers:

a. $I^2t = 4200 \times 4200 \times 0.016 = 282,240$ ampere squared seconds.

b. $I^2t = 40,000 \times 40,000 \times 0.016 = 25,600,000$ ampere squared seconds.

c. $I^2t = 4600 \times 4600 \times 0.004 = 84,640$ ampere squared seconds.

d. Comparing the I^2t withstand rating of the conductor to the I^2t let-through values of the breaker (b) and fuse (c), the proper choice of protection for the conductor is (c).

The use of peak let-through, current-limiting charts is discussed in Unit 17. Peak let-through charts are available from all manufacturers of current-limiting fuses and current-limiting circuit breakers.

Conductors are particularly vulnerable to damage under fault conditions. Circuit conductors can be heated to a point where the insulation is damaged or completely destroyed. The conductor can actually burn off. In the case of equipment grounding conductors, if the conductor burns off under fault conditions, the equipment can become "hot," creating an electrical-shock hazard.

Even if the equipment grounding conductor does not burn off, it can become so hot that it melts the insulation on the other conductors, shorting out the other circuit conductors in the raceway or cable. This results in further fault current conditions, damage, and hazards.

Of extreme importance are bonding jumpers, particularly the main bonding jumpers in service equipment. These main bonding jumpers must be capable of handling extremely high values of fault current.

CALCULATING A BARE COPPER CONDUCTOR AND/OR ITS BOLTED SHORT-CIRCUIT WITHSTAND RATING

A bare conductor can withstand higher levels of current than an insulated conductor of the same cross-sectional area. Do not exceed:

- one ampere (rms current)
- for five seconds
- for every 29.1 circular mils of cross-sectional area of the conductor.

Because bare equipment grounding conductors are often installed in the same raceway or cable as the insulated circuit conductors, the "weakest link" of the system would be the insulation on the circuit conductors. Therefore, the conservative approach when considering conductor safe withstand ratings is to apply the ONE AMPERE — FOR FIVE SECONDS — FOR EVERY 42.25 CIRCULAR MILS formula, or simply refer to a conductor short-circuit withstand rating chart, such as Figure 19-3.

CALCULATING THE MELTING POINT OF A COPPER CONDUCTOR

The melting point of a copper conductor can be calculated by using these values:

- one ampere (rms current)

- for five seconds

- for every 16.19 circular mils of cross-sectional area of the conductor.

NEC® Section 250-2 establishes a performance standard to be attained by the grounding systems. ▶ A note to *NEC® Table 250-122* warns that an equipment grounding conductor may have to be sized larger than shown in the table in order to achieve the desired standard. ◀ Under fault-current conditions, the possible burning off of an equipment grounding conductor, a bonding jumper, or any conductor that is dependent upon safely carrying fault current until the overcurrent protective device can clear the fault results in a hazard to life, safety, and equipment.

USING CHARTS TO DETERMINE A CONDUCTOR'S SHORT TIME WITHSTAND RATING

The Insulated Cable Engineers Association, Inc. publishes much data on this subject. The graph in Figure 19-3 shows the withstand rating of 75° thermoplastic insulated copper conductors. Many engineers will use these tables for bare grounding conductors because, in most cases, the bare equipment grounding conductor is in the same raceway as the phase conductors. An extremely hot equipment grounding conductor in contact with the phase conductors would damage the insulation on the phase conductors. For example, when non-metallic conduit is used, the equipment grounding conductor and the phase conductors are in the same raceway.

To use the table, enter on the left side, at the amount of fault current available. Then draw a line to the right, to the time of opening of the overcurrent protective device. Draw a line downward to the bottom of the chart to determine the conductor size.

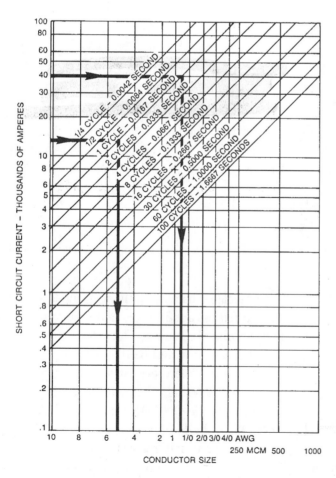

CONDUCTOR–COPPER

INSULATION–THERMOPLASTIC

CURVES BASED ON FORMULA

$$\left[\frac{I}{A}\right]^2 t = 0.0297 \log\left[\frac{T_2 + 234}{T_1 + 234}\right]$$

WHERE

I = SHORT CIRCUIT CURRENT—AMPERES

A = CONDUCTOR AREA—CIRCULAR MILS

t = TIME OF SHORT CIRCUIT—SECONDS

T_1 = MAXIMUM OPERATING TEMPERATURE—75°

T_2 = MAXIMUM SHORT CIRCUIT TEMPERATURE—150°

Fig. 19-3 Available short-circuit currents for insulated copper conductors.

Example:

A circuit is protected by a 60-ampere overcurrent device. Determine the minimum size equipment grounding conductor for available fault currents of 14,000 amperes and 40,000 amperes. Refer to *Table 250-122* of the *NEC*, the fuse peak let-through chart, and the chart showing allowable short-circuit currents for insulated copper conductors, Figure 19-3.

You can see from Table 19-1 that if the conductor size as determined from the allowable short-circuit current chart is larger than the size given in *NEC® Table 250-122*, then you must install the larger size. Installing a conductor too small to handle the available fault current could result in insulation damage or, in the worst case, the burning off of the equipment grounding conductor, leaving the protected equipment "hot."

If the conductor size as determined from the allowable short-circuit current chart is smaller than the size given in *NEC® Table 250-122*, then install an equipment grounding conductor *not smaller* than the minimum size required by *NEC® Table 250-122*.

There is another common situation in which it is necessary to install equipment grounding conductors larger than shown in *NEC® Table 250-122*. This is the case in which circuit conductors have been increased in size because of a voltage-drop calculation, in which case the equipment grounding conductor size shall be increased proportionately according to circular mil area. Voltage-drop calculations are covered in Unit 3.

MAGNETIC FORCES

Magnetic forces acting upon electrical equipment (bus bars, contacts, conductors, and so on) are *proportional to the square of the PEAK current*. This relationship is expressed as I_p^2.

Refer to Figure 19-4 for the case in which there is no fuse in the circuit. The peak current, I_p (available short-circuit current), is indicated as 30,000 amperes. The peak let-through current, I_p, resulting from the current-limiting effect of a fuse is indicated as 10,000 amperes.

Example:

Because magnetic forces are *proportional to the square of the peak current*, the magnetic forces (stresses) on the electrical equipment subjected to the full 30,000-ampere peak current are *nine times* that of the 10,000-ampere peak current let-through by the fuse. Stated another way, a current-limiting fuse or circuit breaker that can reduce the available short-circuit peak current from 30,000 amperes to only 10,000 amperes will subject the electrical equipment to *only one-ninth* the magnetic forces.

Visual signs indicating that too much current was permitted to flow for too long a time include conductor insulation burning, melting and bending of bus bars, arcing damage, exploded overload elements in motor controllers, and welded contacts in controllers.

A current-limiting fuse or current-limiting circuit breaker must be selected carefully. The fuse or breaker must have not only an adequate interrupting

TABLE 19-1 Short-Circuit Currents for Insulated Copper Conductors		
Available Fault Current	**Overcurrent Device**	**Conductor Size**
40,000 amperes	Typical one-cycle breaker	1/0 copper conductor
40,000 amperes	Typical RK1 fuse. Clearing time ¼ cycle or less.	Using a 60-ampere RK1 current-limiting fuse, the apparent rms let-through current is approximately 3000 amperes. *NEC® Table 250-122* shows a minimum #8 equipment grounding conductor. The allowable short-circuit current chart shows a conductor smaller than a #10. Therefore, #8 is the minimum size EGC permitted.
14,000 amperes	Typical one-cycle breaker	#4 copper conductor
14,000 amperes	Typical RK1 fuse. Clearing time ¼ cycle or less.	Using a 60-ampere RK1 current-limiting fuse, the apparent rms let-through current is approximately 2200 amperes. *NEC® Table 250-95* shows a minimum #8 equipment grounding conductor. The allowable short-circuit current chart shows a conductor smaller than a #10. Therefore, #8 is the minimum size EGC permitted.

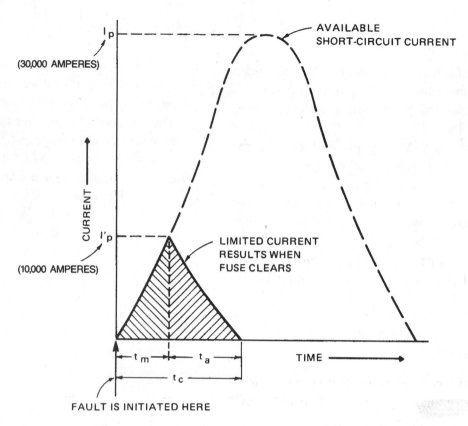

Fig. 19-4 Current-limiting effect of fuses.

rating to clear a fault safely without damage to itself, but it also must be capable of limiting the let-through current (I_p) and the value of I^2t to the withstand rating of the equipment it is to protect.

The graph shown in Figure 19-4 illustrates the current-limiting effect of fuses. The square of the area under the dashed line is energy (I^2t). I_p is the available peak short-circuit current that flows if there is no fuse in the circuit, or if the overcurrent device is not a current-limiting type. For a current-limiting fuse, the square of the shaded area of the graph represents the energy (I^2t), and the peak let-through current I_p. The melting time of the fuse element is t_m and the square of the shaded area corresponding to this time is the melting energy. The arcing time is shown as t_a; similarly, the square of the shaded area corresponding to this time is the arcing energy. The total clearing time, t_c, is the sum of the melting time and the arcing time. The square of the shaded area for time, t_c, is the total energy to which the circuit is subjected when the fuse has cleared. For the graph in figure 19-4, the area under the dashed line is six times greater than the shaded area. Since energy is equal to the area squared, then $6 \times 6 = 36$; that is, the circuit is subjected to 36 times as much energy when it is protected by non-current-limiting overcurrent devices.

Summary of Conductor Short-Circuit Protection

When selecting equipment grounding conductors, bonding jumpers, or other current-carrying conductors, the amount of short-circuit or ground fault current available and the clearing time of the overcurrent protective device must be taken into consideration so as to minimize damage to the conductor and associated equipment. The conductor must not become a fuse. The conductor must remain intact under any values of fault current.

As stated earlier, the important issues are how much current will flow and how long the current will flow.

The choices are:

1. *Limit the current.*
   Install current-limiting overcurrent devices that

will limit the let-through fault current and will reduce the time it takes to clear the fault. Then refer to *NEC® Table 250-122* to select the minimum size equipment grounding conductor permitted by the *National Electrical Code,* and refer to the other tables such as *NEC® Table 310-16* for the selection of circuit conductors.

2. *Do not limit the current.*

   Install conductors that are large enough to handle the full amount of available fault current for the time it takes a non-current-limiting overcurrent device to clear the short-circuit or ground fault. Withstand ratings of conductors can be calculated or determined by referring to conductor withstand rating charts.

TAP CONDUCTORS

Branch-circuit conductors must have an ampacity not less than the maximum load served, *NEC® Section 210-19.*

Feeder conductors must have sufficient ampacity to supply the load served, *NEC® Section 220-10.*

Service-entrance conductors shall be of sufficient size to carry the loads as computed, *NEC® Section 230-42.*

The basic overcurrent protection rule in *NEC® Section 240-3* is that conductors shall be protected at the conductor's ampacity. There are seven subsections to this *Section*, each pertaining to specific applications.

NEC® Section 240-21 supports *Section 240-3* in that it states: "An overcurrent device shall be connected at the point where the conductor to be protected receives its supply." There are many subsections to *NEC® Section 240-21(b)*, each pertaining to specific types of conductor taps.

These exceptions must be read and understood.

Of particular interest are feeder taps that are not over ten feet, and feeder taps that are not over 25 feet. A number of requirements must be met when taps of this sort are made.

There are instances where a smaller conductor must be tapped from a larger conductor. For example, Figure 19-5 shows a large feeder protected by a 400-ampere fuse. The minimum size tap in the wireway or gutter shall not be less than one-tenth the

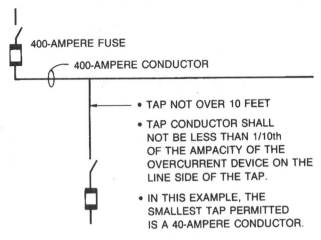

400-AMPERE FUSE

400-AMPERE CONDUCTOR

- TAP NOT OVER 10 FEET
- TAP CONDUCTOR SHALL NOT BE LESS THAN 1/10th OF THE AMPACITY OF THE OVERCURRENT DEVICE ON THE LINE SIDE OF THE TAP.
- IN THIS EXAMPLE, THE SMALLEST TAP PERMITTED IS A 40-AMPERE CONDUCTOR.

Fig. 19-5 *NEC® Section 240-21(b)(1)(d)* requires that the overcurrent device not be greater than 1000% of the tap conductor's ampacity.

ampacity of the 400-ampere fuse. This simply means that the tap conductor must have an ampere rating of not less than:

$$\frac{400}{10} = 40 \text{ amperes}$$

The above example of tapping a small conductor from a larger conductor that is protected by an overcurrent device much larger than the small tap conductor is an indication that the Code is concerned with the protection of the smaller conductor.

Here again, the electrician must properly size the conductors for the load to be served, take into consideration the possibility of voltage drop, then check the short-circuit withstand rating of the conductor to be sure that a severe fault will not cause damage to the conductor's insulation or, in the worst case, vaporize the conductor.

Another common example of a tap is for "fixture whips" that connect a fixture to a branch circuit. The size of taps to branch circuits is found in *NEC® Table 210-24.* For instance, for a 20-ampere branch circuit, the minimum size tap is a No. 14 AWG copper conductor. The tap conductors in a fixture whip shall be suitable for the temperature to be encountered. In most cases, this will call for a 90°C conductor, THHN or equivalent.

Additional Code rules for taps and panelboard protection are found in Unit 12.

REVIEW QUESTIONS

Refer to the *National Electrical Code*® or the working drawings when necessary. Where applicable, responses should be written in complete sentences. Write units using unit names, use no abbreviations or symbols (1 foot, not 1' or 1 ft).

1. Define *withstand rating*. Amount of current a device can see and how long it can withstand it

2. Define an *effective grounding path*. To keep to a min the SCA

3. Define *I²t*. The amount of energy generated during a fault multiplies the current by the time

4. Define a *current-limiting overcurrent device*. will limate the let through fault current and reduces the time it takes to clear

5. Define an *equipment grounding conductor*. A means of proection that has a current fault to go to ground

6. What table in the *National Electrical Code*® shows the minimum size equipment grounding conductor? *Section* 250-122

Respond to the following questions. Show all required calculations.

7. What is the appropriate action if the label on a motor controller, or the nameplate on other electrical equipment is marked "Maximum Fuse Size"? Cite the Code section that supports your proposed action. ~~240-6b~~
    110-3b

8. What is the maximum current that a No. 12 AWG, 75°C thermoplastic insulated conductor can safely carry for five seconds? _____ amperes. Show your calculations.

$$\frac{6530}{42.25} = \frac{154^2}{5} = 4777.5$$

9. What is the maximum current that a No. 8 AWG, 75°C thermoplastic insulated conductor can safely carry for one second? _____ amperes. Show your calculations.

$$\frac{16510}{42.25} = \frac{390^2}{1} = 152,700$$

10. What is the maximum current that a No. 12 AWG, 75°C thermoplastic insulated conductor can safely carry for one-quarter cycle (0.004 second)? _____ amperes. Show your calculations.

11. Referring to Figure 19-3, if the available fault current is 10,000 amperes, and the overcurrent device has a total clearing time of one cycle, the minimum permitted copper 75°C thermoplastic insulated conductor would be a _____ AWG.

12. Referring to Figure 19-2, if the available fault current is 40,000 amperes, a 200-ampere 600-volt fuse of the type represented by the chart will have an instantaneous peak let-through current of approximately _____ amperes.

13. Referring to Figure 19-2, if the available fault current is 40,000 amperes, a 200-ampere 600-volt fuse of the type represented by the chart will have an apparent rms let-through current of approximately _____ amperes.

14. A tap conductor that complies with appropriate Code requirements has an ampacity of 30 amperes and is nine feet long. This is a field installation where the tap conductor leaves the enclosure where the tap is made (a large junction box). The maximum allowable overcurrent protection for the feeder conductor would be _____ amperes.

UNIT 20

Low-Voltage Remote-Control Lighting*

OBJECTIVES

After studying this unit, the student will be able to

- list the components of a low-voltage remote-control wiring system.
- select the appropriate *NEC® Sections* governing the installation of a low-voltage remote-control wiring system.
- demonstrate the correct connections for wiring a low-voltage remote-control system.

LOW-VOLTAGE REMOTE-CONTROL LIGHTING

Conventional general lighting control is used in the major portion of the commercial building. For the drugstore, however, a method known as *low-voltage remote-control lighting* is selected because of the number of switches required and because it is desired to have extensive control of the lighting.

Relays

A low-voltage remote-control wiring system is relay operated. The relay is controlled by a low-voltage switching system and in turn controls the power circuit connected to it, Figure 20-1. The low-voltage, split coil relay is the heart of the low-voltage remote-control system, Figure 20-2. When the *On* coil of the relay is energized, the solenoid mechanism causes the contacts to move into the *On* position to complete the power circuit. The contacts stay in this position until the *Off* coil is energized. When this occurs, the contacts are withdrawn and the power circuit is opened. The red wire is the *On* wire; the black wire is *Off*, and the blue wire is common to the transformer.

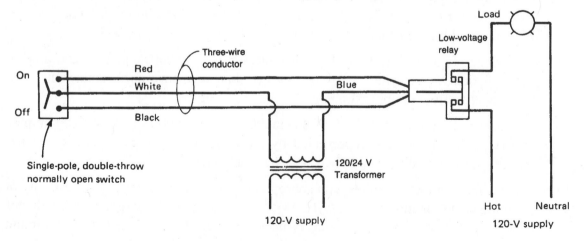

Fig. 20-1   Connection diagram for low-voltage remote-control.

---

*Although General Electric is marketing new hardware items to enhance this wiring system, illustrations were not available at the time of this printing. The user should contact GE directly.

The low-voltage relay is available in two mounting styles. One style of relay is designed to mount through a ½-inch knockout opening, Figure 20-3. For a ¾-inch knockout, a rubber grommet is inserted to isolate the relay from the metal. This practice should insure quieter relay operation. The second relay mounting style is the *plug-in relay*. This type of relay is used in an installation where several relays are mounted in one enclosure. The advantage of the plug-in relay is that it plugs directly into a bus bar. As a result, it is not necessary to splice the line voltage leads.

Single Switch

The switch used in the low-voltage remote-control system is a normally open, single-pole, double-throw, momentary contact switch, Figure 20-4. This switch is approximately one-third the size of a standard single-pole switch. In general, this type of switch has short lead wires for easy connections, Figure 20-5.

To make connections to the low-voltage switches, the white wire is common and is connected to the 24-volt transformer source. The red wire connects to the *On* circuit and the black wire connects to the *Off* circuit.

Master Control

It is often desirable to control several circuits from a single location. Up to eight low-voltage switches, Figure 20-6, can be located in the same area that is required by two conventional switches. This 8-switch master can be mounted on a 4¹¹/₁₆ box using an adaptor provided with the master. Directory strips, identifying the switches functions, can be prepared and inserted in the switch cover. An 8-switch master is used in the commercial building drugstore. The connection diagram is shown in Figure 20-11.

If the control requirements are very complex, or extensive, a Master sequencer, Figure 20-7, can be installed. There is practically no limit to the number of relays that can be controlled, almost instanta-

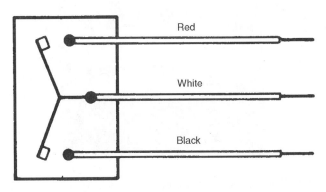

Fig. 20-4 Single-pole, double-throw, normally open switch.

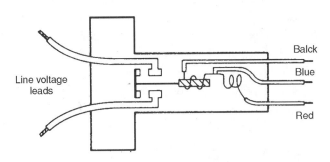

Fig. 20-2 Schematic of low-voltage relay.

Fig. 20-3 Low-voltage relay. (*Courtesy* of General Electric Wiring Devices.)

Fig. 20-5 Low-voltage switch. (*Courtesy* of General Electric Wiring Devices.)

Fig. 20-6   Eight-switch master. (*Courtesy* of General Electric Wiring Devices.)

Fig. 20-7   Master sequences. (*Courtesy* of General Electric Wiring Devices.)

neously, with this microprocessor controlled electronic switch.

Master Control with Rectifiers

Several relays can be operated individually or from a single switch through the use of *rectifiers*. The principle of operation of a master control with rectifiers is based on the fact that a rectifier permits current in only one direction, Figure 20-8.

For example, if two rectifiers are connected as shown in Figure 20-9, current cannot exist from A to B, or from B to A; however, current can exist from A to C, or from B to C. Thus, if a rectifier is placed in

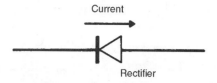

Fig. 20-8   Rectifier showing direction of current.

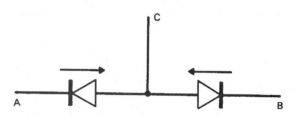

Fig. 20-9   Rectifiers connected in opposition.

one lead of the low-voltage side of the transformer, and additional rectifiers are used to isolate the switches, then a master switching arrangement is achieved, Figure 20-10. This method of master control is used in the drugstore. Although a switching schedule is included in the specifications, the electrician may find it necessary to prepare a connection diagram similar to that shown in Figure 20-11. The relays and rectifiers, Figure 20-12, for the drugstore master control are located in the low-voltage control panel.

WIRING METHODS

NEC® Article 725 governs the installation of remote-control and signal circuits such as is installed in the drugstore. The provisions of this article apply to remote-control circuits, low-voltage relay switching, low-energy power circuits, and low-voltage circuits.

The drugstore low-voltage wiring is classified as a Class 2 circuit, by *NEC® Section 725-2*. Since the power source of the circuit is limited (by the definition of a low-voltage circuit), overcurrent protection is not required.

▶ Power source limitations for alternating current Class 2 circuits are found in *NEC®* Chapter 9, *Tables 11(a)* and *11(b)*. ◀

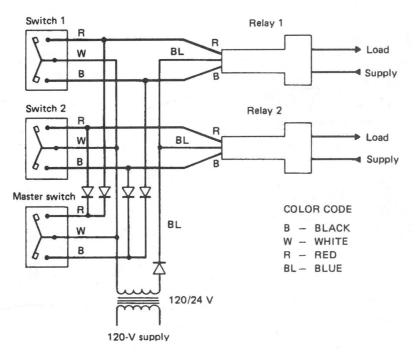

Fig. 20-10 Master control with rectifiers.

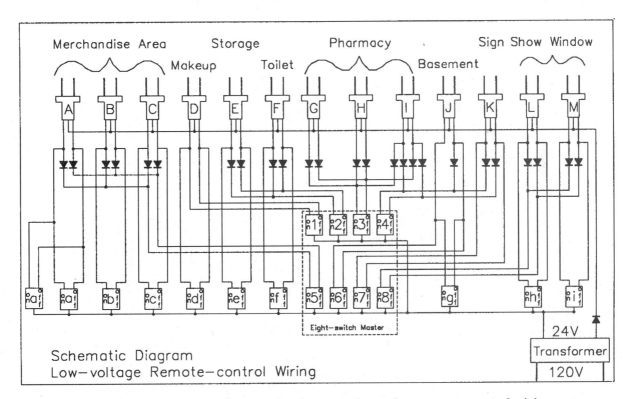

Fig. 20-11 Connection diagram for drugstore low-voltage remote-control wiring.

The circuit transformer, Figure 20-13, is designed so that in the event of an overload, the output voltage decreases and there is less current output. Any overload can be counteracted by these energy-limiting characteristics through the use of a specially designed transformer core. If the transformer is not self-protected, a thermal device may be used to open the primary side to protect the transformer

Fig. 20-12   Transformer.

from overheating. This thermal device resets automatically as soon as the transformer cools. Other transformers are protected with nonresetting fuse links or externally mounted fuses.

Fig. 20-13   Rectifier.

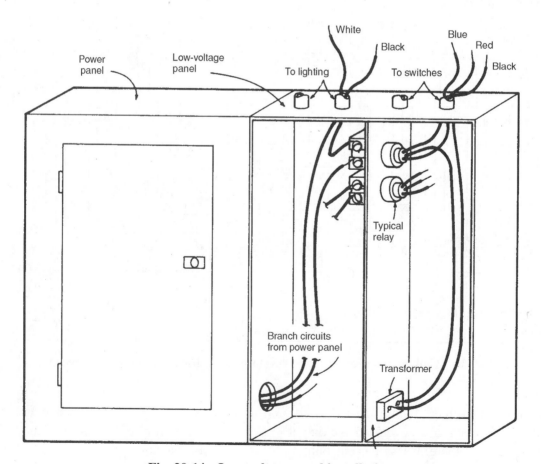

Fig. 20-14   Low-voltage panel installation.

Although the *NEC®* does not require that the low-voltage wiring be installed in a raceway, the specifications for the commercial building do contain this requirement. The advantage of using a raceway for the installation is that it provides a means for making future additions at a minimum cost. A disadvantage of this approach is the initial higher construction cost.

Conductors

No. 18 AWG conductors are used for low-voltage, remote-control systems. Larger conductors should be used for long runs to minimize the voltage drop. The cables for the installation usually contain two or three conductors. The insulation on the individual conductors may be either a double-cotton covering or plastic. Regardless of the type of insulation used, the individual conductors in the cables can be identified readily. To simplify connections, low-voltage cables are available in various color combinations, such as blue-white, red-black, or black-white-red. To install these color-coded cables correctly throughout an entire installation, the wires are connected like-color to like-color. A cable suitable for outdoor use, either overhead or underground, is available for low-voltage, remote-control systems.

Low-Voltage Panel

The low-voltage relays in the drugstore installation are to be mounted in an enclosure next to the power panel, Figure 20-14. A barrier in this low-voltage panel separates the 120-volt power lines from the low-voltage control circuits, in compliance with *NEC® Section 725-54*.

REVIEW QUESTIONS

Refer to the *National Electrical Code®* or the working drawings when necessary. Where applicable, responses should be written in complete sentences. Write units using unit names, use no abbreviations or symbols (1 foot, not 1' or 1 ft).

Refer to figure 20-1 and respond to the following statements.

1. Describe the action of the low-voltage relay. _____
   _____
   _____
   _____

2. Describe the function of the red conductor. _____
   _____
   _____
   _____

3. Describe the function of the white conductor. _____
   _____
   _____
   _____

4. Describe the function of the black conductor. _____
   _____
   _____
   _____

5. Describe the function of the blue conductor. _____

_____

_____

_____

Respond to the following statements.

6. The *NEC®* Article that governs the installation of the low-voltage remote-control system is _____.

7. Complete the diagram below according to the switching schedule that is given at the bottom of the diagram. Indicate the color of the conductors.

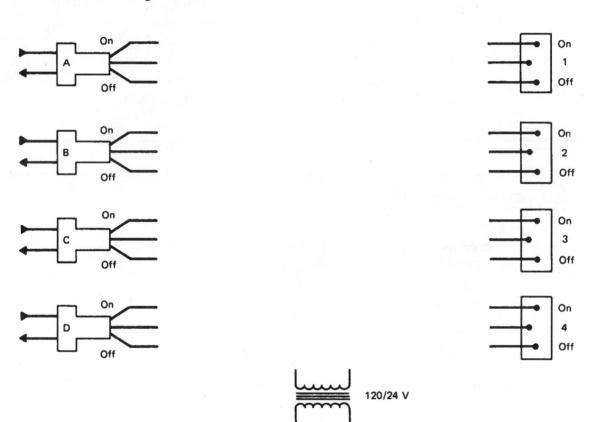

Switching schedule

Switch	Control relay
1	A
2	B
3	C & D
4	A & B

UNIT 21

The Cooling System

OBJECTIVES

After studying this unit, the student will be able to

- list the parts of a cooling system.
- describe the function of each part of the cooling system.
- make the necessary calculations to obtain the sizes of the electrical components.
- read a typical wiring diagram that shows the operation of a cooling unit.

The electrician working on commercial construction is expected to install the wiring of cooling systems and troubleshoot electrical problems in these systems. Therefore, it is recommended that the electrician know the basic theory of refrigeration and the terms associated with it.

REFRIGERATION

Refrigeration is a method of removing energy in the form of heat from an object. When the heat is removed, the object is colder. An energy balance is maintained, which means that the heat must go somewhere. As long as the locations where the heat is discharged and where it is absorbed are remote from each other, it can be said that the space where the heat was absorbed is cooled. The inside of the household refrigerator or freezer is cold to the touch, but this cold cannot be used to cool the kitchen by having the refrigerator door open. Actually, leaving the door open causes the kitchen to become hotter. This situation demonstrates an important principle of mechanical refrigeration: to remove heat energy, it is necessary to add energy or power to it.

Mechanical refrigeration relies primarily on the process of evaporation. This process is responsible for the cool sensation that results when rubbing

alcohol is applied to the skin or when gasoline is spilled on the skin. Body heat supplies the energy required to vaporize the alcohol or gasoline. It is the removal of this energy from the body that causes the sensation of cold. In refrigeration systems such as those used in the commercial building, the evaporation process is controlled in a closed system. The purpose of this arrangement is to preserve the refrigerant so that it can be reused many times in what is known as the *refrigerant cycle*. As shown in Figure 21-1, the four main components of the refrigerant cycle are:

- *Evaporator*
  The refrigerant evaporates here as it absorbs energy from the removal of heat.

- *Compressor*
  This device raises the energy level of the refrigerant so that it can be condensed readily to a liquid.

- *Condenser*
  The compressed refrigerant condenses here as the heat is removed.

- *Expansion valve*
  This metering device maintains a sufficient unbalance in the system so that there is a point of low pressure where the refrigerant can expand and evaporate.

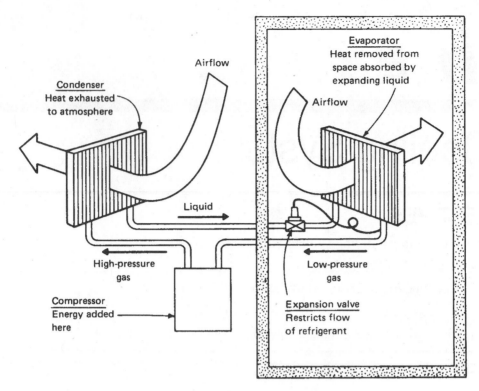

Fig. 21-1   The refrigeration cycle.

EVAPORATOR

The evaporator in a commercial installation normally consists of a fin tube coil, Figure 21-2, through which the building air is circulated by a motor-driven fan. A typical evaporator unit is shown in Figure 21-3. The evaporator may be located inside or outside the building. In either case, the function of the evaporator is to remove heat from the interior of the building or the enclosed space. The air is usually circulated through pipes or ductwork to insure a more even distribution. The window-type air conditioner, however, discharges the air directly from the evaporator coil. In general, the cooling air from the evaporator is recirculated within the space to be cooled and is passed again across the cooling coil. A certain percentage of outside air is added to the circulating air to replace air lost through exhaust systems and fume hoods, or because of the gradual leakage of air through walls, doors, and windows.

COMPRESSOR

The compressor serves as a pump to draw the expanded refrigerant gas from the evaporator. In addition, the compressor boosts the pressure of the

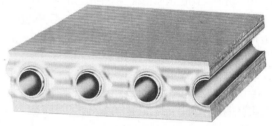

(A) Fin tube coil

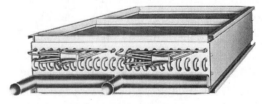

(B) Fin tube evaporator coil

Fig. 21-2   Fin tube coils.

Fig. 21-3   An evaporator with dual fans.

gas and sends it to the condenser, Figure 21-4. The compression of the gas is necessary since this process adds the heat necessary to cause the gas to be condensed to a liquid. When the temperature of the air or water surrounding the condenser is relatively warm, the gas temperature must be increased to insure that the temperature around the condenser will liquefy the refrigerant.

Direct-drive compressors are usually used in large installations. For smaller installations, however, the trend is toward the use of hermetically sealed compressors. Due to several built-in electrical characteristics, these hermetic units cannot be used on all installations. (Restrictions to the use of hermetically sealed compressors are covered later in this text.)

CONDENSER

Condensers are generally available in three types: as air-cooled units, Figure 21-5, as water-cooled units, Figure 21-6, or as evaporative cooling units, Figure 21-7, in which water from a pump is sprayed on air-cooled coils to increase their capacity. The function of the condenser in the refrigerant cycle is to remove the heat taken from the evaporator, plus the heat of compression. Thus, it can be seen that keeping the refrigerator door open causes the kitchen to become hotter because the condenser rejects the combined heat load to the condensing medium. In the case of the refrigerator in a residence, the condensing medium is the room air.

Air-cooled condensers use a motor-driven fan to drive air across the condensing coil. Water-cooled condensers require a pump to circulate the water. Once the refrigerant gas is condensed to a liquid state, it is ready to be used again as a coolant.

EXPANSION VALVE

It was stated previously that the refrigerant must evaporate or boil if it is to absorb heat. The process of boiling at ordinary temperatures can occur in the evaporator only if the pressure is reduced. The task of reducing the pressure is simplified since the compressor draws gas away from the evaporator and tends to evacuate it. In addition, a restricted flow of liquid refrigerant is allowed to enter the high side of the evaporator. As

Fig. 21-4   A motor-driven reciprocating compressor.

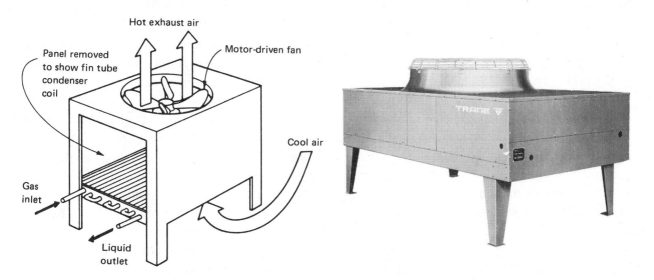

Fig. 21-5   An air-cooled condenser.

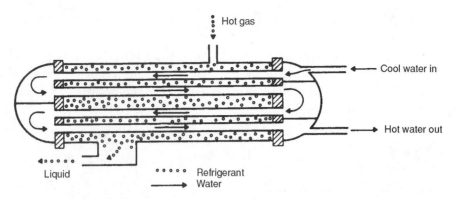

Fig. 21-6   A water-cooled condenser.

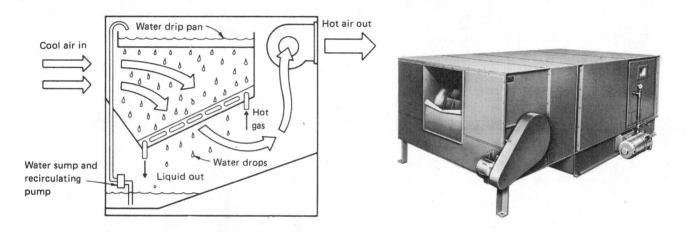

Fig. 21-7   An evaporative condenser.

a result, the pressure remains fairly low in the evaporator coil so that the liquid refrigerant entering through the restriction flashes into a vapor and fills the evaporator as a boiling refrigerant.

The restriction to the evaporator may be a simple orifice. In commercial systems, however, a form of automatic expansion valve is generally used because it is responsive to changes in the heat load. The expansion valve is located in the liquid refrigerant line at the inlet to the evaporator. The valve controls the rate at which the liquid flows into the evaporator. The liquid flow rate is determined by installing a temperature-sensitive bulb at the outlet of the evaporator to sense the heat gained by the refrigerant as it passes through the evaporator. The bulb is filled with a volatile liquid (which may be similar to the refrigerant). This liquid expands and passes through a capillary tube connected to a spring-loaded diaphragm to cause the expansion valve to meter more or less refrigerant to the coil. The delivery of various amounts of refrigerant

compensates for changes in the heat load of the evaporator coil.

HERMETIC COMPRESSORS

The tremendous popularity of mechanical refrigeration for household use in the 1930s and 1940s stimulated the development of a new series of nonexplosive and nontoxic refrigerants. These refrigerants are known as chlorinated hydrofluorides of the halogen family and, at the time of their initial production, were relatively expensive. The expense of these refrigerants was great enough so that it was no longer possible to permit the normal leakage that occurred around the shaft seals of reciprocating, belt-driven units. As a result, the hermetic compressor was developed, Figure 21-8. This unit consists of a motor-compressor completely sealed in a gastight, steel casing. The refrigerant gas is circulated through the compressor and over the motor windings, rotor, bearings, and shaft. The circulation of the expanded gas through the motor helps to cool the motor.

The initial demand for hermetic compressors was for use on residential-type refrigerators. Therefore, most of the hermetic compressors were constructed for single-phase service. In other words, it was necessary to provide auxiliary winding and starting devices for the compressor installation. Because the refrigerant gas surrounded and filled the motor cavity, it was necessary to remove the centrifugal switch commonly provided to disconnect the starting winding at approximately 85% of the full speed. The switch was removed because any arcing in the presence of the refrigerant gas caused the formation of an acid from the hydrocarbons in the gas. This acid attacked and etched the finished surfaces of the shafts, bearings, and valves. The acid also carbonized the organic material used to insulate the motor winding and caused the eventual breakdown of the insulation. To overcome the problem of the switch, the relatively heavy magnetic winding of a relay was connected in series with the main motor winding. The initial heavy inrush of current caused the relay to lift a magnetic core and energize the starting winding. As the motor speed increased, the main winding current decreased and allowed the relay to remove the starting or auxiliary winding from the circuit.

A later refinement of this arrangement was the use of a voltage-sensitive relay that was wound to permit pickup at values of voltage greater than the line voltage. The coil of this voltage-sensitive relay was connected across the starting winding. This connection scheme was based on the principle that the rotor of a single-phase induction motor induces in its own starting winding a voltage that is approximately in quadrature (phase) with the main winding voltage and has a value greater than the main voltage. The voltage-sensitive relay broke the circuit to the starting winding and remained in a sealed position until the main winding was de-energized.

Since the major maintenance problems in hermetic compressor systems were due generally to starting relays and capacitors, it was desirable to eliminate as many as possible of these devices. It was soon realized that small- and medium-sized systems could make use of a different form of refrigerant metering device, thus eliminating the automatic expansion valve. Recall that this valve has a positive shutoff characteristic, which means that the refrigerant is restricted when the operation cycle is finished as indicated by the evaporator reaching the design temperature. As a result, when a new refrigeration cycle begins, the compressor must start against a head of pressure. If a small-bore, open capillary tube is substituted for the expansion valve, the refrigerant is still metered to the evaporator coil, but the gas continues to flow after the compressor stops until the system pressure is equalized. Therefore, the motor is required only to start the compressor against the same pressure on each side of the pistons. This ability to decrease the load led to the development of a new series of motors that contained only a running capacitor (no relay was installed). This type of motor furnished sufficient torque to start the unloaded compressor and, at the same time, greatly improved the overall power factor.

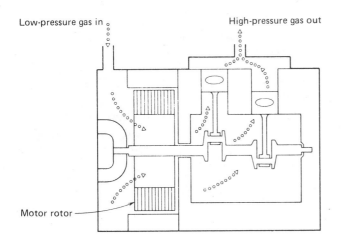
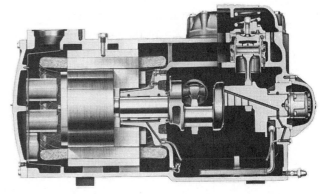

Fig. 21-8   The hermetic compressor.

COOLING SYSTEM CONTROL

Figure 21-9 shows a wiring diagram that is representative of standard cooling units.

When the selector switch in the heating-cooling control is set to COOL and the fan switch is placed on AUTO, any rise in temperature above the set point causes the cooling contacts (TC) to close and complete two circuits. One circuit through the fan switch energizes the evaporator fan relay (EFR), which, in turn, closes EFR contacts to complete the 208-volt circuit to the evaporator fan motor (EFM). The second circuit is to the control relay (CR) and causes the CR1 contacts to open. When the CR1 contacts open, the crankcase heater (CH) is de-energized. (This heater is installed to keep the compressor oil warm and dry when the unit is not running.) In addition, CR2 contacts close to complete a circuit to the condenser fan motor (CFM) and another circuit through low-pressure switch 2 (LP2), the high-pressure switch (HP), the thermal contacts (T), and the overload contacts (OL) to the compressor motor starter coil (CS). When the CR2 contacts close, the three CS contacts in the power circuit to the compressor motor (CM) also close.

Contacts LP1 and LP2 open when the refrigerant pressure drops below a set point. When contacts LP2 open, CS is de-energized. When contacts LP1 open, CR is de-energized, the circuit to CFM opens, and the circuit to CH is completed. The high-pressure switch (HP) contacts open and de-energize the compressor motor starter (CS) when the refrigerant pressure is above the set point. The low-pressure control (LP1) is the normal operating control and the high-pressure control (HP) and LP2 act as safety devices. The T contacts shown in Figure 21-9 are located in the compressor motor and open when the winding temperature of the motor is too high. The OL contacts are controlled by the OL elements installed in the power leads to the motor. The OL elements are sized to correspond to the current draw of the motor. That is, a high current causes the overload elements to overheat and open the OL contacts. As a result, the compressor starter is de-energized and the compressor stops. The evaporator fan motor (EFM) used to circulate air in the store area can be run continuously if the fan switch is turned to FAN. Actually, in many situations, it is recommended that the fan motor run continuously to keep the air in motion.

COOLING SYSTEM INSTALLATION

The owner of the commercial building leases the various office and shop areas on the condition that heat will be furnished to each area. However, tenants agree to pay the cost of operating the refrigeration system to provide cooling in their areas. Only four cooling systems are indicated in the plans for the commercial building since the bakery does not use a cooling system. The cooling equipment for the insurance office, the doctor's office, and the beauty salon are single-package cooling units located on the roof. The compressor, condenser, and evaporator for each of these units are constructed within a single enclosure. The system for the drugstore is a split system with the compressor and condenser located on the roof and the evaporator located in the basement, Figure 21-10. Regardless of the type of cooling system installed, a duct system must be provided to connect the evaporator with the area to be cooled. The duct system shown diagrammatically in Figure 21-10 does not represent the actual duct system, which will be installed by another contractor.

The electrician is expected to provide a power circuit to each air-conditioning unit as shown on the plans. In addition, it is necessary to provide wiring to the thermostat in each area, and, in the case of the drugstore, wiring must be provided to the evaporator located in the basement.

ELECTRICAL REQUIREMENTS FOR AIR-CONDITIONING AND REFRIGERATION EQUIPMENT

NEC® Article 440 provides the requirements for installing air-conditioning and refrigeration equipment that involves one or more hermetic refrigerant motor-compressors. Where these compressors are not involved *NEC® Section 440-3* directs the reader to *NEC® Articles 422, 424*, and *430* and gives other references for special provisions.

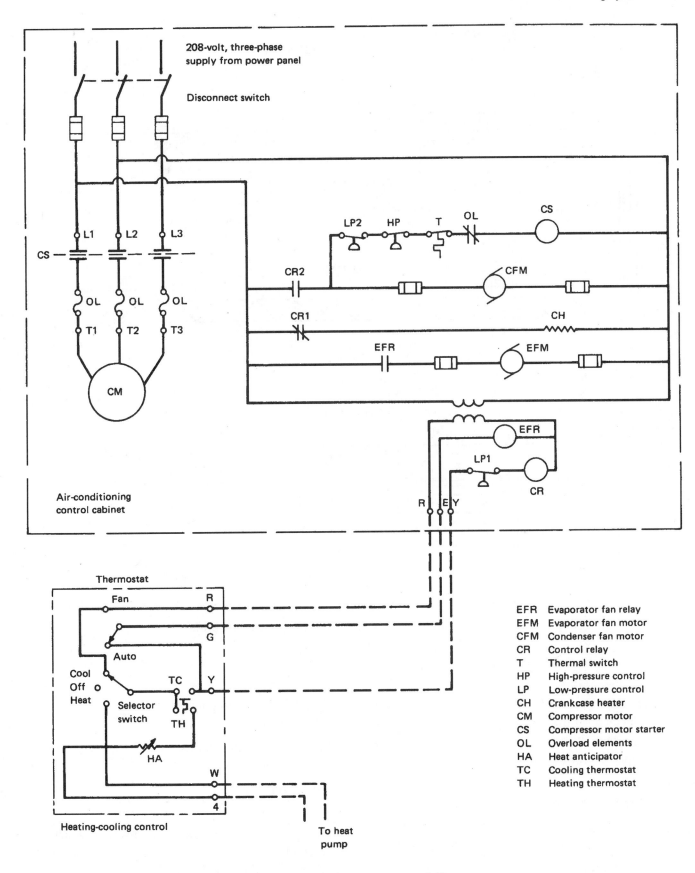

Fig. 21-9 A cooling system control diagram.

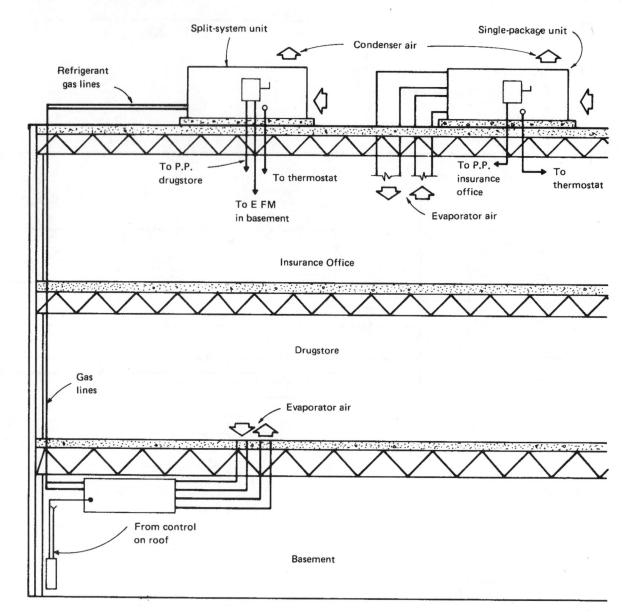

Fig. 21-10   Single-package and split-system cooling units.

For the insurance office, an air-conditioning unit is located on the roof of the commercial building. The data furnished with this air-conditioning unit supply the following information:

Voltage:
208 volts, three-phase, 3-wire, 60 hertz

Hermetic refrigerant compressor-motor:
Rated load amperes 20.2 at 208 volts three-phase

Evaporator motor:
Full load amperes 3.2 at 208 volts single-phase

Condenser motor:
Full load amperes 3.2 at 208 volts single-phase

Minimum circuit ampacity:[1]
31.65 amperes

Maximum overcurrent protection:[2]
50 amperes, time-delay fuse

Locked rotor current:
176 amperes

As discussed elsewhere in this text, when the nameplate on equipment specifies fuses, then only uses are permitted to be used as the branch-circuit

---

[1] $(20.2 \times 1.25) + 3.2 + 3.2 = 31.65$ amperes.
[2] $(20.2 \times 2.25) + 3.2 + 3.2 = 51.85$ amperes. The next lower standard size from *NEC® Section 246-6* is 50 amperes.

overcurrent protection. Circuit breakers would not be permitted. See *NEC® Section 110-3(b)*. This has been covered in Unit 17.

Air-conditioning and refrigeration equipment is almost always marked with the minimum supply circuit ampacity and maximum ampere rating and type of overcurrent protective device. This marking is required by *NEC® Section 440-4*.

SPECIAL TERMINOLOGY

Understanding the following definitions is important when installing air conditioners, heat pumps, and other equipment utilizing hermetic refrigerant motor-compressors. These terms are found throughout *NEC® Article 440* and *UL Standard 1995*.

Rated Load Current (RLA): The RLA is determined by the manufacturer of the hermetic refrigerant motor-compressor through testing at rated refrigerant pressure, temperature conditions, and voltage.

In most instances, the RLA is at least equal to 64.1% of the hermetic refrigerant motor-compressor's maximum continuous current (MCC).

Example:

The nameplate on an air-conditioning unit is marked:

Compressor RLA . . . 17.8 amperes

Branch-Circuit Selection Current (BCSC): Some hermetic refrigerant motor-compressors are designed to operate continuously at currents greater than 156% of the RLA. In such cases, the unit's nameplate is marked with "Branch-Circuit Selection Current." The BCSC will be no less than at least 64.1% of the maximum continuous current rating (MCC) of the hermetic refrigerant motor-compressor.

Note: 156% and 64.1% have an inverse relationship:
$$1/_{1.56} = 0.641 \text{ and } 1/_{0.641} = 1.56$$

Example:

The maximum continuous current of a hermetic refrigerant motor-compressor is 31 amperes. The BCSC will be no less than:

$$31 \times 0.641 = 19.9 \text{ amperes}$$

Since the BCSC value, when marked, is always equal to or greater than the unit nameplate marked RLA value, the manufacturer of the air-conditioning unit must use the BCSC value, instead of the RLA value, to determine the minimum circuit ampacity (MCA), and the maximum overcurrent protection (MOP). For installation of individual hermetic refrigerant motor-compressors, where the electrician must select the conductors, the controller, the disconnecting means, and the short-circuit and ground-fault protection, the electrician must use the BCSC, if given, instead of the RLA; see *NEC® Section 440-4(c)*.

Maximum Continuous Current (MCC): The MCC is determined by the manufacturer of the hermetic refrigerant motor-compressor under specific test conditions. The MCC is needed to properly design the end-use product. The installing electrician is not directly involved with the MCC.

The MCC is not on the nameplate of the packaged air-conditioning unit. The MCC is established by the *NEC®* as being no greater than 156% of the RLA or the BCSC, for the hermetic refrigerant motor-compressor.

Except in special conditions, the overload protective system must operate for current in excess of 156%.

Example:

A hermetic refrigerant motor-compressor is marked:

Maximum Continuous Current . . . 31 amperes

Minimum Circuit Ampacity (MCA): The manufacturer of an air-conditioning unit is required to mark the nameplate with this value. This is what the electrician needs to know. The manufacturer determines the MCA by multiplying the RLA, or the BCSC, of the hermetic refrigerant motor-compressor by 125%. The current ratings of all other concurrent loads, such as fan motors, transformers, relay coils, etc., are then added to this value.

Example:

An air-conditioning unit's nameplate is marked:
Minimum Circuit Ampacity . . . 26.4 amperes

This was derived as follows:

BCSC of 19.9 × 1.25	=	24.875 amperes
plus ¼ HP fan motor	@	1.5 amperes
Therefore, the MCA	=	26.4 amperes

The electrician must install the conductors, the disconnect switch, and the overcurrent protection based upon the MCA value. No further calculations are needed; the manufacturer of the equipment has done it all.

Maximum Overcurrent Protective Device (MOP): The electrician should always check the nameplate of an air-conditioning unit for this information.

The manufacturer is required to mark this value on the nameplate. This value is determined by multiplying the RLA, or the BCSC, of the hermetic refrigerant motor-compressor by 225%, then adding all concurrent loads such as electric heaters, motors, etc.

Example:

The nameplate of an air-conditioning unit is marked:

Maximum Time Delay Fuse . . . 45 amperes
This was derived as follows:

BCSC of 19.9×2.25 = 44.775 amperes
plus ¼ HP fan motor   @   1.5 amperes
     Therefore, the MOP   = 46.275 amperes

Since 46.275 is the maximum, the next lower standard size time-delay fuse (that is, 45 amperes) must be used. See *NEC® Section 240-6* and *Article 440 Part C*.

The electrician installs branch-circuit overcurrent devices with a rating **not to exceed** the MOP as marked on the nameplate of the air-conditioning unit. The electrician makes no further calculations; all calculations have been done by the manufacturer of the air-conditioning unit.

Overcurrent Protection Device Selection: How does an electrician know if fuses or circuit breakers are to be used for the branch-circuit overcurrent device? The electrician reads the nameplate and the instructions carefully. *NEC® Section 110-3(b)* requires that "Listed or labeled equipment be used or installed in accordance with any instructions included in the listing or labeling."

Example:

If the nameplate of an air-conditioning unit reads "Maximum Size Time-Delay Fuse: 45 amperes," fuses are required for the branch-circuit protection. To install a circuit breaker would be a violation of *NEC® Section 110-3(b)*.

If the nameplate reads "Maximum Fuse or HACR Type Breaker . . . ," then either fuses or a HACR type circuit breaker is permitted. See figures 17-33, 34, and 35.

Disconnecting Means Rating: The Code rules for determining the size of disconnecting means required for HVAC equipment is covered in *NEC® Article 440 Part B*.

The horsepower rating of the disconnecting means must be at least equal to the sum of all of the individual loads within the equipment . . . at rated load conditions . . . and at locked rotor conditions. See *NEC® Section 440-12(b)(1)*.

The ampere rating of the disconnecting means must be at least 115% of the sum of all of the individual loads within the equipment. Refer to *NEC® Sections 440-12(a)(1)* and *-12(b)(2)* for details.

Because disconnect switches are horsepower rated, it is sometimes necessary to convert the locked rotor current information found on the nameplate of air-conditioning equipment to an equivalent horsepower rating; *NEC® Article 440 Part B* sets forth the conversion procedure.

Air-Conditioning and Refrigeration Equipment Disconnecting Means

- The disconnecting means for an individual hermetic motor/compressor must not be less than 115% of the rated load current or the branch-circuit selection current, whichever is greater. See *NEC® Section 440-12(a)(1)*.

- The disconnecting means for an individual hermetic motor/compressor shall have a horsepower rating equivalent to the horsepower rating shown in *NEC® Tables 430-151(A)* and *-151(B)*. This table is a conversion table to convert locked rotor current to an equivalent horsepower. See *NEC® Section 440-12(a)(2)*.

- For equipment that has at least one hermetic motor/compressor and other loads, such as fans, heaters, solenoids, and coils, the disconnecting means must not be less than 115% of the sum of the currents of all of the components. See *NEC® Section 440-12(b)(2)*.

- For equipment that has at least one hermetic motor/compressor and other loads, such as fans, heaters, solenoids, and coils, the disconnecting means is determined by adding up all of the individual currents. This total is then considered to be a single motor for the purpose of determining the disconnecting means horsepower rating. See *NEC® Section 440-12(b)(1)*.

▶ In our example, the full-load current equals 31.65 amperes. Checking *NEC® Table 430-150*, will reveal that a full-load current rating of 31.65 amperes falls between a 10-horsepower and a 15-horsepower motor. Therefore, we must consider our example to be equivalent to a 15-horsepower motor. ◀

NEC® Table 430-148 lists the full-load current ratings for single-phase motors.

- The total locked-rotor current must also be checked to be sure that the horsepower rating of the disconnecting means is capable of safely disconnecting the full amount of locked-rotor current, see *NEC® Section 440-12(b)(1)*.

▶ In our example, the locked-rotor current equals 176 amperes. Checking *NEC® Table 430-151B* in the 230-volt column (this table does not have a 208-volt column), for the conversion of locked-rotor current to horsepower, we find that an LRA of 176 amperes falls between a 10-horsepower and a 15-horsepower motor. Therefore, we again consider our example to be equivalent to a 15-horsepower motor, see *NEC® Sections 440-12(a)* and *(b)*. ◀

▶ If one of the selection methods (FLA or LRA) indicates a larger size disconnect switch than the other method does, then the larger switch shall be installed. See *NEC® Sections 440-12(a)* and *-12(b)*. ◀

- The disconnecting means must be within sight of and readily accessible from, the air-conditioning and refrigeration equipment. See *NEC® Section 440-14*.

- The disconnecting means may be mounted inside the equipment, or it may be mounted on the equipment. Very large equipment often has the disconnect switch mounted inside the enclosure. See *NEC® Section 440-14*.

Air-Conditioning and Refrigeration Equipment Branch-Circuit Conductors

- For individual motor/compressor equipment, the branch-circuit conductors must have an ampacity of not less than 125% of the rated load current or branch-circuit selection current, whichever is greater. See *NEC® Section 440-32*.

- For motor/compressor(s), plus other loads such as a typical air-conditioning unit that contains a motor/compressor, fan, heater, coils, etc., add all of the individual current values, plus 25% of that of the largest motor or motor/compressor. See *NEC® Section 440-33*.

Table 430-148. Full-Load Currents in Amperes, Single-Phase Alternating-Current Motors

The following values of full-load currents are for motors running at usual speeds and motors with normal torque characteristics. Motors built for especially low speeds or high torques may have higher full-load currents, and multispeed motors will have full-load current varying with speed, in which case the namplate current ratings shall be used.

The voltages listed are rated motor voltages. The currents listed shall be permitted for system voltage ranges of 110 to 120 and 220 to 240 volts.

Horsepower	115 Volts	200 Volts	208 Volts	230 Volts
⅙	4.4	2.5	2.4	2.2
¼	5.8	3.3	3.2	2.9
⅓	7.2	4.1	4.0	3.6
½	9.8	5.6	5.4	4.9
¾	13.8	7.9	7.6	6.9
1	16	9.2	8.8	8.0
1½	20	11.5	11.0	10
2	24	13.8	13.2	12
3	34	19.6	18.7	17
5	56	32.2	30.8	28
7½	80	46.0	44.0	40
10	100	57.5	55.0	50

Reprinted with permission from NFPA 70-1999.

Table 430-150. Full-Load Current Three-Phase Alternating-Current Motors

The following values of full-load currents are typical for motors running at speeds usual for belted motors and motors with normal torque characteristics.

Motors built for low speeds (1200 rpm or less) or high torques may require more running current, and multispeed motors will have full-load current varying with speed. In these cases, the nameplate current rating shall be used.

The voltages listed are rated motor voltages. The currents listed shall be permitted for system voltage ranges of 110 to 120, 220 to 240, 440 to 480, and 550 to 600 volts.

Horsepower	Induction Type Squirrel Cage and Wound Rotor (Amperes)							Synchronous-Type Unity Power Factor* (Amperes)			
	115 Volts	200 Volts	208 Volts	230 Volts	460 Volts	575 Volts	2300 Volts	230 Volts	460 Volts	575 Volts	2300 Volts
½	4.4	2.5	2.4	2.2	1.1	0.9	—	—	—	—	—
¾	6.4	3.7	3.5	3.2	1.6	1.3	—	—	—	—	—
1	8.4	4.8	4.6	4.2	2.1	1.7	—	—	—	—	—
1½	12.0	6.9	6.6	6.0	3.0	2.4	—	—	—	—	—
2	13.6	7.8	7.5	6.8	3.4	2.7	—	—	—	—	—
3	—	11.0	10.6	9.6	4.8	3.9	—	—	—	—	—
5	—	17.5	16.7	15.2	7.6	6.1	—	—	—	—	—
7½	—	25.3	24.2	22	11	9	—	—	—	—	—
10	—	32.2	30.8	28	14	11	—	—	—	—	—
15	—	48.3	46.2	42	21	17	—	—	—	—	—
20	—	62.1	59.4	54	27	22	—	—	—	—	—
25	—	78.2	74.8	68	34	27	—	53	26	21	—
30	—	92	88	80	40	32	—	63	32	26	—
40	—	120	114	104	52	41	—	83	41	33	—
50	—	150	143	130	65	52	—	104	52	42	—
60	—	177	169	154	77	62	16	123	61	49	12
75	—	221	211	192	96	77	20	155	78	62	15
100	—	285	273	248	124	99	26	202	101	81	20
125	—	359	343	312	156	125	31	253	126	101	25
150	—	414	396	360	180	144	37	302	151	121	30
200	—	552	528	480	240	192	49	400	201	161	40
250	—	—	—	—	302	242	60	—	—	—	—
300	—	—	—	—	361	289	72	—	—	—	—
350	—	—	—	—	414	336	83	—	—	—	—
400	—	—	—	—	477	382	95	—	—	—	—
450	—	—	—	—	515	412	103	—	—	—	—
500	—	—	—	—	590	472	118	—	—	—	—

*For 90 and 80 percent power factor, the figures shall be multiplied by 1.1 and 1.25, respectively.

Reprinted with permission from NFPA 70-1999.

One can readily see the difficulty encountered when attempting to compute the actual currents flowing in phases A, B, and C for a typical three-phase unit, Figure 21-11.

Rather than attempting to add the currents vectorally, for simplicity, one generally uses the arithmetic sum of the current. However, the student must realize that an ammeter reading taken while the air-conditioning unit is operating will not be the same as the calculated current values.

For the air-conditioning unit in the insurance office, the calculation is:

$$20.2 + 3.2 + 3.2 = 26.60 \text{ amperes}$$
$$\text{plus 25\% of } 20.2 = \underline{5.05} \text{ amperes}$$
$$\text{Total} = 31.65 \text{ amperes}$$

The minimum ampacity for the conductors serving the above air-conditioning unit is 31.65 amperes, with the calculation done as above.

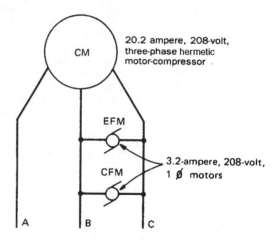

20.2 ampere, 208-volt, three-phase hermetic motor-compressor

3.2-ampere, 208-volt, 1 Ø motors

Fig. 21-11   Load connections.

However, the nameplate of the air-conditioning unit calls for a minimum circuit ampacity of 31.65 amperes. Referring to *NEC® Table 310-16* it is determined that a No. 8 AWG, Type THHN conductor has an allowable ampacity of 40-ampere.

Table 430-151(A). Conversion Table of Single-Phase Locked-Rotor Currents for Selection of Disconnecting Means and Controllers as Determined from Horsepower and Voltage Rating

For use only with Sections 430-110, 440-12, 440-41, and 455-8(c).

Rated Horsepower	Maximum Locked-Rotor Current in Amperes, Single Phase		
	115 Volts	208 Volts	230 Volts
½	58.8	32.5	29.4
¾	82.8	45.8	41.4
1	96	53	48
1½	120	66	60
2	144	80	72
3	204	113	102
5	336	186	168
7½	480	265	240
10	600	332	300

Reprinted with permission from NFPA 70-1999.

Section 440-35 references *NEC® Section 440-4(b)*, which sets forth the requirement that the manufacturers mark the equipment with the minimum supply circuit ampacity and the maximum branch-circuit overcurrent protective device ampere rating.

Air-Conditioning and Refrigeration Equipment Branch-Circuit Short-Circuit and Ground-Fault Protection

- For individual units, the branch-circuit overcurrent protection shall not exceed 175% of the rated load current or the branch-circuit selection current, whichever is greater. But if the 175% sizing will not allow the equipment to start, then *NEC® Section 440-22(a)* permits a further increase in size not to exceed 225%.

- Where the circuit has more than one hermetic refrigerant motor/compressor, or one hermetic refrigerant motor/compressor and other motors or other loads, *NEC® Section 440-22(b)* references *NEC® Section 430-53*, which addresses the situation where more than one motor is fed by a single circuit.

The branch-circuit overcurrent protection must also meet the requirements of *NEC® Section 440-22(b)*. Simply stated, this means that overcurrent protection must be provided

Table 430-151(B). Conversion Table of Polyphase Design B, C, D, and E Maximum Locked-Rotor Currents for Selection of Disconnecting Means and Controllers as Determined from Horsepower and Voltage Rating and Design Letter

For use only with Sections 430-110, 440-12,* 440-41,* and 455-8(c).

Rated Horsepower	Maximum Motor Locked-Rotor Current in Amperes, Two- and Three-Phase, Design B, C, D and E											
	115 Volts		200 Volts		208 Volts		230 Volts		460 Volts		575 Volts	
	B, C, D	E	B, C, D	E	B, C, D	E	B, C, D	E	B, C, D	E	B, C, D	E
½	40	40	23	23	22.1	22.1	20	20	10	10	8	8
¾	50	50	28.8	28.8	27.6	27.6	25	25	12.5	12.5	10	10
1	60	60	34.5	34.5	33	33	30	30	15	15	12	12
1½	80	80	46	46	44	44	40	40	20	20	16	16
2	100	100	57.5	57.5	55	55	50	50	25	25	20	20
3	—	—	73.6	84	71	81	64	73	32	36.5	25.6	29.2
5	—	—	105.8	140	102	135	92	122	46	61	36.8	48.8
7½	—	—	146	210	140	202	127	183	63.5	91.5	50.8	73.2
10	—	—	186.3	259	179	249	162	225	81	113	64.8	90
15	—	—	267	388	257	373	232	337	116	169	93	135
20	—	—	334	516	321	497	290	449	145	225	116	180
25	—	—	420	646	404	621	365	562	183	281	146	225
30	—	—	500	775	481	745	435	674	218	337	174	270
40	—	—	667	948	641	911	580	824	290	412	232	330
50	—	—	834	1185	802	1139	725	1030	363	515	290	412
60	—	—	1001	1421	962	1367	870	1236	435	618	348	494
75	—	—	1248	1777	1200	1708	1085	1545	543	773	434	618
100	—	—	1668	2154	1603	2071	1450	1873	725	937	580	749
125	—	—	2087	2692	2007	2589	1815	2341	908	1171	726	936
150	—	—	2496	3230	2400	3106	2170	2809	1085	1405	868	1124
200	—	—	3335	4307	3207	4141	2900	3745	1450	1873	1160	1498
250	—	—	—	—	—	—	—	—	1825	2344	1460	1875
300	—	—	—	—	—	—	—	—	2200	2809	1760	2247
350	—	—	—	—	—	—	—	—	2550	3277	2040	2622
400	—	—	—	—	—	—	—	—	2900	3745	2320	2996
450	—	—	—	—	—	—	—	—	3250	4214	2600	3371
500	—	—	—	—	—	—	—	—	3625	4682	2900	3746

*In determining compliance with Sections 440-12 and 440-41, the values in the B, C, D columns shall be used.

Reprinted with permission from NFPA 70-1999.

that does not exceed the protection selected for the motor/compressor that draws the largest amount of current, plus the sum of the rated load current of branch-circuit selection currents of other motor/compressors, plus the ampere ratings of any other loads.

In our example:

$$(20.2 \times 1.75) + 3.2 + 3.2 = 41.75 \text{ amperes}$$

It is important to know that *NEC®* Section, because *Section 440-22(c)* states that the over-current protection shall not exceed the value indicated on the nameplate as determined by the manufacturer of the unit.

Underwriters Laboratories Standard No. 1995 allows the manufacturer of hermetic motor-compressor equipment to determine the maximum size overcurrent protection as follows:

$$\begin{aligned}
\text{Largest load} \times 2.25 = \\
20.2 \times 2.25 = 45.45 \text{ amperes} \\
\text{Plus other loads} \quad 3.20 \text{ amperes} \\
\underline{3.20 \text{ amperes}} \\
51.85 \text{ amperes} \\
\text{maximum}
\end{aligned}$$

This value is "rounded down" to the next lower standard fuse rating, which is 50-ampere (*NEC®* Section 240-6). This size requires a 60-ampere disconnect switch. Because the 50-ampere value is maximum, the use of a lower rated time-delay fuse, such as a 45-ampere or 40-ampere fuse, is permitted. The question is, will the lower ratings open unnecessarily on repetitive start-ups of the air-conditioning unit? The ampere rating of the maximum size fuse or circuit breaker as determined by the manufacturer is generally the size to install.

The nameplate for the air-conditioning unit supplying the cooling for the drugstore calls for maximum 50-ampere, dual-element, time-delay fuses. This value must not be exceeded.

NEC® Section 110-3(b) requires that "Listed or labeled equipment shall be installed and used in accordance with any instructions included in the listing or labeling." It is this section that prohibits the use of a circuit breaker to protect this particular air-conditioning unit. If the feeder or branch-circuit originates from a circuit-breaker panel, then the requirements of this Section can be met by installing a fusible disconnect switch near, on, or within the air-conditioning equipment.

Air-Conditioning and Refrigerator Equipment Motor/Compressor and Branch-Circuit Overload Protection

This topic is covered in *NEC® Article 440, Part F.*

The manufacturer of this type of equipment provides the overload protection for the various internal components that make up the complete air-conditioning unit. The manufacturer submits the equipment to Underwriters Laboratories for testing and listing of the product. Overload protection of the internal components is covered in the Underwriters Standards. This protection can be in the form of fuses, thermal protectors, certain types of inverse-time circuit breakers, overload relays, etc.

NEC® Section 440-52 discusses overload protection for motor/compressor appliances. This section requires that the motor/compressor be protected from overloads and failure to start by:

- a separate overload relay that is responsive to current flow and trips at not more than 140% of the rated load current, or

- an integral thermal protector arranged so that its action interrupts the current flow to the motor/compressor, or

- a time-delay fuse or inverse-time circuit breaker rated at not more than 125% of the rated load current.

It is rare for the electrician to have to provide additional overload protection for these components. The electrician's job is to make sure that he or she supplies the equipment with the proper size conductors, the proper size and type of branch-circuit overcurrent protection, and the proper horse-power-rated disconnecting means.

REVIEW QUESTIONS

Refer to the *National Electrical Code*® or the working drawings when necessary. Where applicable, responses should be written in complete sentences. Write units using unit names, use no abbreviations or symbols (1 foot, not 1' or 1 ft).

Complete the following sentences by writing in a component used in a refrigeration system.

1. The _____ raises the energy level of the refrigerant.

2. The heat is removed from the refrigerant in the _____.

3. The refrigerant absorbs heat energy in the _____.

4. The _____ is a metering device.

5. In the typical residential heating/cooling system, the evaporator would be located _____.

6. In the typical residential heating/cooling system, the condenser would be located _____.

Respond to the following statements.

7. Name three types of condensers. (1) _____ , (2) _____ and (3) _____

8. A 230-volt single-phase hermetic compressor with an FLA of 21 and an LRA of 250 requires a disconnect switch with a rating of _____ hp.

9. A motor-compressor unit with a rating of 32 amperes is protected from overloads by a separate overload relay selected to trip at not more than _____ ampere(s).

10. Does the Code permit the disconnect switch for a large rooftop air conditioner to be mounted inside the unit? Give the appropriate Code reference.

_____

11. The nameplate of an air-conditioning unit specifies 85 amperes as the minimum ampacity for the supply conductors. Select the conductor size and type.

_____

12. Where the nameplate of an air-conditioner unit specifies the minimum ampacity required for the supply conductors, must the electrician multiply this ampacity by 1.25 to determine the correct size conductors? Explain.

_____

_____

_____

_____

13. Which current rating of an air-conditioning unit is greater? Check the correct answer.

   a. _____ rate load current

   b. _____ branch-circuit selection current

APPENDIX

ELECTRICAL SPECIFICATIONS

General Clauses and Conditions

The General Clauses and Conditions and the Supplementary General Conditions are hereby made a part of the Electrical Specifications.

Scope

The electrical contractor shall furnish all labor and material to install the electrical equipment shown on the drawings herein specified, or both. The contractor shall secure all necessary permits and licenses required and in accepting the contract agrees to have all equipment in working order at the completion of the project.

Materials

All materials used shall be new and shall bear the label of the Underwriters Laboratories, Inc. and shall meet the requirements of the drawings and specifications.

Workmanship

All electrical work shall be in accordance with the requirements of the *National Electrical Code,* shall be executed in a workmanlike manner, and shall present a neat and symmetrical appearance when completed.

Show Drawings

The contractor shall submit for approval descriptive literature for all equipment installed as part of this contract.

Motors

All motors shall be installed by the contractor furnishing the motor. All wiring to the motors and the connection of motors will be completed by the electrical contractor. All control wiring will be the responsibility of the contractor furnishing the equipment to be controlled.

Wiring

The wiring shall meet the following requirements:

a. The minimum size conductors will be No. 12 AWG.

b. In the doctor's office, all branch-circuit conductors will be installed in electrical metallic tubing. In all other areas of the building, branch-circuit conductors may be installed in either electrical metallic tubing or electrical nonmetallic tubing.

c. Three spare ¾-inch conduits shall be installed from each flush-mounted panel into the ceiling joist space above the panel.

d. In the doctor's office, branch-circuit raceways shall be electrical metallic tubing. In all other areas of the building, raceways may be either electrical metallic tubing or electrical nonmetallic conduit.

e. Feeders shall be installed in either rigid metal conduit or rigid nonmetallic conduit. The supply to the boiler shall be rigid metal conduit or Schedule 80 rigid nonmetallic conduit.

f. Conductors shall have a 90°C rating, Types THHN, THHW, or XHHW. All conductors shall be copper.

Switches

Switches shall be ac general-use, snap switches, specification grade, 20 amperes, 120–277 volts, either single-pole, three-way, or four-way, as shown on the plans.

Time Clock

A time clock shall be installed to control the

lighting in the front and rear entries. The clock will be connected to Panel EM circuit No. 1. The clock will be 120 volts, one circuit, with astronomic control and a spring-wound carry-over mechanism.

Receptacles

The electrical contractor shall furnish and install, as indicated in blueprint E-4 (Electrical Symbol Schedule), receptacles meeting NEMA standards and listed by Underwriters Laboratories, Inc.

Panelboards

The electrical contractor shall furnish panelboards, as shown on the plans and detailed in the panelboard schedules. Panelboards shall be listed by the Underwriters Laboratories, Inc. All interiors will have 225-ampere bus with lugs, rated at 75°C, for feed conductors. Boxes will be of galvanized sheet steel and will provide wiring gutters, as required by the *National Electrical Code.* Fronts will be suitable for either flush or surface installation and shall be equipped with a keyed lock and a directory card holder. All panelboards shall have an equipment grounding bus bonded to the cabinet. See Sheet E4 for a listing of the required poles and overcurrent devices.

Molded Case Circuit Breakers

Molded case circuit breakers shall be installed in branch-circuit panelboards as indicated on the panelboard schedules following the Specifications. Each breaker shall provide inverse time delay under overload conditions and magnetic tripping for short circuits. The breaker operating mechanism shall be trip-free and multipole units will open all poles if one pole trips. Breakers shall have sufficient interrupting rating to interrupt 10,000 RMS symmetrical amperes.

Motor-Generator

The electrical contractor will furnish and install a motor-generator plant capable of delivering 12 kVA at 208Y/120 volts, three phase. Motor shall be for use with diesel fuel, liquid cooled, complete with 12-volt batteries and battery charger, mounted on antivibration mounts with all necessary accessories,

including mufflers, exhaust piping, fuel tanks, fuel lines, and remote derangement annunciator.

Transfer Switch

The electrical contractor will furnish and install a complete automatic load transfer switch capable of handling 12 kVA at 208Y/120 volts, three phase. The switch control will sense a loss of power on any phase and signal the motor-generator to start. When emergency power reaches 80% of voltage and frequency, the switch will automatically transfer to the generator source. When the normal power has been restored for a minimum of five minutes, the switch will reconnect the load to the regular power and shut off the motor-generator. Switch shall be in a NEMA 1 enclosure.

Luminaires

The electrical contractor will furnish and install luminaires, as described in the schedule shown in the plans. Luminaires shall be complete with diffusers and lamps where indicated. Ballast, when required, shall be Class P and have a sound rating of A. Each ballast is to be individually fused; fuse size and type shall be as recommended by the ballast manufacturer. All rapid-start ballasts shall be Class E.

Service Entrance and Equipment

The electrical contractor will furnish and install a service entrance as shown on the Electrical Power Distribution Riser Diagram, which is found in the plans. Transformers and all primary service will be installed by the power company. The electrical contractor will provide a concrete pad and conduit rough-in as required by the power company.

The electrical contractor will install a ground field adjacent to the transformer pad consisting of three 8-foot-long copperweld rods connected to the grounded conductor. In addition, a connection shall be made to the main water pipe.

The electrical contractor will furnish and install a service lateral consisting of three sets of three 500 kcmil and one 1/0 AWG conductors, each set in a three-inch rigid metallic or nonmetallic conduit.

The electrical contractor will furnish and install service-entrance equipment, as shown in the plans and detailed herein. The equipment will consist of nine switches, and the metering equipment will be fabricated in three type NEMA 1 sections. A continuous neutral bus will be furnished for the length of the equipment and shall be isolated except for the main bonding jumper to the grounding bus, which shall also be connected to each section of the service-entrance equipment and to the water main. The switchboard shall be braced for 50,000 RMS symmetrical amperes.

The metering will be located in one section and shall consist of seven meters. Five of these meters shall be for the occupants of the building and two meters shall serve the owner's equipment.

The switches shall be as follows:

1. Bolted pressure switch, three-pole, 600 amperes with three 600-ampere fuses.

2. Bolted pressure switch, three-pole, 800 amperes with three 700-ampere fuses.

3. Quick make-quick break switch, three-pole, 100 amperes, with three 100-ampere fuses.

4. Quick make-quick break switch, three-pole, 200 amperes with three 200-ampere fuses.

5. Quick make-quick break switch, three-pole, 200 amperes with three 150-ampere fuses.

6. Quick make-quick break switch, two-pole, 200-ampere switch with two 125-ampere fuses.

7. Quick make-quick break switch, three-pole, 100-ampere with three 90-ampere fuses.

8. Quick make-quick break switch, three-pole, 100-ampere with three 60-ampere fuses.

9. Quick make-quick break switch, three-pole, 100-ampere switch.

The bolted pressure switches shall be knife-type switches constructed with a mechanism that automatically applies a high pressure to the blade when the switch is closed. Switch shall be rated to interrupt 200,000 symmetrical RMS amperes when used with current-limiting fuses having an equal rating.

The quick make-quick break switches shall be constructed with a device that assists the operator in opening or closing the switch to minimize arcing. Switches shall be rated to interrupt 200,000 symmetrical RMS amperes when used with fuses having an equal rating. They shall have rejection type R fuse clips for Class R fuses.

Fuses

a. Fuses 601 amperes and larger shall have an interrupting rating of 200,000 symmetrical RMS amperes. They shall provide time-delay of not less than 4 seconds at 500% of their ampere rating. They shall be current limiting and of silver-sand Class L construction.

b. Fuses 600 amperes and less shall have an interrupting rating of 200,000 symmetrical RMS amperes. They shall be of the rejection Class RK1 type, having a time delay of not less than 10 seconds at 500% of their ampere rating.

c. Plug fuses shall be dual-element, time-delay type with Type S base.

d. All fuses shall be so selected to assure positive selective coordination as indicated on the schedule.

e. All lighting ballasts (fluorescent, mercury vapor, or others) shall be protected on the supply side with an appropriate fuse in a suitable approved fuseholder mounted within or on the fixture. Fuses shall be sized as recommended by the ballast manufacturer.

f. Spare fuses shall be provided in the amount of 20% of each size and type installed, but in no case shall less than three spares of a specific size and type be supplied. These spare fuses shall be delivered to the owner at the time of acceptance of the project, and shall be placed in a spare fuse cabinet mounted on the wall adjacent to, or located in, the switchboard.

g. Fuse identification labels, showing the size and type of fuses installed, shall be placed inside the cover of each switch.

Low-Voltage Remote-Control Switching

A low-voltage, remote-control switch system shall be installed in the drugstore as shown on the plans and detailed herein. All components shall be specification grade and constructed to operate on 24-volt control power. The transformer shall be a 120/24-volt, energy-limiting type for use on a Class II signal system.

Cabinet. A metal cabinet matching the panelboard cabinets shall be installed for the installation of relays and other components. A barrier will separate the control section from the power wiring.

Relay.   A 24-volt ac, split-coil design relay rated to control 20 amperes of tungsten or fluorescent lamp loads shall be provided.

Switches.   The switches shall be complete with wall plate and mounting bracket. They shall be normally open, single-pole, double-throw, momentary contact switches with on-off identification.

Rectifiers.   A heavy-duty silicon rectifier with 7.5-ampere continuous duty rating shall be provided.

Wire.   The wiring shall be in two- or three-conductor, color coded, No. 20 AWG wire.

Rubber grommet.   An adapter will be installed on all relays to isolate the relay from the metal cabinet to reduce noise.

Switching schedule.   Connections will be made to accomplish the lighting control as shown in the switching schedule, figure S1.

HEATING AND AIR-CONDITIONING SPECIFICATIONS

Only those sections that pertain to the electrical work are listed here.

Boiler

The heating contractor will furnish and install a 200-kW electric hot water heating boiler completely equipped with safety, operating, and sequencing controls for 208-volt, three-phase electric power.

Hot Water Circulating Pumps

The heating contractor will furnish and install five circulating pumps. Each pump will serve a separate rental area and be controlled from a thermostat located in that area as indicated on the electrical plans. Pumps will be $1/6$ horsepower, 120 volts, single phase. Overload protection will be provided by a manual motor starter.

TABLE A-1
Drugstore Low-Voltage, Remote-Control Switching Schedule

Switch	Relay	Area Served	Branch Circuit #
RCa	A	Main area lighting	1
RCb	B	" " "	3
RCc	C	" " "	3
RCd	D	Makeup area	5
RCe	E	Storage	5
RCf	F	Toilet	5
	G	Pharmacy	7
	H	"	7
	I	"	7
RCg	J	Stairway and Basement	9
RCh, RCi	L,M	Show window	11,13
	K	Sign	15
RCM-1	D		
RCM-2	E,F		
RCM-3	H,I,J		
RCM-4	I,J,K		
RCM-5	A,B,C		
RCM-6	J		
RCM-7	K		
RCM-8	L,M		

Air-Conditioning Equipment

Air-conditioning equipment will be furnished and installed by the heating contractor in four of the rental areas indicated as follows:

Drugstore. A split system packaged unit will be installed with a rooftop compressor-condenser and a remote evaporator located in the basement. The electrical characteristics of this system are:

Voltage:
   208 volts, three-phase, 3 wire, 60 hertz
Hermetic refrigerant compressor-motor:
   Rated load current 20.2 amperes, 208 volts three-phase
Evaporator motor:
   Full load current 3.2 amperes, 208 volts single-phase
Condenser motor:
   Full load current 3.2 amperes, 208 volts single-phase
Minimum circuit ampacity:
   31.65 amperes
Maximum overcurrent protection:
   50 amperes, time-delay fuse

Insurance Office. A single package unit will be installed on the roof with electrical data identical to that of the unit specified for the drugstore.

Beauty Salon. A single package unit will be installed on the roof and will have the following electrical characteristics:
Voltage:
   208 volts, three-phase, 3 wire, 60 hertz
Hermetic refrigerant compressor-motor:
   Rated load amperes 14.1, 208 volts three-phase
Condenser-evaporator motor:
   Full load amperes 3.3, 208 volts single-phase
Minimum circuit ampacity:
   20.92 amperes
Maximum overcurrent protection:
   35 amperes, time-delay fuse

Doctor's Office. A single package unit will be installed on the roof and will have the following electrical characteristics:
Voltage:
   208 volts, single-phase, 2 wire, 60 hertz
Hermetic refrigerant compressor-motor:
   Rated load amperes 16.8 at 208 volts single-phase
Condenser-evaporator motor:
   Full load amperes 3.7 at 208 volts single-phase
Minimum circuit ampacity:
   24.7 amperes
Maximum overcurrent protection:
   40 amperes, time-delay fuse

Heating Control

In cooled areas, the heating and cooling will be controlled by a combination thermostat located in the proper area as shown on the electrical plans.

PLUMBING SPECIFICATIONS

Only those sections that pertain to the electrical work are listed here.

Motors

All motors will be installed by the contractor furnishing the motor. All electrical power wiring to the motors and the connection of the motors shall be made by the electrical contractor. All control wiring and control devices will be the responsibility of the contractor furnishing the equipment to be controlled.

Sump Pump

The plumbing contractor will furnish and install an electric motor-driven, fully automatic sump pump. Motor will be 1/2 horsepower, 120 volts, single phase. Overload protection will be provided by a manual motor starter.

TABLE A-2
Aluminum Conductors — Ratings and Volt Loss†

Conduit	Wire Size	Ampacity: Type T, TW, UF (60°C Wire)	Type RH, RHW, THHW (wet), THW, THWN, XHHW (wet), USE (75°C Wire)	Type RHHW, THHW (dry), XHHW (dry) (90°C Wire)	Direct Current	Volt Loss Three Phase (60 Cycle, Lagging Power Factor.) 100%	90%	80%	70%	60%	Single Phase (60 Cycle, Lagging Power Factor.) 100%	90%	80%	70%	60%
Steel Conduit	12	20*	20*	25*	6360	5542	5039	4504	3963	3419	6400	5819	5201	4577	3948
	10	25	30*	35*	4000	3464	3165	2836	2502	2165	4000	3654	3275	2889	2500
	8	30	40	45	2520	2251	2075	1868	1656	1441	2600	2396	2158	1912	1663
	6	40	50	60	1616	1402	1310	1188	1061	930	1620	1513	1372	1225	1074
	4	55	65	75	1016	883	840	769	692	613	1020	970	888	799	708
	3	65	75	85	796	692	668	615	557	497	800	771	710	644	574
	2	75	90	110	638	554	541	502	458	411	640	625	580	529	475
	1	85	100	115	506	433	432	405	373	338	500	499	468	431	391
	0	100	120	135	402	346	353	334	310	284	400	407	386	358	328
	00	115	135	150	318	277	290	277	260	241	320	335	320	301	278
	000	130	155	175	252	225	241	234	221	207	260	279	270	256	239
	0000	155	180	205	200	173	194	191	184	174	200	224	221	212	201
	250	170	205	230	169	148	173	173	168	161	172	200	200	194	186
	300	190	230	255	141	124	150	152	150	145	144	174	176	173	168
	350	210	250	280	121	109	135	139	138	134	126	156	160	159	155
	400	225	270	305	106	95	122	127	127	125	110	141	146	146	144
	500	260	310	350	85	77	106	112	113	113	90	122	129	131	130
	600	285	340	385	71	65	95	102	105	106	76	110	118	121	122
	750	320	385	435	56	53	84	92	96	98	62	97	107	111	114
	1000	375	445	500	42	43	73	82	87	89	50	85	95	100	103
Non-Magnetic Conduit (Lead Covered Cables Or Installation In Fibre Or Other Non-Magnetic Conduit. Etc.)	12	20*	20*	25*	6360	5542	5029	4490	3946	3400	6400	5807	5184	4557	3926
	10	25	30*	35*	4000	3464	3155	2823	2486	2147	4000	3643	3260	2871	2480
	8	30	40	45	2520	2251	2065	1855	1640	1423	2600	2385	2142	1894	1643
	6	40	50	60	1616	1402	1301	1175	1045	912	1620	1502	1357	1206	1053
	4	55	65	75	1016	883	831	756	677	596	1020	959	873	782	688
	3	65	75	85	796	692	659	603	543	480	800	760	696	627	555
	2	75	90	100	638	554	532	490	443	394	640	615	566	512	456
	1	85	100	115	506	433	424	394	360	323	500	490	455	415	373
	0	100	120	135	402	346	344	322	296	268	400	398	372	342	310
	00	115	135	150	318	277	281	266	247	225	320	325	307	285	260
	000	130	155	175	252	225	234	223	209	193	260	270	258	241	223
	0000	155	180	205	200	173	186	181	171	160	200	215	209	198	185
	250	170	205	230	169	147	163	160	153	145	170	188	185	177	167
	300	190	230	255	141	122	141	140	136	130	142	163	162	157	150
	350	210	250	280	121	105	125	125	123	118	122	144	145	142	137
	400	225	270	305	106	93	114	116	114	111	108	132	134	132	128
	500	260	310	350	85	74	96	100	100	98	86	111	115	115	114
	600	285	340	385	71	62	85	90	91	91	72	98	104	106	105
	750	320	385	435	56	50	73	79	82	82	58	85	92	94	95
	1000	375	445	500	42	39	63	70	73	75	46	73	81	85	86

* The overcurrent protection for conductor types marked with an (*) shall not exceed 15 amperes for 14 AWG, 20 amperes for 12 AWG, and 30 amperes for 10 AWG copper; or 15 amperes for 12 AWG and 25 amperes for 10 AWG aluminum and copper-clad aluminum after any correction factors for ambient temperature and number of conductors have been applied.

† Figures are L-L for both single phase and three phase. Three phase figures are average for the three phase.

TABLE A-3
Copper Conductors — Ratings and Volt Loss†

Conduit	Wire Size	Ampacity Type T, TW, UF (60°C Wire)	Type RH, RHW, THHW (wet), THW, THWN, XHHW (wet), USE (75°C Wire)	Type RHHW, THHW (dry), XHHW (dry) (90°C Wire)	Direct Current	Volt Loss (See Explanation Preface.) Three Phase (60 Cycle, Lagging Power Factor.) 100%	90%	80%	70%	60%	Single Phase (60 Cycle, Lagging Power Factor.) 100%	90%	80%	70%	60%
Steel Conduit	14	20*	20*	25*	6140	5369	4887	4371	3848	3322	6200	5643	5047	4444	3836
	12	25*	25*	30*	3860	3464	3169	2841	2508	2172	4000	3659	3281	2897	2508
	10	30	35*	40*	2420	2078	1918	1728	1532	1334	2400	2214	1995	1769	1540
	8	40	50	55	1528	1350	1264	1148	1026	900	1560	1460	1326	1184	1040
	6	55	65	75	982	848	812	745	673	597	980	937	860	777	690
	4	70	85	95	616	536	528	491	450	405	620	610	568	519	468
	3	85	100	110	490	433	434	407	376	341	500	501	470	434	394
	2	95	115	130	388	346	354	336	312	286	400	409	388	361	331
	1	110	130	150	308	277	292	280	264	245	320	337	324	305	283
	0	125	150	170	244	207	228	223	213	200	240	263	258	246	232
	00	145	175	195	193	173	196	194	188	178	200	227	224	217	206
	000	165	200	225	153	136	162	163	160	154	158	187	188	184	178
	0000	195	230	260	122	109	136	140	139	136	126	157	162	161	157
	250	215	255	290	103	93	123	128	129	128	108	142	148	149	148
	300	240	285	320	86	77	108	115	117	117	90	125	133	135	135
	350	260	310	350	73	67	98	106	109	109	78	113	122	126	126
	400	280	335	380	64	60	91	99	103	104	70	105	114	118	120
	500	320	380	430	52	50	81	90	94	96	58	94	104	109	111
	600	355	420	475	43	43	75	84	89	92	50	86	97	103	106
	750	400	475	535	34	36	68	78	84	88	42	79	91	97	102
	1000	455	545	615	26	31	62	72	78	82	36	72	84	90	95
Non-Magnetic Conduit (Lead Covered Cables Or Installation In Fibre Or Other Non-Magnetic Conduit, Etc.)	14	20*	20*	25*	6140	5369	4876	4355	3830	3301	6200	5630	5029	4422	3812
	12	25*	25*	30*	3860	3464	3158	2827	2491	2153	4000	3647	3264	2877	2486
	10	30	35*	40*	2420	2078	1908	1714	1516	1316	2400	2203	1980	1751	1520
	8	40	50	55	1528	1350	1255	1134	1010	882	1560	1449	1310	1166	1019
	6	55	65	75	982	848	802	731	657	579	980	926	845	758	669
	4	70	85	95	616	536	519	479	435	388	620	599	553	502	448
	3	85	100	110	470	433	425	395	361	324	500	490	456	417	375
	2	95	115	130	388	329	330	310	286	259	380	381	358	330	300
	1	110	130	150	308	259	268	255	238	219	300	310	295	275	253
	0	125	150	170	244	207	220	212	199	185	240	254	244	230	214
	00	145	175	195	193	173	188	183	174	163	200	217	211	201	188
	000	165	200	225	153	133	151	150	145	138	154	175	173	167	159
	0000	195	230	260	122	107	127	128	125	121	124	147	148	145	140
	250	215	255	290	103	90	112	114	113	110	104	129	132	131	128
	300	240	285	320	86	76	99	103	104	102	88	114	119	120	118
	350	260	310	350	73	65	89	94	95	94	76	103	108	110	109
	400	280	335	380	64	57	81	87	89	89	66	94	100	103	103
	500	320	380	430	52	46	71	77	80	82	54	82	90	93	94
	600	355	420	475	43	39	65	72	76	77	46	75	83	87	90
	750	400	475	535	34	32	58	65	70	72	38	67	76	80	83
	1000	455	545	615	26	25	51	59	63	66	30	59	68	73	77

* The overcurrent protection for conductor types marked with an (*) shall not exceed 15 amperes for 14 AWG, 20 amperes for 12 AWG, and 30 amperes for 10 AWG copper; or 15 amperes for 12 AWG and 25 amperes for 10 AWG aluminum and copper-clad aluminum after any correction factors for ambient temperature and number of conductors have been applied.

† Figures are L-L for both single phase and three phase. Three phase figures are average for the three phase.

TABLE A-4
Bakery – Loading Schedule

	Count	VA/Unit	NEC®	Actual	Computed	Balanced	Nonlinear
General Lighting:							
Style C luminaire	15	87		1305			1305
Style B luminaire	4	87		348			348
Style E luminaire	2	144		288			288
Style N luminaire	2	60		120			
Totals:				2061	2061		
Storage:							
NEC® Section 220-2	1200	0.25	300				
Style L luminaire	12	87		1044			1044
Totals:			300	1044	1044		
Show Window:							
NEC® Section 220-12	14	200	2800				
Style C luminaire	3	87		261			261
Receptacle outlets	4	650		2600			
Totals:			2800	2861	2861		
Other Loads:							
Receptacle outlets	21	180		3780			
Sign outlet	1	1200		1200			1200
Bake oven outlet	1	16000		16000		16000	
Totals:				20980	20980		
Motors & Appliances:	**Amperes**						
Exhaust fan	2.9	120		348			
Multi-mixer	3.96	360		1426		1426	
Multi-mixer	7.48	360		2693		2693	
Dough divider	2.2	360		792		792	
Doughnut machine	2.2	360		792		792	
Doughnut heater		2000		2000		2000	
Dishwasher	23.9	360		8604		8604	
Disposer	7.44	360		2678		2678	
Totals:				19333	19333		
TOTAL LOADS:					46279	34985	4446

TABLE A-5
Bakery – Panelboard Worksheet

Phase	Circuit Number	Load/Area Served	Computed Volt-Amperes	Computed Amperes	Load Type	Load Modifier	OCPD Selection Amperes	OCPD Rating	Minimum Conductor Size AWG	Minimum Derated Ampacity	Ambient Temp. C	Correction Factor	Current Carrying Conductors	Adjustment Factor	Minimum Allowable Ampacity	Conductor Size AWG	Allowable Ampacity	Derated Ampacity
A	1	15 Style C, Bake Area Exhaust fan	1305	11	C	1.25	17	20	12	18	30	1		1	17	12	30	30
			348	2.9	C	1.25												
A	2	4 Rec., Sales Area	720	6	R	1	6	20	12	21	30	1		1	21	12	30	30
B	3	4 Style B, 3 Style C	609	5.1	C	1.25	7	20	12	16	30	1		1	7	12	30	30
B	4	2 Style N, 1 E, Toilet	264	2.2	C	1.25	3	20	12	16	30	1		1	3	12	30	30
C	5	Sign Outlet	1200	10	C	1.25	13	20	12	16	30	1		1	13	12	30	30
C	6	12 Style L, 1 E	1188	9.9	C	1.25	13	20	12	16	30	1		1	13	12	30	30
A	7	Multi-mixer	1426	4			14	20	12	20	30	1		1	20	12	30	30
B		Multi-mixer	2693	7.5	R	1					30	1						
C		Dough Divider	792	2.2							30							
A	8	Motor	792	2.2			8	20	12	20	30	1		1	20	12	30	30
B		Doughnut Machine	2000	5.6	R	1					30	1						
C											30							
A	9	Recp. Show Window	1300	11	R	1	11	20	12	16	30	1		1	21	12	30	30
B	10	Oven	16000	44	C	1.25	56	60	4	56	30	1		1	56	4	95	95
C											30							
A	11	Recp. Show Window	1300	11	R	1	11	20	12	21	30	1		1	21	12	30	30
B	12	Dishwasher	8604	24	R	1	24	50	8	50	30	1		1	50	8	55	55
C											30							
A	13	Recp. Basement North	900	7.5	R	1	8	20	12	21	30	1		1	21	12	30	30
B	14	Disposer	2678	22	R	1	22	50	8	50	30	1		1	50	8	55	55
C											30							
A	15	Recp. North Wall Bake	540	4.5	R	1	5	20	12	21	30	1		1	21	12	30	30
B	17	Recp. Basement South	720	6	R	1	6	20	12	21	30	1		1	21	12	30	30
C	19	Recp. S & W Walls Bake	900	7.5	R	1	8	20	12	21	30	1		1	21	12	30	30

TABLE A-6
Bakery – Load Summary

Connection Summary:	VA Total
Connected load Phase A	4213
Connected load Phase B	2893
Connected load Phase C	4188
Balanced Loads	34985
Connection Total	46279
Loads Summary:	VA Total
Continuous	22566
Noncontinuous + Receptacle	6380
Highest Motor	8604
Other Motors	8729
Loads Summary Total	46279

TABLE A-7
Bakery – Feeder Selection

Load Summary, 208Y/120, 3 Phase, 4-Wire	Computed	OCPD
Continuous Load	22566	28208
Noncontinuous + Receptacle Load	6380	6380
Highest Motor Load	8604	10755
Other Motor Load	8729	8729
Growth	11570	14463
Computed Load & OCPD Load	57849	68535
OCPD Selection	**Input**	**Output**
OCPD Load Volt-Amperes & Amperes	68535	190
OCPD Rating		200
Minimum Conductor Size		3/0 AWG
Minimum Derated Ampacity		176
Phase Conductor Selection	**Input**	**Output**
Ambient Temperature & Correction Factor	28°C	1
Current Carrying Conductors & Adjustment Factor	3	1
Derating Factor		1
Minimum Allowable Ampacity		176
Conductor Size		3/0 AWG
Conductor Type, Allowable Ampacity	THHN	225
Derated Ampacity		225
Voltage Drop, 0.9 pf, per 100 ft		2.6
Neutral Conductor Selection	**Input**	**Output**
OCPD Load Volt-Amperes	68535	
Balanced Load Volt-Amperes	−34985	
Nonlinear Load Volt-Amperes	4446	
Total Load Volt-Amperes & Amperes	37996	106
Minimum Neutral Size		2 AWG
Minimum Allowable Ampacity		106
Neutral Conductor Type & Size	THHN	2 AWG
Allowable Ampacity & Derated Ampacity	130	130
Raceway Size Determination	**Input**	**Output**
Feeder Conductors Size & Total Area	3/0 AWG	0.8037
Neutral Conductor Size & Area	2 AWG	0.1158
Grounding Conductor Size & Area		0
Total Conductor Area		0.9195
Raceway Type & Size	RMC	2 in

TABLE A-8
Beauty Salon – Loading Schedule

	Count	VA/Unit	NEC®	Actual	Computed	Balanced	Nonlinear
General Lighting:							
NEC® Section 220-2	480	3	1440				
Style A luminaire	5	87		435			
Style M luminaire	9	50		450			
Style E luminaire	5	144		720			720
Totals:			1440	1605	1605		
Other Loads:							
Receptacle outlets	6	180		1080			
Receptacle outlets	3	1500		4500			
Roof receptacle	1	1500		1500			
Washer/dryer	1	4000		4000		4000	
Water heater	1	3800		3800		3800	
Totals:				14880	14880		
Motors & Appliances:	**Amperes**						
Air Conditioning							
Compressor	14.1	360		5076		5076	
Cond-Evap motor	3.3	208		686		686	
Totals:				5762	5762		
TOTAL LOADS:					22247	13562	1155

328   Appendix

TABLE A-9
Beauty Salon – Panelboard Worksheet

Phase	Circuit Number	Load/Area Served	Computed Volt-Amperes	Computed Amperes	Load Type	Load Modifier	OCPD Selection Amperes	OCPD Rating	Minimum Conductor Size AWG	Minimum Derated Ampacity	Ambient Temp. C	Correction Factor	Current Carrying Conductors	Adjustment Factor	Minimum Allowable Ampacity	Conductor Size AWG	Allowable Ampacity	Derated Ampacity
A	1	General use receptacles	1080	9	R	1	9	20	12	21	30	1		1	21	12	30	30
A	2	Receptacle Station A	1500	13	N	1	13	20	12	16	30	1		1	13	12	30	30
B	3	9 Style M Station Spots	450	3.8	C	1.25	5	20	12	16	30	1		1	5	12	30	30
B	4	Receptacle Station B	1500	13	N	1	13	20	12	16	30	1		1	13	12	30	30
C	5	Evaporator/Condenser		3.3		1												
A	5	Cooling System	5762	14	H	2.25	35	35	10	21	38	0.9	5	0.8	35	10	40	29
B																		
C	6	Receptacle Station C	1500	13	N	1	13	20	12	16	30	1		1	13	12	30	30
C	7	4 Style A Station Area	348	2.9	C	1.25	4	20	12	16	30	1		1	4	12	30	30
A	8	5 Style E, 1 A Waiting Area	807	6.7	C	1.25	9	20	12	16	30	1		1	9	12	30	30
A	9	Water Heater	3800	18	M	1.25	23	25	12	23	30	1	5	0.8	23	12	30	24
B																		
B	10	Washer/Dryer	4000	19	C	1.25	25	25	12	25	30	1	5	0.8	25	12	30	24
C																		
C	11	Roof Receptacle	1500	13	N	1	13	20	12	16	38	0.9	5	0.8	18	12	30	21

TABLE A-10
Beauty Salon – Load Summary

Connection Summary:	VA Total
Connected load Phase A	3387
Connected load Phase B	1950
Connected load Phase C	3348
Balanced Loads	13562
Connection Total	22247
Loads Summary:	**VA Total**
Continuous	5605
Noncontinuous + Receptacle	7080
Highest Motor	5762
Other Motors	3800
Loads Summary Total	22247

TABLE A-11
Beauty Salon – Feeder Selection

Load Summary, 208Y/120, 3 Phase, 4-Wire	Computed	OCPD
Continuous Load	5605	7006
Noncontinuous + Receptacle Load	7080	7080
Highest Motor Load	5762	7203
Other Motor Load	3800	3800
Growth	5562	6953
Computed Load & OCPD Load	27809	32042
OCPD Selection	**Input**	**Output**
OCPD Load Volt-Amperes & Amperes	32042	89
OCPD Rating		90
Minimum Conductor Size		2 AWG
Minimum Derated Ampacity		81
Phase Conductor Selection	**Input**	**Output**
Ambient Temperature & Correction Factor	28°C	1
Current Carrying Conductors & Adjustment Factor	3	1
Derating Factor		1
Minimum Allowable Ampacity		81
Conductor Size		2 AWG
Conductor Type, Allowable Ampacity	THHN	130
Derated Ampacity		130
Voltage Drop, 0.9 pf, per 100 ft		2.73
Neutral Conductor Selection	**Input**	**Output**
OCPD Load Volt-Amperes	32042	
Balanced Load Volt-Amperes	−13562	
Nonlinear Load Volt-Amperes	1155	
Total Load Volt-Amperes & Amperes	19635	55
Minimum Neutral Size		6 AWG
Minimum Allowable Ampacity		55
Neutral Conductor Type & Size	THHN	6 AWG
Allowable Ampacity & Derated Ampacity	75	75
Raceway Size Determination	**Input**	**Output**
Feeder Conductors Size & Total Area	2 AWG	0.3474
Neutral Conductor Size & Area	6 AWG	0.0507
Grounding Conductor Size & Area		0
Total Conductor Area		0.3981
Raceway Type & Size	RMC	1¼ in

TABLE A-12
Doctor's Office – Loading Schedule

	Count	VA/Unit	*NEC*®	Actual	Computed	Balanced	Nonlinear
General Lighting:							
NEC® Section 220-2	457	3	1600				
Style F luminaire	3	132		396			396
Style G luminaire	2	132		264			264
Style J luminaire	5	150		750			
Style N luminaire	2	60		120			
Totals:			1600	1530	1600		
Other Loads:							
Receptacle outlets	10	180		1800			
Equipment outlet	1	2000		2000			
Roof receptacle	1	1500		1500			
Water heater	1	3800		3800		3800	
Totals:				9100	9100		
Motors & Appliances:	**Amperes**						
Air Conditioning							
Compressor	16.8	208		3494		3494	
Cond-Evap motor	3.7	208		770		770	
Totals:				4264	4264		
TOTAL LOADS:					14964	8064	660

TABLE A-13
Doctor's Office – Branch Circuit Selection

Phase	Circuit Number	Load/Area Served	Computed Volt-Amperes	Computed Amperes	Load Type	Load Modifier	OCPD Selection Amperes	OCPD Rating	Minimum Conductor Size AWG	Minimum Derated Amperes	Ambient Temp. C	Correction Factor	Current Carrying Conductors	Adjustment Factor	Minimum Allowable Ampacity	Conductor Size AWG	Allowable Ampacity	Derated Ampacity
B	1	5 Style J, 2 N, Waiting &	820	7.3	C	1.25	9	20	12	16	30	1		1	9	12	30	30
B	2	5 Receptacles, Waiting Rm	720	6	R	1	6	20	12	21	30	1		1	21	12	30	30
C	3	2 Style G, 3 F, Examining Rm	660	5.5	C	1.25	7	20	12	16	30	1		1	7	12	30	30
C	4	4 Receptacles, Examining Rm	720	6	R	1	6	20	12	21	30	1		1	21	12	30	30
B	5	Equipment Recp., Exam Rm	2000	17	N	1	17	20	12	16	30	1		1	17	12	30	30
B	6	2 Receptacles, Laboratory	360	3	R	1	3	20	12	21	30	1		1	21	12	30	30
C / B	7	Cooling System	4264	21	H	2.25	46	45	10	26	38	0.9	4	0.8	29	10	40	29
C	8	Roof Receptacle	1500	13	N	1	13	20	12	16	38	0.9	4	0.8	18	12	30	21
C / B	9	Water Heater	3800	32	C	1.25	40	40	8	40	30	1		1	40	8	55	55

TABLE A-14
Doctor's Office – Load Summary

Connection Summary:	VA Total
Connected load Phase A	XXX
Connected load Phase B	3950
Connected load Phase C	2880
Balanced Loads	8064
Connection Total	14894
Loads Summary:	**VA Total**
Continuous	5330
Noncontinuous + Receptacle	5300
Highest Motor	4264
Other Motors	0
Loads Summary Total	14894

TABLE A-15
Doctor's Office – Feeder Selection

Load Summary, 208Y/120, 1 Phase, 3-Wire	Computed	OCPD
Continuous Load	5330	6663
Noncontinuous + Receptacle Load	5330	5330
Highest Motor Load	4264	5330
Other Motor Load	0	0
Growth	3724	4655
Computed Load & OCPD Load	18618	21948
OCPD Selection	**Input**	**Output**
OCPD Load Volt-Amperes & Amperes	21948	106
OCPD Rating		110
Minimum Conductor Size		2 AWG
Minimum Derated Ampacity		101
Phase Conductor Selection	**Input**	**Output**
Ambient Temperature & Correction Factor	28°C	1
Current Carrying Conductors & Adjustment Factor	3	1
Derating Factor		1
Minimum Allowable Ampacity		101
Conductor Size		2 AWG
Conductor Type, Allowable Ampacity	THHN	130
Derated Ampacity		130
Voltage Drop, 0.9 pf, per 100 ft		3.66
Neutral Conductor Selection	**Input**	**Output**
OCPD Load Volt-Amperes	21948	
Balanced Load Volt-Amperes	0	
Nonlinear Load Volt-Amperes	660	
Neutral Load Volt-Amperes & Amperes	22608	109
Minimum Neutral Size		2 AWG
Minimum Allowable Ampacity		109
Neutral Conductor Type & Size	THHN	2 AWG
Allowable Ampacity & Derated Ampacity	130	130
Raceway Size Determination	**Input**	**Output**
Feeder Conductors Size & Total Area	2 AWG	0.2316
Neutral Conductor Size & Area	2 AWG	0.1158
Grounding Conductor Size & Area		0
Total Conductor Area		0.3474
Raceway Type & Size	RMC	1 in

TABLE A-16
Drugstore – Loading Schedule

	Count	VA/Unit	NEC®	Actual	Computed	Balanced	Nonlinear
General Lighting:							
NEC® Section 220-2	1395	3	4185				
Style I luminaire	27	87		2349			2349
Style E luminaire	4	144		576			576
Style D luminaire	15	74		1110			1110
Style N luminaire	2	60		120			
Totals:			4185	4155	4185		
Storage:							
NEC® Section 220-2	1012	0.25	253				
Style L luminaire	9	60		540			540
Totals:			253	540	540		
Show Window:							
NEC® Section 220-12	16	200	3200				
Receptacle outlets	3	500		1500			
Lighting Track	8	150		1200			
Totals:			3200	2700	3200		
Other Loads:							
Receptacle outlets	21	180		3780			
Roof receptacle	1	1500		1500			
Sign Outlet	1	1200		1200			
Totals:				6480	6480		
Motors & Appliances:	**Amperes**						
Air Conditioning:							
Compressor	20.2	360		7272		7272	
Condenser	3.2	208		666		666	
Evaporator	3.2	208		666		666	
Totals:				8604	8604		
TOTAL LOADS:					23009	8604	4575

TABLE A-17
Drugstore – Panelboard Worksheet

Phase	Circuit Number	Load/Area Served	Computed Volt-Amperes	Computed Amperes	Load Type	Load Modifier	OCPD Selection Amperes	OCPD Rating	Minimum Conductor Size AWG	Minimum Derated Amperes	Ambient Temp. C	Correction Factor	Current Carrying Conductors	Adjustment Factor	Minimum Allowable Ampacity	Conductor Size AWG	Allowable Ampacity	Derated Ampacity
A	1	9 Style I,	783	6.5	C	1.25	9	20	12	16	30	1		1	16	12	30	30
A	2	4 Rec., Merchandise	720	6	R	1	6	20	12	21	30	1		1	21	12	30	30
B	3	18 Style I,	1566	13	C	1.25	17	20	12	17	30	1		1	17	12	30	30
B	4	3 Rec., Toilet area	540	4.5	R	1	5	20	12	21	30	1		1	21	12	30	30
C	5	3 Style E, 2 N,	552	4.6	C	1.25	6	20	12	16	30	1		1	16	12	30	30
C	6	5 Rec., Merchandise	900	7.5	R	1	8	20	12	21	30	1		1	21	12	30	30
A	7	2 Rec., Pharmacy	360	3	R	1	3	20	12	21	30	1		1	21	12	30	30
A	8	3 Rec., Show Window	1550	13	C	1.25	16	20	12	16	30	1		1	16	12	30	30
B	9	9 Style L, 1 E	684	5.7	C	1.25	8	20	12	16	30	1		1	16	12	30	30
B	10	15 Style D, Pharmacy	1110	9.3	C	1.25	12	20	12	16	30	1		1	16	12	30	30
C	11	Track Show Window	600	5	C	1.25	7	20	12	16	30	1		1	16	12	30	30
C	12	4 Rec., S. Basement	720	6	R	1	6	20	12	21	30	1		1	21	12	30	30
A	13	Track Show Window	600	5	C	1.25	7	20	12	16	30	1		1	16	12	30	30
A	14	3 Rec., N. Basement	540	4.5	R	1	5	20	12	21	30	1		1	21	12	30	30
B	15	Sign	1200	10	C	1.25	13	20	12	16	30	1		1	16	12	30	30
B	16	Evaporator		3.2		1												
C		Compressor	8604	20	H	2.25	52	50	10	32	38	0.9	5	0.8	44	8	55	40
A		Condenser		3.2		1												
C	17	1 Receptacle, Roof	1500	13	N	1	13	20	12	16	38	0.9	5	0.8	22	12	30	22

TABLE A-18
Drugstore – Load Summary

Connection Summary:	VA Total
Connected load Phase A	4503
Connected load Phase B	5100
Connected load Phase C	4272
Balanced Loads	8604
Connection Total	22479
Loads Summary:	**VA Total**
Continuous	8595
Noncontinuous + Receptacle	5280
Highest Motor	8604
Other Motors	0
Loads Summary Total	22479

TABLE A-19
Drugstore – Feeder Selection

Load Summary, 208Y/120, 3-	Computed	OCPD
Continuous Load	8595	10744
Noncontinuous + Receptacle Load	5280	5280
Highest Motor Load	8604	10755
Other Motor Load	0	0
Growth	5620	7025
Computed Load & OCPD Load	28099	33804
OCPD Selection	**Input**	**Output**
OCPD Load Volt-Amperes & Amperes	33804	94
OCPD Rating		100
Minimum Conductor Size		1 AWG
Minimum Derated Ampacity		94
Phase Conductor Selection	**Input**	**Output**
Ambient Temperature & Correction Factor	28°C	1
Current Carrying Conductors & Adjustment Factor	4	0.8
Derating Factor		0.8
Minimum Allowable Ampacity		118
Conductor Size		1 AWG
Conductor Type, Allowable Ampacity	THHN	150
Derated Ampacity		120
Voltage Drop, 0.9 pf, per 100 ft		2.28
Neutral Conductor Selection	**Input**	**Output**
OCPD Load Volt-Amperes	33804	
Balanced Load Volt-Amperes	−8604	
Nonlinear Load Volt-Amperes	4575	
Total Load Volt-Amperes & Amperes	29775	83
Minimum Neutral Size		4 AWG
Minimum Allowable Ampacity		104
Neutral Conductor Type & Size	THHN	3 AWG
Allowable Ampacity & Derated Ampacity	110	88
Raceway Size Determination	**Input**	**Output**
Feeder Conductors Size & Total Area	1 AWG	0.4686
Neutral Conductor Size & Area	3 AWG	0.0973
Grounding Conductor Size & Area		0
Total Conductor Area		0.5659
Raceway Type & Size	RMC	1¼ in

TABLE A-20
Insurance Office – Loading Schedule

	Count	VA/Unit	NEC®	Actual	Computed	Balanced	Nonlinear
General Lighting:							
NEC® Section 220-2	1374	3	4122				
Style E luminaire	10	144		1440			1440
Style O luminaire	16	143		2288			2288
Style P luminaire	2	87		174			174
Style Q luminaire	4	143		572			572
Totals:			4122	4474	4474		
Other Loads:							
Copier outlets	1	1500		1500			1500
Computer outlets	9	500		4500			4500
Roof receptacle	1	1500		1500			
Totals:				7500	7500		
Receptacle Outlets							
NEC® Section 220-3	15	180		2700			
NEC® Section 220-3	56	180		10080			
NEC® Table 220-13			11390	12780	11390		11390
Motors & Appliances:	**Amperes**						
Air Conditioning:							
Compressor	20.2	360		7272		7272	
Evaporator	3.2	208		666		666	
Condenser	3.2	208		666		666	
Totals:				8604	8604		
TOTAL LOADS:					31968	8604	21864

TABLE A-21
Insurance Office – Panelboard Worksheet

Phase	Circuit Number	Load/Area Served	Computed Volt-Amperes	Computed Amperes	Load Type	Load Modifier	OCPD Selection Amperes	OCPD Rating	Minimum Conductor Size AWG	Minimum Derated Amperes	Ambient Temp. C	Correction Factor	Current Carrying Conductors	Adjustment Factor	Minimum Allowable Ampacity	Conductor Size AWG	Allowable Ampacity	Derated Ampacity
A	1	14 2 Lamp, Style O, Staff	1218	10	C	1.25	13	20	12	16	30	1		1	16	12	30	30
A	2	10 Rec. South Wall Staff	1800	15	R	1	15	20	12	21	30	1		1	21	12	30	30
B	3	14 1 Lamp, Style O; 2 P, Staff	958	8	C	1.25	10	20	12	16	30	1		1	16	12	30	30
B	4	10 Rec. South Wall Staff	1800	15	R	1	15	20	12	21	30	1		1	21	12	30	30
C	5	6 Style E, 2 O, Reception	1150	9.6	C	1.25	12	20	12	16	30	1		1	16	12	30	30
C	6	10 Rec. South Wall Staff	1800	15	R	1	15	20	12	21	30	1		1	21	12	30	30
A	7	4 Style Q, 4 E, Computer	1148	9.6	C	1.25	12	20	12	16	30	1		1	16	12	30	30
A	8	3 Rec. Computer Room	1500	13	C	1.25	16	20	12	16	30	1	6	0.8	20	12	30	24
B	9	13 Rec. East & South Walls	2340	20	R	1	20	20	12	21	30	1		1	21	12	30	30
B	10	3 Rec. Computer Room	1500	13	C	1.25	16	20	12	16	30	1	6	0.8	20	12	30	24
C	11	13 Rec. East & North Walls	2340	20	R	1	20	20	12	21	30	1		1	21	12	30	30
C	12	3 Rec. Computer Room	1500	13	C	1.25	16	20	12	16	30	1	6	0.8	20	12	30	24
A	13	Evaporator		3.2		1												
B		Compressor	8604	20	H	2.25	52	50	10	32	38	0.9	5	0.8	44	8	55	40
C		Condenser		3.2		1												
A	14	4 Floor Rec. Staff Office	720	6	R	1	6	20	12	21	30	1		1	21	12	30	30
A	15	Copy Machine	1500	13	N	1	13	20	12	16	30	1		1	16	12	30	30
B	16	5 Rec. West Wall Staff Office	900	7.5	R	1	8	20	12	21	30	1		1	21	12	30	30
B	17	Roof Receptacle	1500	13	N	1	13	20	12	16	38	0.9	5	0.8	22	12	30	21
C	18	6 Rec. Reception Room	1080	9	R	1	9	20	12	21	30	1		1	21	12	30	30

TABLE A-22
Insurance Office – Load Summary

Connection Summary:	VA Total
Phase A Loads	7886
Phase B Loads	8998
Phase C Loads	7870
Balanced Loads	8604
Connection Total	33358
Loads Summary:	**VA Total**
Continuous	8974
Noncontinuous + Receptacle	15780
Highest Motor	8604
Other Motors	0
Loads Summary Total	33358

TABLE A-23
Insurance Office – Feeder Selection

Load Summary, 208Y/120, 3 Phase, 4-Wire	Computed	OCPD
Continuous Load	8974	11218
Noncontinuous + Receptacle Load	15780	15780
Highest Motor Load	8604	10755
Other Motor Load	0	0
Growth	8340	10425
Computed Load & OCPD Load	41698	48178
OCPD Selection	**Input**	**Output**
OCPD Load Volt-Amperes & Amperes	48178	134
OCPD Rating		150
Minimum Conductor Size		1/0 AWG
Minimum Derated Ampacity		126
Phase Conductor Selection	**Input**	**Output**
Ambient Temperature & Correction Factor	28°C	1
Current Carrying Conductors & Adjustment Factor	4	0.8
Derating Factor		0.8
Minimum Allowable Ampacity		158
Conductor Size		1/0 AWG
Conductor Type & Allowable Ampacity	THHN	170
Derated Ampacity		136
Voltage Drop, 0.9 pf, per 100 ft		3.27
Neutral Conductor Selection	**Input**	**Output**
OCPD Load Volt-Amperes	41698	
Balanced Load Volt-Amperes	−8604	
Nonlinear Load Volt-Amperes	21864	
Total Load Volt-Amperes & Amperes	54958	153
Minimum Neutral Size		2/0 AWG
Minimum Allowable Ampacity		191
Neutral Conductor Type & Size	THHN	2/0 AWG
Allowable Ampacity & Derated Ampacity	195	156
Raceway Size Determination	**Input**	**Output**
Feeder Conductors Size & Total Area	1/0 AWG	0.5565
Neutral Conductor Size & Area	2/0 AWG	0.2223
Grounding Conductor Size & Area	8 AWG	0.0366
Total Conductor Area		0.8154
Raceway Type & Size	RMC	1½ in

TABLE A-24
Owner's – Loading Schedule

	Count	VA/Unit	NEC®	Actual	Computed	Balanced	Nonlinear
General Lighting:							
Style E luminaire	4	144		576			576
Style H luminaire	4	87		348			348
Style J luminaire	3	150		450			
Style K luminaire	3	192		576			576
Style L luminaire	6	87		522			522
Style N luminaire	4	60		240			
Totals:				2712	2712		
Other Loads:							
Receptacle outlets	8	180		1440			
Telephone outlets	2	1250		2500			
Totals:				3940	3940		
Motors & Appliances:	**Amperes**						
Sump pump	9.4	120		1128			
Circulating pump	4.4	120		528			
Circulating pump	4.4	120		528			
Circulating pump	4.4	120		528			
Circulating pump	4.4	120		528			
Circulating pump	4.4	120		528			
Totals:				3768	3768		
TOTAL LOADS:					10420	0	2022

TABLE A-25
Owner's Panelboard Worksheet

Phase	Circuit Number	Load/Area Served	Computed Volt-Amperes	Computed Amperes	Load Type	Load Modifier	OCPD Selection Amperes	OCPD Rating	Minimum Conductor Size AWG	Minimum Derated Amperes	Ambient Temp. C	Correction Factor	Current Carrying Conductors	Adjustment Factor	Minimum Allowable Ampacity	Conductor Size AWG	Allowable Ampacity	Derated Ampacity
A	1	3 Style E. 4 N, 2nd Floor	672	5.6	C	1.25	7	20	12	16	38	0.9		1	18	12	30	27
A	2	3 Style K, Exterior Lighting	576	4.8	C	1.25	6	20	12	16	38	0.9		1	18	12	30	27
B	3	2 Style H, 2nd Floor Corridor	174	1.5	C	1.25	2	20	12	16	38	0.9		1	18	12	30	27
B	4	2 Exterior Receptacles	360	3	R	1	3	20	12	21	38	0.9		1	23	12	30	27
C	5	2 Recep. 2nd Floor Utility	360	3	N	1	3	20	12	16	38	0.9		1	18	12	30	27
C	6																	
A	7	3 Style J, 1 E, East & West	594	5	N	1	5	20	12	16	38	0.9		1	18	12	30	27
A	8	6 Style L, Utility Area	522	4.4	N	1	5	20	12	16	38	0.9		1	18	12	30	27
B	9	1 Recp. Telephone Panel	1250	10	C	1.25	14	20	12	16	38	0.9		1	18	12	30	27
B	10	4 Recp. Utility Area	720	6	R	1	6	20	12	21	38	0.9		1	23	12	30	27
C	11	1 Recp. Telephone Panel	1250	10	C	1.25	14	20	12	16	38	0.9		1	18	12	30	27
C	12	2 Style H, 2nd Floor Corridor	174	1.5	C	1.25	2	20	12	16	38	0.9		1	18	12	30	27
A	13	Drugstore Hot Water Pump	528	4.4	M	1.25	6	20	12	16	38	0.9		1	18	12	30	27
A	14	Bakery Hot Water Pump	528	4.4	M	1.25	6	20	12	16	38	0.9		1	18	12	30	27
B	15	Insurance Office Hot Water	528	4.4	M	1.25	6	20	12	16	38	0.9		1	18	12	30	27
B	16	Beauty Salon Hot Water Pump	528	4.4	M	1.25	6	20	12	16	38	0.9		1	18	12	30	27
C	17	Doctor's Office Hot Water	528	4.4	M	1.25	6	20	12	16	38	0.9		1	18	12	30	27
C	18	Sump Pump	1128	9.4	M	1.25	12	20	12	16	38	0.9		1	18	12	30	27
A	19																	
A	20																	

TABLE A-26
Owner's – Feeder Selection

Load Summary, 208Y/120, 3 Phase, 4-	Computed	OCPD
Continuous Load	4096	5120
Noncontinuous + Receptacle Load	2556	2556
Highest Motor Load	1128	1410
Other Motor Load	2640	2640
Growth	2605	3256
Computed Load & OCPD Load	13025	14982
OCPD Selection	**Input**	**Output**
Selection Amperes	14982	42
OCPD Rating		45
Minimum Conductor Size		6 AWG
Minimum Derated Ampacity		41
Phase Conductor Selection	**Input**	**Output**
Ambient Temperature & Correction Factor	38°C	0.91
Current Carrying Conductors & Adjustment Factor	3	1
Derating Factor		0.91
Minimum Allowable Ampacity		45
Conductor Size		6 AWG
Conductor Type & Allowable Ampacity	THHN	75
Derated Ampacity		68
Voltage Drop, 0.9 pf, 3 ph, per 100 ft		2.94
Neutral Conductor Selection	**Input**	**Output**
Total Load	14982	
Balanced Load	0	
Nonlinear Load	2022	
Total Load Volt-Amperes & Amperes	17004	47
Minimum Neutral Size		6 AWG
Minimum Allowable Ampacity		68
Neutral Conductor Type & Size	THHN	6 AWG
Allowable Ampacity & Derated Ampacity	75	68
Raceway Size Determination	**Input**	**Output**
Feeder Conductors Size & Area	6 AWG	0.1521
Neutral Conductor Size & Area	6 AWG	0.0507
Grounding Conductor Size & Area		0
Total Conductor Area		0.2028
Raceway Type & Size	RMC	¾ in

TABLE A-27
Owner's – Panelboard Summary

Connection Summary:	VA Total
Connected load Phase A	3420
Connected load Phase B	3560
Connected load Phase C	3440
Balanced Loads	0
Connection Total	10420
Loads Summary:	**VA Total**
Continuous	4096
Noncontinuous + Receptacle	2556
Highest Motor	1128
Other Motors	2640
Loads Summary Total	10420

TABLE A-28
Useful Formulas

TO FIND	SINGLE PHASE	THREE PHASE	DIRECT CURRENT
AMPERES when KVA is known	$\dfrac{kVA \times 1000}{E}$	$\dfrac{kVA \times 1000}{E \times 1.73}$	not applicable
AMPERES when horsepower is known	$\dfrac{hp \times 746}{E \times \% \text{ eff.} \times pf}$	$\dfrac{hp \times 746}{E \times 1.73 \times \% \text{ eff.} \times pf}$	$\dfrac{hp \times 746}{E \times \% \text{ eff.}}$
AMPERES when kilowatts are known	$\dfrac{kw \times 1000}{E \times pf}$	$\dfrac{kW \times 1000}{E \times 1.73 \times pf}$	$\dfrac{kW \times 1000}{E}$
KILOWATTS	$\dfrac{I \times E \times pf}{1000}$	$\dfrac{I \times E \times 1.73 \times pf}{1000}$	$\dfrac{I \times E}{1000}$
KILOVOLT-AMPERES	$\dfrac{I \times E}{1000}$	$\dfrac{I \times E \times 1.73}{1000}$	not applicable
HORSEPOWER	$\dfrac{I \times E \times \% \text{ eff.} \times pf}{746}$	$\dfrac{I \times E \times 1.73 \times \% \text{ eff.} \times pf}{746}$	$\dfrac{I \times E \times \% \text{ eff.}}{746}$
WATTS	$E \times I \times pf$	$E \times I \times 1.73 \times pf$	$E \times I$

I = amperes	E = volts	kW = kilowatts	kVA = kilovolt-amperes
hp = horsepower		% eff. = percent efficiency	pf = power factor

METRIC SYSTEM OF MEASUREMENT

The following table provides useful conversions of English customary terms to SI terms, and SI terms to English customary terms. The metric system is known as the International System of Units (SI), taken from the French *Le Système International d'Unites.* Whenever the slant line / is found, say it as "per." Be practical when using the values in the following table. Present day use of calculators and computers provides many "places" beyond the decimal point. You must decide how accurate your results must be when performing a calculation. When converting centimeters to feet, for example, one could be reasonably accurate by rounding the value of 0.03281 to 0.033. When a manufacturer describes a product's measurement using metrics, but does not physically change the product, it is referred to as a

"soft conversion." When a manufacturer actually changes the physical measurements of the product to standard metric size or a rational whole number of metric units, it is referred to as a "hard conversion." When rounding off numbers, be sure to round off in such a manner that the final result does not violate a "maximum" or "minimum" value, such as might be required by the *National Electrical Code®.* The rule for discarding digits is that if the digits to be discarded begin with a 5 or higher, increase the last digit retained by one; for example: 6.3745, if rounded to three digits, would become 6.37. If 6.3745 were rounded to four digits, it would become 6.375. The following information has been developed from the latest U.S. Department of Commerce and the National Institute of Standards and Technology publications covering the metric system.

Multiply This Unit(s)	By This Factor	To Obtain This Unit(s)
acre	4 046.9	square meters (m²)
acre	43 560	square feet (ft²)
ampere hour	3 600	coulombs (C)
angstrom	0.1	nanometers (nm)
atmosphere	101.325	kilopascals (kPa)
atmosphere	33.9	feet of water (at 4°C)
atmosphere	29.92	inches of mercury (at 0°C)
atmosphere	0.76	meters of mercury (at 0°C)
atmosphere	0.007 348	tons per square inch
atmosphere	1.058	tons per square foot
atmosphere	1.0333	kilograms per square centimeter
atmosphere	10 333	kilograms per square meter
atmosphere	14.7	pounds per square inch
bar	100	kilopascals (kPa)
barrel (oil, 42 U.S. gallons)	0.158 987 3	cubic meters (m³)
barrel (oil, 42 U.S. gallons)	158.987 3	liters (L)
board foot	0.002 359 737	cubic meters (m³)
bushel	0.035 239 07	cubic meters (m³)
Btu	778.16	feet-pounds
Btu	252	grams-calories
Btu	0.000 393 1	horsepower-hours
Btu	1 054.8	joules (J)
Btu	1.055 056	kilojoules (kJ)
Btu	0.000 293 1	kilowatt-hours (kWH)
Btu per hour	0.000 393 1	horsepower (hp)
Btu per hour	0.293 071 1	watts (W)
Btu per degree Fahrenheit	1.899 108	kilojoules per kelvin (kJ/K)
Btu per pound	2.326	kilojoules per kilogram (kJ/kg)
Btu per second	1.055 056	kilowatts (kW)
calorie	4.184	joules (J)
calorie, gram	0.003 968 3	Btus
candela per foot squared (cd/ft²)	10.763 9	candelas per meter squared (cd/m²)

Note: The former term *candlepower* has been replaced with the term *candela*.

candela per meter squared (cd/m²)	0.092 903	candelas per foot squared (cd/ft²)
candela per meter squared (cd/m²)	0.291 864	footlambert*
candela per square inch (cd/in²)	1550.003	candelas per square meter (cd/m²)

Celsius = (Fahrenheit − 32) × ⁵⁄₉

Celsius = (Fahrenheit − 32) × 0.555555

Celsius = (0.556 × Fahrenheit) − 17.8

Note: The term *centigrade* was officially discontinued in 1948, and was replaced by the term *Celsius*. The term *centigrade* may still be found in some publications.

centimeter (cm)	0.032 81	feet (ft)
centimeter (cm)	0.393 7	inches (in)
centimeter (cm)	0.01	meters (m)
centimeter (cm)	10	millimeters (mm)
centimeter (cm)	393.7	mils
centimeter (cm)	0.010 94	yards
circular mil	0.000 005 067	square centimeters (cm²)
circular mil	0.785 4	square mils (mil²)
circular mil	0.000 000 785 4	square inches (in²)
cubic centimeter (cm³)	0.061 02	cubic inches (in³)

Multiply This Unit(s)	By This Factor	To Obtain This Unit(s)
cubic foot per second	0.028 316 85	cubic meters per second (m^3/s)
cubic foot per minute	0.000 471 947	cubic meters per second (m^3/s)
cubic foot per minute	0.471 947	liters per second (L/s)
cubic inch (in^3)	16.39	cubic centimeters (cm^3)
cubic inch (in^3)	0.000 578 7	cubic feet (ft^3)
cubic meter (m^3)	35.31	cubic feet (ft^3)
cubic yard per minute	12.742 58	liters per second (L/s)
cup (c)	0.236 56	liters (L)
decimeter	0.1	meters (m)
dekameter	10	meters (m)
Fahrenheit = (⅘ Celsius) + 32		
Fahrenheit = (Celsius × 1.8) + 32		
fathom	1.828 804	meters (m)
fathom	6.0	feet
foot	30.48	centimeters (cm)
foot	12	inches
foot	0.000 304 8	kilometers (km)
foot	0.304 8	meters (m)
foot	0.000 189 4	miles (statute)
foot	304.8	millimeters (mm)
foot	12 000	mils
foot	0.333 33	yard
foot, cubic (ft^3)	0.028 316 85	cubic meters (m^3)
foot, cubic (ft^3)	28.316 85	liters (L)
foot, board	0.002 359 737	cubic meters (m^3)
cubic feet per second (ft^3/s)	0.028 316 85	cubic meters per second (m^3/s)
cubic feet per minute (ft^3/min)	0.000 471 947	cubic meters per second (m^3/s)
cubic feet per minute (ft^3/min)	0.471 947	liters per second (L/s)
foot, square (ft^2)	0.092 903	square meters (m^2)
footcandle	10.763 91	lux (lx)
footlambert*	3.426 259	candelas per square meter (cd/m^2)
foot of water	2.988 98	kilopascals (kPa)
foot pound	0.001 286	Btus
foot pound-force	1 055.06	joules (J)
foot pound-force per second	1.355 818	joules (J)
foot pound-force per second	1.355 818	watts (W)
foot per second	0.304 8	meters per second (m/s)
foot per second squared	0.304 8	meters per second squared (m/s^2)
gallon (U.S. liquid)	3.785 412	liters (L)
gallons per day	3.785 412	liters per day (L/d)
gallons per hour	1.051 50	milliliters per second (mL/s)
gallons per minute	0.063 090 2	liters per second (L/s)
gauss	6.452	lines per square inch
gauss	0.1	millitesla (mT)
gauss	0.000 000 064 52	webers per square inch
grain	64.798 91	milligrams (mg)
gram (g) (a little more than the weight of a paper clip)	0.035 274	ounce (avoirdupois)
gram (g)	0.002 204 6	pound (avoirdupois)
gram per meter (g/m)	3.547 99	pounds per mile (lb/mile)
grams per square meter (g/m^2)	0.003 277 06	ounces per square foot (oz/ft^2)

Multiply This Unit(s)	By This Factor	To Obtain This Unit(s)
grams per square meter (g/m²)	0.029 494	ounces per square yard (oz/yd²)
gravity (standard acceleration)	9.806 65	meters per second squared (m/s²)
quart (U.S. liquid)	0.946 352 9	liters (L)
horsepower (550 ft·lbf/s)	0.745 7	kilowatts (kW)
horsepower	745.7	watts (W)
horsepower hours	2.684 520	megajoules (mJ)
inch per second squared (in/s²)	0.025 4	meters per second squared (m/s²)
inch	2.54	centimeters (cm)
inch	0.254	decimeters (dm)
inch	0.025 4	meters (m)
inch	25.4	millimeters (mm)
inch	1 000	mils
inch	0.027 78	yards
inch, cubic (in³)	16 387.1	cubic millimeters (mm³)
inch, cubic (in³)	16.387 06	cubic centimeters (cm³)
inch, cubic (in³)	645.16	square millimeters (mm²)
inches of mercury	3.386 38	kilopascals (kPa)
inches of mercury	0.033 42	atmospheres
inches of mercury	1.133	feet of water
inches of water	0.248 84	kilopascals (kPa)
inches of water	0.073 55	inches of mercury
joule (J)	0.737 562	foot pound-force (ft·lbf)
kilocandela per meter squared (kcd/m²)	0.314 159	lambert*
kilogram (kg)	2.204 62	pounds (avoirdupois)
kilogram (kg)	35.274	ounces (avoirdupois)
kilogram per meter (kg/m)	0.671 969	pounds per foot (lb/ft)
kilogram per square meter (kg/m²)	0.204 816	pounds per square foot (lb/ft²)
kilogram meter squared (kg·m²)	23.730 4	pounds foot squared (lb·ft²)
kilogram meter squared (kg·m²)	3 417.17	pounds inch squared (lb·in²)
kilogram per cubic meter (kg/m³)	0.062 428	pounds per cubic foot (lb/ft³)
kilogram per cubic meter (kg/m³)	1.685 56	pounds per cubic yard (lb/yd³)
kilogram per second (kg/s)	2.204 62	pounds per second (lb/s)
kilojoule (kJ)	0.947 817	Btus
kilometer (km)	1 000	meters (m)
kilometer (km)	0.621 371	miles (statute)
kilometer (km)	1 000 000	millimeters (mm)
kilometer (km)	1 093.6	yards
kilometer per hour (kg/hr)	0.621 371	miles per hour (mph)
kilometer squared (km²)	0.386 101	square miles (mile²)
kilopound-force per square inch	6.894 757	megapascals (MPa)
kilowatts (kW)	56.921	Btus per minute
kilowatts (kW)	1.341 02	horsepower (hp)
kilowatts (kW)	1 000	watts (W)
kilowatt-hour (kWh)	3 413	Btus
kilowatt-hour (kWh)	3.6	megajoules (MJ)
knots	1.852	kilometers per hour (km/h)
lamberts*	3 183.099	candelas per square meter (cd/m²)
lamberts*	3.183 01	kilocandelas per square meter (kcd/m²)
liter (L)	0.035 314 7	cubic feet (ft³)
liter (L)	0.264 172	gallons (U.S. liquid)
liter (L)	2.113	pints (U.S. liquid)

Multiply This Unit(s)	By This Factor	To Obtain This Unit(s)
liter (L)	1.056 69	quarts (U.S. liquid)
liter per second (L/s)	2.118 88	cubic feet per minute (ft³/min)
liter per second (L/s)	15.850 3	gallons per minute (gal/min)
liter per second (L/s)	951.022	gallons per hour (gal/hr)
lumen per square foot (lm/ft²)	10.763 9	lux (lx); plural: luces
lumen per square foot (lm/ft²)	1.0	footcandles
lumen per square foot (lm/ft²)	10.763	lumens per square meter (lm/m²)
lumen per square meter (lm/m²)	1.0	lux (lx)
lux (lx)	0.092 903	lumens per square foot (footcandle)
maxwell	10	nanowebers (nWb)
megajoule (MJ)	0.277 778	kilowatt hours (kWh)
meter (m)	100	centimeters (cm)
meter (m)	0.546 81	fathoms
meter (m)	3.280 9	feet
meter (m)	39.37	inches
meter (m)	0.001	kilometers (km)
meter (m)	0.000 621 4	miles (statute)
meter (m)	1 000	millimeters (mm)
meter (m)	1.093 61	yards
meter, cubic (m³)	1.307 95	cubic yards (yd³)
meter, cubic (m³)	35.314 7	cubic feet (ft³)
meter, cubic (m³)	423.776	board feet
meter per second (m/s)	3.280 84	feet per second (ft/s)
meter per second (m/s)	2.236 94	miles per hour (mph)
meter, squared (m²)	1.195 99	square yards (yd²)
meter, squared (m²)	10.763 9	square feet (ft²)
mho per centimeter (mho/cm)	100	siemens per meter (S/m)

Note: The older term "mho" has been replaced with "siemens." The term "mho" may still be found in some publications.

micro inch	0.025 4	micrometers (μm)
mil	25.4	micrometers (μm)
mil	0.025 4	millimeters (mm)
mil	$2.540 \times 10{-3}$	centimeters (cm)
mil	$8.333 \times 10{-5}$	feet
mil	0.001	inches
mil	$2.540 \times 10{-8}$	kilometers
mil	$2.778 \times 10{-5}$	yards
miles per hour	1.609 344	kilometers per hour (km/h)
miles per hour	0.447 04	meters per second (m/s)
miles per gallon	0.425 143 7	kilometers per liter (km/L)
miles	1.609 344	kilometers (km)
miles	5 280	feet
miles	1 609	meters (m)
miles	1 760	yards
miles (nautical)	1.852	kilometers (km)
miles squared	2.590 000	kilometers squared (km²)
millibar	0.1	kilopascals (kPa)
milliliter (mL)	0.061 023 7	cubic inches (in³)
milliliter (mL)	0.033 814	fluid ounces (U.S.)
millimeter (about the thickness of a dime)	0.1	centimeters (cm)

Multiply This Unit(s)	By This Factor	To Obtain This Unit(s)
millimeter (mm)	0.003 280 8	feet
millimeter (mm)	0.039 370 1	inches
millimeter (mm)	0.001	meters (m)
millimeter (mm)	39.37	mils
millimeter (mm)	0.001 094	yards
millimeter squared (mm²)	0.001 550	square inches (in²)
millimeter cubed (mm³)	0.000 061 023 7	cubic inches (in³)
millimeter of mercury	0.133 322 4	kilopascals (kPa)
ohm	0.000 001	megohms
ohm	1 000 000	micro ohms
ohm circular mil per foot	1.662 426	nano ohms meter (nΩ·m)
oersted	79.577 47	amperes per meter (A/m)
ounce (avoirdupois)	28.349 52	grams (g)
ounce (avoirdupois)	0.062 5	pounds (avoirdupois)
ounce, fluid	29.573 53	milliliters (mL)
ounce (troy)	31.103 48	grams (g)
ounce per foot squared (oz/ft²)	305.152	grams per meter squared (g/m²)
ounces per gallon (U.S. liquid)	7.489 152	grams per liter (g/L)
ounce per yard squared (oz/yd²)	33.905 7	grams per meter squared (g/m²)
pica	4.217 5	millimeters (mm)
pint (U.S. liquid)	0.473 176 5	liters (L)
pint (U.S. liquid)	473.177	milliliters (mL)
pound (avoirdupois)	453.592	grams (g)
pound (avoirdupois)	0.453 592	kilograms (kg)
pound (avoirdupois)	16	ounces (avoirdupois)
poundal	0.138 255	newtons (N)
pound foot (lb·ft)	0.138 255	kilograms meter (kg·m)
pound foot per second	0.138 255	kilograms meter per second (kg·m/s)
pound foot squared (lb·ft²)	0.042 140 1	kilograms meter squared (kg·m²)
pound-force	4.448 222	newtons (N)
pound-force foot	1.355 818	newton meters (N·m)
pound-force inch	0.112 984 8	newton meters (N·m)
pound-force per square inch	6.894 757	kilopascals (kPa)
pound-force per square foot	0.047 880 26	kilopascals (kPa)
pound per cubic foot (lb/ft³)	16.018 46	kilograms per cubic meter (kg/m³)
pound per foot (lb/ft)	1.488 16	kilograms per meter (kg/m)
pound per foot squared (lb/ft²)	4.882 43	kilograms per meter squared (kg/m²)
pound per foot cubed (lb/ft³)	16.018 5	kilograms per meter cubed (kg/m³)
pound per gallon (U.S. liquid)	119.826 4	grams per liter (g/L)
pound per second (lb/s)	0.453 592	kilograms per second (kg/s)
pound inch squared (lb·in²)	292.640	kilograms millimeter squared (kg·mm²)
pound per mile	0.281 849	grams per meter (g/m)
pound square foot (lb·ft²)	0.042 140 11	kilograms square meter (kg/m²)
pound per cubic yard (lb/yd³)	0.593 276	kilograms per cubic meter (kg/m³)
quart (U.S. liquid)	946.353	milliliters (mL)
square centimeter (cm²)	197 300	circular mils
square centimeter (cm²)	0.001 076	square feet (ft²)
square centimeter (cm²)	0.155	square inches (in²)

Multiply This Unit(s)	By This Factor	To Obtain This Unit(s)
square centimeter (cm²)	0.000 1	square meters (m²)
square centimeter (cm²)	0.000 119 6	square yards (yd²)
square foot (ft²)	144	square inches (in²)
square foot (ft²)	0.092 903 04	square meters (m²)
square foot (ft²)	0.111 1	square yards (yd²)
square inch (in²)	1 273 000	circular mils
square inch (in²)	6.451 6	square centimeters (cm²)
square inch (in²)	0.006 944	square feet (ft²)
square inch (in²)	645.16	square millimeters (mm²)
square inch (in²)	1 000 000	square mils
square meter (m²)	10.764	square feet (ft²)
square meter (m²)	1 550	square inches (in²)
square meter (m²)	0.000 000 386 1	square miles
square meter (m²)	1.196	square yards (yd²)
square mil	1.273	circular mils
square mil	0.000 001	square inches (in²)
square mile	2.589 988	square kilometers (km²)
square millimeter (mm²)	1 973	circular mils
square yard (yd²)	0.836 127 4	square meters (m²)
tablespoon (tbsp)	14.786 75	milliliters (mL)
teaspoon (tsp)	4.928 916 7	milliliters (mL)
therm	105.480 4	megajoules (MJ)
ton (long) (2,240 lb)	1 016.047	kilograms (kg)
ton (long) (2,240 lb)	1.016 047	metric tons (t)
ton, metric	2 204.62	pounds (avoirdupois)
ton, metric	1.102 31	tons, short (2,000 lb)
ton, refrigeration	12 000	Btus per hour
ton, refrigeration	4.716 095 9	horsepower-hours
ton, refrigeration	3.516 85	kilowatts (kW)
ton (short) (2,000 lb)	907.185	kilograms (kg)
ton (short) (2,000 lb)	0.907 185	metric tons (t)
ton per cubic meter (t/m³)	0.842 778	tons per cubic yard (ton/yd³)
ton per cubic yard (ton/yd³)	1.186 55	tons per cubic meter (ton/m³)
torr	133.322 4	pascals (Pa)
watt (W)	3.412 14	Btus per hour (Btu/hr)
watt (W)	0.001 341	horsepower
watt (W)	0.001	kilowatts (kW)
watt-hour (Wh)	3.413	Btus
watt-hour (Wh)	0.001 341	horsepower-hours
watt-hour (Wh)	0.001	kilowatt-hours (kW/hr)
yard	91.44	centimeters (cm)
yard	3	feet
yard	36	inches
yard	0.000 914 4	kilometers (km)
yard	0.914 4	meters (m)
yard	914.4	millimeters
yard, cubic (yd³)	0.764 55	cubic meters (m³)
yard squared (yd²)	0.836 127	meters squared (m²)

---

*These terms are no longer used, but may still be found in some publications.

GLOSSARY

ELECTRICAL TERMS COMMONLY USED IN THE APPLICATION OF THE *NEC*®

The following terms are used extensively in the application of the *National Electrical Code.*® When added to the terms related to electric phenomena they constitute a "culture literacy" for the electrical construction industry. Many of these definitions are quotes or paraphrases of definitions given by the *NEC*®; these have been marked with an *.

Accessible (wiring methods)*: Capable of being removed or exposed without damaging the building structure or finish, or not permanently closed in by the structure or finish of the building.

Accessible (equipment)*: Admitting close approach: not guarded by locked doors, elevation, or other effective means.

Accessible, Readily*: Capable of being reached quickly for operation, renewal, or inspections, without requiring those to whom ready access is requisite to climb over or remove obstacles or to resort to portable ladders, chairs, etc.

Addendum: Modification (change) made to the construction documents (plans and specifications) during the bidding period.

AL/CU: Terminal marking on switches and receptacles rated 30 amperes and greater suitable for use with aluminum, copper, and copper-clad aluminum conductors. If not marked, suitable for copper conductors only.

American National Standards Institute (ANSI): An organization that identifies industrial and public requirements for national consensus standards and coordinates and manages their development, resolves national standards problems, and ensures effective participation in international standardization. ANSI does not itself develop standards. Rather, it facilitates development by establishing consensus among qualified groups. ANSI ensures that the guiding principles—consensus, due process, and openness are followed.

Ampacity*: The current in amperes that a conductor can carry continuously under the conditions of use without exceeding its temperature rating.

Ampere: The measurement of intensity of rate of flow of electrons in an electric circuit. An ampere is the amount of current that will flow through a resistance of one ohm under a pressure of one volt.

Appliance*: Utilization equipment, generally other than industrial, normally built in standardized sizes or types, that is installed or connected as a unit to perform one or more functions such as clothes washing, air conditioning, food mixing, deep frying, etc.

Approved*: Acceptable to the authority having jurisdiction.

Arc-Fault Circuit Interrupter*: A device intended to provide protection from the effects of arc faults by recognizing characteristics unique to arcing and by functioning to de-energize a circuit when an arcing fault is detected.

As-Built Plans: When required, a modified set of working drawings that are prepared for a construction project that includes all variances from the original working drawings that occurred during the project construction.

Authority Having Jurisdiction: A governmental body that has legal jurisdiction over electrical installations. The Authority Having Jurisdiction has the responsibility for making interpretations of the rules, for deciding upon the approval of equipment and materials, and for granting the special permission where required. See *Section 90-4* of the *NEC*®.

Ballast: Used for energizing fluorescent lamps. It is constructed of a laminated core and coil windings or solid-state electronic components.

Bonding*: The permanent joining of metallic parts to form an electrically conductive path that will ensure electrical continuity and the capacity to conduct safely any current likely to be imposed.

Bonding Jumper*: A reliable conductor to ensure the required electrical conductivity between metal parts required to be electrically connected.

Bonding Jumper, Circuit*: The connection

between portions of a conductor in a circuit to maintain required ampacity of the circuit.

Bonding Jumper, Equipment*: The connection between two or more portions of the equipment grounding conductor.

Bonding Jumper, Main*: The connection between the grounded circuit conductor and the equipment grounding conductor at the service.

Branch-Circuit*: The circuit conductors between the final overcurrent device protecting the circuit and the outlet(s).

Branch-Circuit, Appliance*: A branch-circuit supplying energy to one or more outlets to which appliances are to be connected; such circuits are to have no permanently connected lighting fixtures not a part of an appliance.

Branch-Circuit, General Purpose*: A branch-circuit that supplies a number of outlets for lighting and appliances.

Branch-Circuit, Individual*: A branch-circuit that supplies only one utilization equipment.

Branch-Circuit, Multiwire*: A branch-circuit consisting of two or more ungrounded conductors having a potential difference between them, and a grounded conductor having equal potential difference between it and each ungrounded conductor of the circuit and that is connected to the neutral or grounded conductor of the system.

Branch-Circuit Selection Current*: Branch-circuit selection current is the value in amperes to be used instead of the rated-load current in determining the ratings of motor branch-circuit conductors, disconnecting means, controllers, and branch-circuit short-circuit and ground-fault protective devices wherever the running overload protective device permits a sustained current greater than the specified percentage of the rated-load current. The value of branch-circuit selection current will always be equal to or greater than the marked rated-load current.

Circuit Breaker*: A device designed to open and close a circuit by nonautomatic means and to open the circuit automatically on a predetermined overcurrent without damage to itself when properly applied within its rating.

CC: A marking on wire connectors and soldering lugs indicating they are suitable for use with copper-clad aluminum conductors only.

CC/CU: A marking on wire connectors and soldering lugs indicating they are suitable for use with copper or copper-clad aluminum conductors only.

CU: A marking on wire connectors and soldering lugs indicating they are suitable for use with copper conductors only.

CO/ALR: Terminal marking on switches and receptacles rated 15 and 20 amperes that are suitable for use with aluminum, copper, and copper-clad aluminum conductors. If not marked, they are suitable for copper or copper-clad conductors only.

Conductor*:

Bare: A conductor having no covering or electrical insulation whatsoever. (See "Conductor, Covered.")

Covered: A conductor encased within material of composition or thickness that is not recognized by this Code as electrical insulation. (See "Conductor, Bare.")

Insulated: A conductor encased within material of composition and thickness that is recognized by this Code as electrical insulation.

Continuous Load*: A load where the maximum current is expected to continue for three hours or more.

Contract Modifications: After the agreements has been signed, any additions, deletions, or modifications of the work to be done are accomplished by change order, supplemental instruction, and field order. They can be issued at any time during the contract period.

Correction Factor: A multiplier (penalty) that is applied to conductors when the ambient temperature is greater than 30°C (86°F) . These multipliers are found at the bottom of *Table 310-16* of the *NEC.*®

CSA: This is the Canadian Standards Association that develops safety and performance standards in Canada for electrical products similar to but not always identical to those of UL (Underwriters Laboratories) in the United States.

Current: The flow of electrons through an electrical circuit, measured in amperes.

Derating Factor: A multiplier (penalty) that is applied to conductors when there are more than three current-carrying conductors in a raceway or cable. These multipliers are found in *Table 310-15(b)(2)*. A *derating factor* is also referred to as an *adjustment factor*.

Details: Plans, elevations, or sections that provide more specific information about a portion of a project component or element than smaller scale drawings.

Diagrams: Nonscaled views showing arrangements of special system components and connections not possible to clearly show in scaled views. A *schematic diagram* shows circuit components and their electrical connections without regard to actual physical locations. A *wiring diagram* shows circuit components and the actual electrical connections.

Dimmer: A switch with components that permits variable control of lighting intensity. Some dimmers have electronic components; others have core and coil (transformer) components.

Disconnecting Means*: A device, or group of devices, or other means by which the conductors of a circuit can be disconnected from their source of supply.

Drawings: A graphic representation of the work to be done. They show the relationship of the materials to each other, including sizes, shapes, locations, and connections. The drawings may include schematic diagrams showing such things as mechanical and electrical systems. They may also include schedules of structural elements, equipment, finishes, and other similar items.

Dry Niche Lighting Fixture*: A lighting fixture intended for installation in the wall of a pool or fountain in a niche that is sealed against the entry of pool water.

Dwelling Unit*: One or more rooms for the use of one or more persons as a housekeeping unit with space for eating, living, and sleeping, and permanent provisions for cooking and sanitation.

Elevations: Views of vertical planes, showing components in their vertical relationship, viewed perpendicularly from a selected vertical plane.

Feeder*: All circuit conductors between the service equipment or the source of a separately derived system and the final branch-circuit overcurrent device.

Fire Rating: The classification indicating in time (hours) the ability of a structure or component to withstand fire conditions.

Fine Print Note (FPN)*: Explanatory material is in the form of Fine Print Notes (FPN). Fine Print Notes are informational only, and are not enforceable as a requirement by the Code.

Flame Detector: A radiant energy-sensing fire detector that detects the radiant energy emitted by a flame.

Fully Rated System: Panelboards are marked with their short circuit rating in RMS symmetrical amperes. In a fully-rated system, the panelboard short circuit current rating will be equal to the lowest interrupting rating of any branch-circuit breaker or fuse installed. All devices installed shall have an interrupting rating greater than or equal to the specified available fault current. (See "Series-Rated System").

Fuse: An overcurrent protective device with a fusible link that operates and opens the circuit on an overcurrent condition.

Ground-Fault Protection of Equipment (GFPE)*: A system intended to provide protection of equipment from damaging line-to-ground fault currents by operating to cause a disconnecting means to open all ungrounded conductors of the faulted circuit. This protection is provided at current levels less than those required to protect conductors from damage through the operation of a supply circuit overcurrent device.

Ground*: A conducting connection, whether intentional or accidental, between an electrical circuit or equipment and the earth, or to some conducting body that serves in place of the earth.

Ground-Fault: A condition where an ungrounded conductor comes in contact with a grounded object such as but not limited to a grounded metal water pipe. A ground-fault is often referred to as a "line-to-ground" fault. See *Section 230-95* of the *NEC*.®

Grounded*: Connected to earth or to some conducting body that serves in place of the earth.

Grounded Conductor*: A system or circuit conductor that is intentionally grounded.

Grounding Conductor*: A conductor used to connect equipment or the grounded circuit of a wiring system to a grounding electrode or electrodes.

Grounded, Effectively*: Intentionally connected to earth through a ground connection or connections of sufficiently low impedance and having sufficient current-carrying capacity to prevent the buildup of voltages that may result in undue hazards to connected equipment or to persons.

Ground-Fault Circuit-Interrupter (GFCI)*: A device intended for the protection of personnel that functions to de-energize a circuit or portion thereof within an established period of time when a current to ground exceeds some predetermined value that is less than that required to operate the overcurrent protective device of the supply circuit.

Grounding Conductor, Equipment*: The conductor used to connect the noncurrent-carrying metal parts of equipment, raceways, and other enclosures to the system grounded conductor, the grounding electrode conductor, or both, at the service equipment or at the source of a separately derived system.

Grounding Electrode Conductor*: The conductor used to connect the grounding electrode to the equipment grounding conductor, to the grounded conductor, or to both, of the circuit at the service equipment or at the source of a separately derived system.

HACR: A circuit breaker that has been tested and found suitable for use on heating, air-conditioning, and refrigeration equipment. The circuit breaker is marked with the letters HACR. Equipment that is suitable for use with HACR circuit breakers will be marked for use with HACR-type circuit breakers.

Heat Alarm: A single or multiple station alarm responsive to heat.

Hermetic Refrigerant Motor-Compressor: A combination consisting of a compressor and motor, both of which are enclosed in the same housing, with no external shaft or shaft seals; the motor operating in the refrigerant.

Household Fire Alarm System: A system of devices that produces an alarm signal in the household for the purpose of notifying the occupants of the presence of a fire so that they will evacuate the premises.

Identified*: (as applied to equipment) Recognizable as suitable for the specific purpose, function, use, environment, application, etc., where described in a particular Code requirement.

Identified* (conductor): The identified conductor is the insulated grounded conductor.
- For sizes No. 6 or smaller, the insulated grounded conductor shall be identified by a continuous white or natural gray outer finish or by three continuous white stripes on other than green insulation along its entire length.
- For sizes larger than No. 6, the insulated grounded conductor shall be identified either by a continuous white or natural gray outer finish or by three continuous white stripes on other than green insulation along its entire length, or at the time of installation by a distinctive white marking at its terminations.

Identified* (terminal): The identification of terminals to which a grounded conductor is to be connected shall be substantially white in color. The identification of other terminals shall be of a readily distinguishable different color.

IEC: The International Electrotechnical Commission is a worldwide standards organization. These standards differ from those of Underwriters Laboratories. Some electrical equipment might conform to a specific IEC Standard, but may not conform to the U.L. Standard for the same item.

Immersion Detection Circuit Interrupter: A device integral with grooming appliances that will shut off the appliance when the appliance is dropped in water.

In Sight: Where this Code specifies that one equipment shall be "in sight from," "within sight from," or "within sight," etc., of another equipment, the specified equipment is to be visible and not more than 50 ft (15.24 m) from the other.

Inductive Load: A load that is made up of coiled or wound wire that creates a magnetic field when energized. Transformers, core and coil ballasts, motors, and solenoids are examples of inductive loads.

International Association of Electrical Inspectors: A not-for-profit and educational organization cooperating in the formulation and uniform application of standards for the safe installation and use of electricity, and collecting and disseminating information relative thereto. The IAEI is made up of electrical inspectors, electrical contractors, electrical apprentices, manufacturers, and governmental agencies.

Interrupting Rating*: The highest current at rated voltage that a device is intended to interrupt under standard test conditions.

Isolated Ground Receptacle: A grounding type device in which the equipment ground contact and terminal are electrically isolated from the receptacle mounting means.

Labeled*: Equipment or materials to which has been attached a label, symbol, or other identifying mark of an organization that is acceptable to the authority having jurisdiction and concerned with product evaluation that maintains periodic inspection of production of labeled equipment or materials and by whose labeling the manufacturer indicates compliance with appropriate standards or performance in a specified manner.

Lighting Outlet*: An outlet intended for the direct connection of a lampholder, a lighting fixture, or a pendant cord terminating in a lampholder.

Listed*: Equipment or materials included in a list published by an organization acceptable to the authority having jurisdiction and concerned with product evaluation, that maintains periodic inspection of production of listed equipment or materials, and whose listing states that either the equipment or materia service meets appropriate designated standards or has been tested and found suitable for use in a specified manner.

Load: The electric power used by devices connected to an electrical system. Loads can be figured in amperes, volt-amperes, kilovolt-amperes, or kilowatts. Loads can be intermittent, continuous intermittent, periodic, short-time, or varying. See the definition of "Duty" in the *NEC*.

Load Center: A common name for residential panelboards. A load center may not be as deep as a panelboard, and generally does not contain relays or other accessories as are available for panelboards. Circuit breakers "plug-in" as opposed to "bolt-in" types used in panelboards. Manufacturers' catalogs will show both load centers and panelboards. The UL standards do not differentiate.

Locked Rotor Current: The steady-state current taken from the line with the rotor locked and with rated voltage and frequency applied to the motor.

Luminaire: A complete lighting unit consisting of lamp or lamps together with the parts designed to distribute the light, to position and protect the lamps, and to connect the lamps to the power supply. This is a lighting fixture.

Maximum Continuous Current (MCC): A value that is determined by the manufacturer of hermetic refrigerant motor compressors under high load (high refrigerant pressure) conditions. This value is established in the *NEC* as being no greater than 156% of the marked rated amperes or the branch-circuit selection current.

Maximum Overcurrent Protection (MOP): A term used with equipment that has a hermetic motor compressor(s). MOP is the maximum ampere rating for the equipment's branch-circuit overcurrent protective device. The ampere rating is determined by the manufacturer, and is marked on the nameplate of the equipment. The nameplate will also indicate if the overcurrent device can be a HACR-type circuit breaker, a fuse, or either.

Minimum Circuit Ampacity (MCA): A term used with equipment that has a hermetic motor compressor(s). MCA is the minimum ampere rating value used to determine the proper size conductors, disconnecting means, and controllers. MCA is determined by the manufacturer, and is marked on the nameplate of the equipment.

National Electrical Code (*NEC*): The electrical code published by the National Fire Protection Association. This Code provides for practical safeguarding of persons and property from hazards arising from the use of electricity. It does not become law until adopted by federal, state, and local laws and regulations. The *NEC* is not intended as a design specification nor an instruction manual for untrained persons.

National Electrical Manufacturers Association (NEMA): NEMA is a trade organization made up of many manufacturers of electrical equipment. They develop and promote standards for electrical equipment.

National Fire Protection Association (NFPA): Located in Quincy, MA. The NFPA is an international Standards Making Organization dedicated to the protection of people from the ravages of fire and electric shock. The NFPA is responsible for developing and writing the *National Electrical Code*, The Sprinkler Code, The Life Safety Code, The National Fire Alarm Code, and over 295 other codes, standards, and recommended practices. The NFPA may be contacted at (800) 344-3555 or on their web site at www.nfpa.org.

Nationally Recognized Testing Laboratory (NRTL): The term used to define a testing laboratory that has been recognized by OSHA. For example, Underwriters Laboratories.

Neutral: In residential wiring, the neutral conductor is the grounded conductor in a circuit consisting of three or more conductors. There is no neutral conductor in a two-wire circuit, although electricians many times refer to the white grounded conductor as the neutral.

The voltages from the ungrounded conductors to the grounded conductor are of equal magnitude.

See the definition of "Multiwire Branch-Circuit" in the *NEC*.

No Niche Lighting Fixture*: A lighting fixture intended for installation above or below the water without a niche.

Noncoincidental Loads: Loads that are not likely to be on at the same time. Heating and cooling loads would not operate at the same time. See *Section 220-21* of the *NEC*.

Notations: Words found on plans to describe something.

Occupational Safety and Health Act (OSHA): This is the Code of Federal Regulations, developed by the Occupational Safety and Health Administration, U.S. Department of Labor. The electrical regulations are covered in *Part 1910*,

Subpart S. So as not to shorten and simplify the regulations, *Part 1920, Subpart S* contains only the most common performance requirements of the *NEC*.® The *NEC*® must still be referred to, in conjunction with OSHA regulations.

Ohm: A unit of measure for electric resistance. An ohm is the amount of resistance that will allow one ampere to flow under a pressure of one volt.

Outlet*: A point on the wiring system at which current is taken to supply utilization equipment.

Overcurrent*: Any current in excess of the rated current of equipment or the ampacity of a conductor. It may result from overload, short circuit, or ground fault.

Overload*: Operation of equipment in excess of normal, full-load rating, or of a conductor in excess of rated ampacity that, when it persists for a sufficient length of time, would cause damage or dangerous overheating. The current flow is contained in its normal intended path. A fault, such as a short circuit or ground fault, is not an overload.

Overcurrent Device: Also referred to as an overcurrent protective device. A form of protection that operates when current exceeds a predetermined value. Common forms of overcurrent devices are circuit breakers, fuses, and thermal overload elements found in motor controllers.

Panelboard*: A single panel or group of panel units designed for assembly in the form of a single panel; including buses, automatic overcurrent devices, and equipped with or without switches for the control of light, heat, or power circuits; designed to be placed in a cabinet or cutout box placed in or against a wall or partition and accessible only from the front.

Plans: Views of horizontal planes, showing components in their horizontal relationship. A set of construction drawings.

Power Supply: A source of electrical operating power including the circuits and terminations connecting it to the dependent system components.

Qualified Person*: One familiar with the construction and operation of the equipment and the hazards involved.

Raceway*: An enclosed channel of metal or nonmetallic materials designed expressly for holding wires, cables, or busbars, with additional functions as permitted in this Code. Raceways include, but are not limited to, rigid metal conduit, rigid nonmetallic conduit, intermediate metal conduit, liquidtight flexible conduit, flexible metallic tubing, flexible metal conduit, electrical nonmetallic tubing, electrical metallic tubing, underfloor raceways, cellular concrete floor raceways, cellular metal floor raceways, surface raceways, wireways, and busways.

Rated-Load Current: The rated-load current for a hermetic refrigerant motor-compressor is the current resulting when the motor-compressor is operated at the rated load, rated voltage, and rated frequency of the equipment it serves.

Rate of Rise Detector: A device that responds when the temperature rises at a rate exceeding a predetermined value.

Receptacle*: A receptacle is a contact device installed at the outlet for the connection of a single contact device. A single receptacle is a single contact device with no other contact device on the same yoke. ▶A multiple receptacle is two or more contact devices on the same yoke. ◀

Receptacle Outlet*: An outlet where one or more receptacles are installed.

Resistive Load: An electric load that opposes the flow of current. A resistive load does not contain cores or coils of wire. Some examples of resistive loads are the electric heating elements in an electric range, ceiling heat cables, and electric baseboard heaters.

Root-Mean-Square (RMS): The square root of the average of the square of the instantaneous values of current or voltage. For example, the RMS value of voltage line to neutral in a home is 120 volts. During each electrical cycle, the voltage rises from zero to a peak value $(120 \times 1.4142 = 169.7$ volts), back through zero to a negative peak value, then back to zero. The RMS value is 0.707 of the peak value $(169.7 \times 0.707 = 120$ volts). The root-mean-square value is what an electrician reads on an ammeter or voltmeter.

Schedules: Tables or charts that include data about materials, products, and equipment.

Sections: Views of vertical cuts through and perpendicular to components, showing their detailed arrangement.

Series-Rated System: Panelboards are marked with their short-circuit rating in RMS symmetrical amperes. A series-rated panelboard will be determined by the main circuit breaker or fuse, and branch-circuit breaker combination tested in accordance to UL *Standard 489.* The series-rating will be less than or equal to the interrupting rating of the main overcurrent device, and greater than the interrupting rating of the branch-circuit overcurrent devices. (See "Fully-Rated System.")

Service*: ▶The conductors and equipment for delivering energy from the serving utility to the wiring system of the premises served.◀

Service Conductors*: ▶The conductors from the service point to the service disconnecting means.◀

Service Drop*: The overhead service conductors from the last pole or other aerial support to and including the splices, if any, connecting to the service-entrance conductors at the building or other structure.

Service Equipment*: The necessary equipment, usually consisting of a circuit breaker or switch and fuses, and their accessories, located near the point of entrance of supply conductors to a building or other structure, or an otherwise defined area, and intended to constitute the main control and means of cutoff of the supply.

Service Lateral*: The underground service conductors between the street main, including any risers at a pole or other structure or from transformers, and the first point of connection to the service-entrance conductors in a terminal box or meter or other enclosure with adequate space, inside or outside the building wall. Where there is no terminal box, meter, or other enclosure with adequate space, the point of connection shall be considered to be the point of entrance of the service conductors into the building.

Service Point*: Service point is the point of connection between the facilities of the serving utility and the premises wiring.

Shall: Indicates a mandatory requirement.

Short Circuit: A connection between any two or more conductors of an electrical system in such a way as to significantly reduce the impedance of the circuit. The current flow is outside of its intended path, thus the term "short circuit." A short circuit is also referred to as a fault.

Should: Indicates a recommendation or that which is advised but not required.

Smoke Alarm: A single or multiple station alarm responsive to smoke.

Smoke Detector: A device that detects visible or invisible particles of combustion.

Specifications: Text setting forth details such as description, size, quality, performance, and workmanship, etc. Specifications that pertain to all of the construction trades involved might be subdivided into "General Conditions" and "Supplemental General Conditions." Further subdividing of the Specifications might be specific requirements for the various contractors such as electrical, plumbing, heating, and masonry, etc. Typically, the electrical specifications are found in Division 16.

Split Circuit Receptacle: A receptacle that can be connected to two branch-circuits. These receptacles may also be used so that one receptacle is live at all times, and the other receptacle is controlled by a switch. The terminals on these receptacles usually have breakaway tabs so the receptacle can be used either as a split circuit receptacle or as a standard receptacle.

Standard Network Interface (SNI): A device usually installed by the telephone company at the demarcation point where their service leaves off and the customers service takes over. This is similar to the service point for electrical systems.

Surge Suppressor: A device that limits peak voltage to a predetermined value when voltage spikes or surges appear on the connected line.

Switches*:

   General-Use Switch: A switch intended for use in general distribution and branch-circuits. It is rated in amperes, and it is capable of interrupting its rated current at its rated voltage.

   Motor-Circuit Switch: A switch, rated in horsepower, capable of interrupting the maximum operating overload current of a motor of the same horsepower rating as the switch at the rated voltage.

Symbols: A graphic representation that stands for or represents another thing. A symbol is a simple way to show such things as lighting outlets, switches and receptacles on an electrical Plan. The American Institute of Architects has developed a very comprehensive set of symbols that represent just about everything used by all building trades. When an item cannot be shown using a symbol, then a more detailed explanation using a notation or inclusion in the Specifications is necessary.

Terminal: A screw or a quick-connect device where a conductor(s) is intended to be connected.

Thermocouple: A pair of dissimilar conductors so joined at two points that an electromotive force is developed by the thermoelectric effects when the junctions are at different temperatures.

Thermopile: More than one thermocouple connected together. The connections may be series, parallel, or both.

Transient Voltage Surge Suppressors (TVSS): A device that clamps transient voltages by absorbing the major portion of the energy (joules) created by the surge, allowing only a small, safe amount of energy to enter the actual connected load.

UL: Underwriters Laboratories is an independent not-for-profit organization that develops standards, and tests electrical equipment to these standards.

UL Listed: Indicates that an item has been tested and approved to the standards established by UL for that particular item. The UL Listing Mark may appear in various forms, such as the letters UL in a circle. If the product is too small for the marking to be applied to the product, the marking must appear on the smallest unit container in which the product is packaged.

UL Recognized: Refers to a product that is incomplete in construction features or limited in performance capabilities. A "Recognized" product is intended to be used as a component part of equipment that has been "listed." A "Recognized" product must not be used by itself. A UL product may contain a number of components that have been "Recognized."

Ungrounded Conductor: The conductor of an electrical system that is not intentionally connected to ground. This conductor is referred to as the "hot" or "live" conductor.

Volt: The difference of electric potential between two points of a conductor carrying a constant current of one ampere, when the power dissipated between these points is equal to one watt. A voltage of one volt can push one ampere through a resistance of one ohm.

Voltage (of a circuit)*: The greatest root-mean-square (effective) difference of potential between any two conductors of the circuit concerned.

Voltage (nominal)*: A nominal value assigned to a circuit or system for the purpose of conveniently designating its voltage class (e.g., 120/240 volts, 480Y/277 volts, 600 volts).

The actual voltage at which a circuit operates can vary from the nominal within a range that permits satisfactory operation of equipment.

Voltage to Ground*: For grounded circuits, the voltage between the given conductor and that point or conductor of the circuit that is grounded; for ungrounded circuits, the greatest voltage between the given conductor and any other conductor of the circuit.

Voltage Drop: Also referred to as IR drop. Voltage drop is most commonly associated with conductors. A conductor has resistance. When current is flowing through the conductor, a voltage drop will be experienced across the conductor. Voltage drop across a conductor can be calculated using Ohm's Law: $E = IR$.

Volt-ampere: A unit of power determined by multiplying the voltage and current in a circuit. A 120-volt circuit carrying 1 ampere is 120 volt-amperes.

Watt: A measure of true power. A watt is the power required to do work at the rate of 1 joule per second. Wattage is determined by multiplying voltage times amperes times the power factor of the circuit: $W = E \times I \times PF$.

Watertight: Constructed so that moisture will not enter the enclosure under specified test conditions.

Weatherproof*: Constructed or protected so that exposure to the weather will not interfere with successful operation.

(FPN): Rainproof, raintight, or watertight equipment can fulfill the requirements for weatherproof where varying weather conditions other than wetness, such as snow, ice, dust, or temperature extremes, are not a factor.

Wet Niche Lighting Fixture*: A lighting fixture intended for installation in a forming shell mounted in a pool.

WEB SITES

The following list of Internet World Wide Web sites has been included for your convenience. These are current as of time of printing. Web sites are "moving targets" with continual changes. We will do our utmost to keep these web sites current. If you are aware of Web sites that should be added, deleted, or that are different than shown, please let us know. Most manufacturers will provide catalogs and technical literature upon request. You will be amazed at the amount of electrical equipment information that is available on these web sites.

Electrical Industry (*This search engine links to over 800 sites related to the electrical industry.*)	www.electricalsearch.com
ACCUBID	www.accubid.com
Advance Transformer Co.	www.advancetransformer.com
AFC Cable Systems	www.afcweb.com
Alflex	www.alflex.com
Allen-Bradley	www.ab.com
Allied Moulded Products, Inc.	www.alliedmoulded.com
Allied Tube & Conduit	www.industry.net/allied.tube.conduit
Alpha Wire	www.alphawire.com
American Lighting Association	www.americanlightingassoc.com
American Insulated Wire Corp.	www.aiwc.com
American National Standards Institute	www.ansi.org
American Society of Heating, Refrigerating, and Air-Conditioning Engineers	www.ashrae.org
American National Standards Institute	www.ansi.org
AMP	www.amp.com
Amprobe	www.amprobe.com
Angelo	www.angelobrothers.com
Appleton Electric Co.	www.appletonelec.com
Arrow Hart	www.crouse-hinds.com
Associated Builders & Contractors	www.abc.org
Automatic Switch Co. (ASCO)	www.asco.com
Baldor Motors & Drives	www.baldor.com
Belden	www.belden.com
Best Power	www.bestpower.com
BOCA	www.bocai.org
Brady	www.whbrady.com
Bridgeport	www.bpfittings.com
Broan	www.broan.com
Burndy	www.burndy.com
Bussmann	www.bussmann.com
CABO	www.cabo.org/
Canadian Standards Association	www.csa.ca

CEBus Industry Council	www.cebus.org
CEE News magazine	www.ceenews.com/
Certified Ballast Manufacturers	www.certbal.org
Chromalox	www.chromalox.com
Construction Specifications Institute	www.csinet.org
Continental Automated Building Association	www.caba.org
Cooper Hand Tools	www.coopertools.com
Cooper Power Systems	www.cooperpower.com
Copper Development Association Inc.	www.copper.org
Copper information	www.copper.org
Crouse-Hinds	www.crouse-hinds.com
Crouse-Hinds (Hazardous locations)	www.zonequest.com
Daniel Woodhead Company	www.danielwoodhead.com
Delmar Publishers	www.thomson.com/delmar/index.htm
Dual-Lite	www.dual-lite.com
Eagle Electric	www.eagle-electric.com
Eaton/Cutler-Hammer	www.cutlerhammer.eaton.com
EC&M magazine	www.ecmweb.com
Edwards	www.edwards-signals.com
Electrical Contractor magazine	www.ecmag.com
Erico, Inc.	www.erico.com
Estimation Inc.	www.estimation.com
Factory Mutual Insurance Co.	www.factorymutual.com
Federal Pacific Co.	www.electro-mechanical.com/fpc.html
Ferraz	www.ferraz.com
First Alert	www.FirstAlert.com
Fluke Corporation	www.fluke.com
F. W. Dodge	www.fwdodge.com
Gardner Bender	gardnerbender.com
GE Electrical Distribution & Control	www.ge.com/edc
GE Lighting	www.ge.com/lighting/business
Generac Power Systems	www.generac.com
GE Total Lighting Control	www.ge.com/tlc
Gould Shawmut	www.gouldshawmut.com
Greenlee Textron Inc.	www.greenlee.textron.com
Halo (Cooper Lighting)	www.cooperlighting.com
Heyco Molded Products, Inc.	www.heyco.com
Hilti	www.hilti.com
Home Automation Association	www.creator.hometeam.com/haa
HomeStar Wiring Systems	www.lucent/netsys/homestar
Home Toys (Home automation)	www.hometoys.com
Honeywell	www.Honeywell.com
Housing Urban Development	www.HUD.gov
Houston Wire & Cable Company	www.houwire.com
Hubbell Incorporated	www.hubbell.com
Hubbell Incorporated (Lighting)	www.hubbell-ltg.com
Hubbell Incorporated (Premise)	www.hubbell-premise.com
Hubbell Incorporated (Wiring)	www.hubbell-wiring.com
Hunt Dimmers	www.huntdimming.com

Husky	www.mphusky.com
ICBO	www.icbo.org
Ideal Industries, Inc.	www.idealindustries.com
ILSCO	www.ilsco.com
Independent Electrical Contractors	www.ieci.org
Institute of Electrical and Electronic Engineers, Inc.	www.ieee.org
inter.Light, Inc.	www.lightsearch.com
Intermatic	www.intermatic.com
International Association of Electrical Inspectors (IAEI)	www.iaei.com
Juno Lighting, Inc.	www.junolighting.com
Klein Tools	www.klein-tools.com
Kohler Generators	www.kohlergenerators.com
Leviton Manufacturing Co.	www.leviton.com
Lightolier	www.lightolier.com
Lithonia Lighting	www.lithonia.com
Littelfuse, Inc.	www.littelfuse.com
Lutron Electronics Co., Inc.	www.lutron.com
MagneTek Lighting Products	www.magnetek.com/ballast
Manhattan/CDT	www.manhattancdt.com
McGill	www.mcgillelectrical.com
Megger	www.avointl.com
Midwest Electric Products, Inc.	www.midwestelectric.com
Milbank Manufacturing Co.	www.milbankkmfg.com
Minerallac	www.minerallac.com
Motorola Lighting	www.motorola.com/aegg/mli
MYTECH	www.lightswitch.com
National Association of Electrical Distributors	www.naed.org
National Electrical Contractors Association (NECA)	www.necanet.org
National Electrical Manufacturers Association (NEMA)	www.nema.org
National Fire Protection Association	www.nfpa.org
National Joint Apprenticeship and Training Committee	www.njatc.org
NuTone	www.nutone.com
Occupational Safety & Health Act	www.osha.gov
Onan Corporation	www.onan.com
Optical Cable Corporation	www.occfiber.com
Osram Sylvania, Inc.	www.sylvania.com
Panasonic Lighting	www.panasonic.com/lighting
Panduit Corporation	www.panduit.com
Paragon Electric Company, Inc.	www.paragonelectric.com
Penn-Union	www.penn-union.com
Philips Lamps	www.ALTOlamp.com
Philips Lighting Company	www.lighting.philips.com/nam
Progress Lighting	www.progresslighting.com
RACO	www.racoinc.com
Russelectric	www.russelectric.com

SBCCI	www.sbcci.org
S&C Electric Company	www.sandc.com
Siemens-Furnas Controls	www.furnas.com
Simplex	www.simplexnet.com
Sola/Hevi Duty	www.sola-hevi-duty.com
Southwire Co.	www.southwire.com
Square D Company	www.squared.com
Sylvania	www.sylvania.com
Thomas & Betts Corporation	www.tnb.com
Trade Service	www.tra-ser.com
Underwriters Laboratories	www.ul.com
Underwriters Laboratories (Code Authority)	www.ul.com/auth/index.html
Underwriters Laboratories (Contact)	www.ul.com/contact.html
Waukesha Electric Systems	www.waukeshaelectric.com
Wheatland Tube Company	www.wheatland.com
Wiremold	www.wiremold.com
X-10 Pro	www.x10pro.com

CODE REFERENCE INDEX

Note, page numbers in **bold type** reference non-text material.

INDEX

Note, page numbers in **bold type** reference non-text material.